SECONDARY METABOLITES: THEIR FUNCTION AND EVOLUTION

The Ciba Foundation is an international scientific and educational charity. It was established in 1947 by the Swiss chemical and pharmaceutical company of CIBA Limited—now Ciba-Geigy Limited. The Foundation operates independently in London under English trust law.

The Ciba Foundation exists to promote international cooperation in biological, medical and chemical research. It organizes about eight international multidisciplinary symposia each year on topics that seem ready for discussion by a small group of research workers. The papers and discussions are published in the Ciba Foundation symposium series. The Foundation also holds many shorter meetings (not published), organized by the Foundation itself or by outside scientific organizations. The staff always welcome suggestions for future meetings.

The Foundation's house at 41 Portland Place, London W1N 4BN, provides facilities for meetings of all kinds. Its Media Resource Service supplies information to journalists on all scientific and technological topics. The library, open five days a week to any graduate in science or medicine, also provides information on scientific meetings throughout the world and answers general enquiries on biomedical and chemical subjects. Scientists from any part of the world may stay in the house during working visits to London.

Ciba Foundation Symposium 171

SECONDARY METABOLITES: THEIR FUNCTION AND EVOLUTION

A Wiley-Interscience Publication

1992

JOHN WILEY & SONS

Chichester · New York · Brisbane · Toronto · Singapore

Published in 1992 by John Wiley & Sons Ltd
Baffins Lane, Chichester
West Sussex PO19 1UD, England

Other Wiley Editorial Offices

John Wiley & Sons, Inc., 605 Third Avenue,
New York, NY 10158-0012, USA

Jacaranda Wiley Ltd, G.P.O. Box 859, Brisbane,
Queensland 4001, Australia

John Wiley & Sons (Canada) Ltd, 22 Worcester Road,
Rexdale, Ontario M9W 1L1, Canada

John Wiley & Sons (SEA) Pte Ltd, 37 Jalan Pemimpin #05-04,
Block B, Union Industrial Building, Singapore 2057

Suggested series entry for library catalogues:
Ciba Foundation Symposia

Ciba Foundation Symposium 171
x+318 pages, 49 figures, 26 tables, 20 structures

Library of Congress Cataloging-in-Publication Data

Secondary metabolites: their function and evolution/Derek J.
Chadwick and Julie Whelan, editors.
p. cm. – (Ciba Foundation symposium: 171)
Includes bibliographical references and index.
ISBN 0 471 93447 X
1. Metabolism, Secondary—Congresses. I. Chadwick, Derek.
II. Whelan, Julie. III. Series.
QH521.S43 1992
574.19′24—dc20 92-28934
CIP

British Library Cataloguing in Publication Data

A catalogue record for this book is
available from the British Library

ISBN 0 471 93447 X

Phototypeset by Dobbie Typesetting Limited, Tavistock, Devon.
Printed and bound in Great Britain by Biddles Ltd, Guildford.

Contents

Symposium on Secondary Metabolites: their Function and Evolution, held at the Ciba Foundation, London, 18–20 February 1992

The topic of the symposium was proposed by Professor Julian Davies and Dr Dudley Williams

Editors: Derek J. Chadwick (Organizer) and Julie Whelan

Participants

J. E. Baldwin Department of Organic Chemistry, University of Oxford, Dyson Perrins Laboratory, South Parks Road, Oxford OX1 3QY, UK

T. Beppu Department of Agricultural Chemistry, Faculty of Agriculture, University of Tokyo, Bunkyo-ku, Tokyo 113, Japan

M. Brandl Ciba-Geigy Ltd, Research Services, Natural Products PP 2.21, R-1040.P.68, CH-4002 Basel, Switzerland

B. Brückner Department of General Microbiology, Friedrich Schiller University, Neugasse 24, D-6900 Jena, Germany

J. D. Bu'Lock Weizmann Microbial Chemistry Laboratory, Department of Chemistry, University of Manchester, Oxford Road, Manchester M13 9PL, UK

D. E. Cane Department of Chemistry, Brown University, Providence, RI 02912, USA

T. Cavalier-Smith Department of Botany, University of British Columbia, 3529-6270 University Boulevard, Vancouver BC, Canada V6T 1Z4

K. F. Chater Department of Genetics, John Innes Institute, John Innes Centre, Colney Lane, Norwich NR4 7UH, UK

E. Cundliffe Department of Biochemistry, University of Leicester, Leicester LE1 7RH, UK

J. Davies (*Chairman*) Department of Microbiology, University of British Columbia, 300-6174 University Boulevard, Vancouver BC, Canada V6T 1Z3

A. L. Demain Department of Biology, Massachusetts Institute of Technology, Cambridge, MA 02139, USA

P. Escoubas (*Bursar*) Mizutani Plant Ecochemicals Project, Research Development Corporation of Japan, Eniwa RBP, Eniwa-shi, Megumino kita 3-1-1, Hokkaido 061-13, Japan

E. Haslam Department of Chemistry, University of Sheffield, Sheffield S3 7HF, UK

D. A. Hopwood Department of Genetics, John Innes Institute, John Innes Centre, Colney Lane, Norwich NR4 7UH, UK

I. S. Hunter Robertson Institute for Biotechnology, University of Glasgow, Glasgow G11 5JS, UK

P. F. Leadlay Department of Biochemistry, University of Cambridge, Tennis Court Road, Cambridge CB2 1QW, UK

R. I. Lehrer Department of Medicine, UCLA School of Medicine, Center for Health Sciences, 10833 Le Conte Avenue, Los Angeles, CA 90024-1678, USA

L. J. Nisbet Xenova Ltd, 240 Bath Road, Slough, Berks SL1 4EF, UK

L. E. Orgel The Salk Institute for Biological Sciences, PO Box 85800, San Diego, CA 92186-8500, USA

W. Piepersberg Chemische Mikrobiologie, Bergische Universität-Gesamthochschule Wuppertal, Gauss-Strasse 20, D-5600 Wuppertal 1, Germany

K. L. Rinehart 454 Roger Adams Laboratory, Box 45-5, Department of Chemistry, University of Illinois, 1209 West California Street, Urbana, IL 61801, USA

G. Turner Department of Molecular Biology & Biotechnology and Krebs Institute for Biomolecular Research, University of Sheffield, Sheffield S10 2TN, UK

L. C. Vining Department of Biology, Dalhousie University, Halifax, Nova Scotia, Canada B3H 4J1

G. Wächtershäuser Tal 29, D-8000 Munich 2, Germany

P. G. Waterman Phytochemistry Research Laboratories, Department of Pharmaceutical Sciences, University of Strathclyde, 204 George Street, Glasgow G1 1XW, UK

D. H. Williams Department of Chemistry, University of Cambridge, Lensfield Road, Cambridge CB2 1EW, UK

H. Zähner Institute of Biology 2, University of Tübingen, D-7400 Tübingen, Germany

Introduction

Julian Davies

Department of Microbiology, University of British Columbia, 300-6174 University Boulevard, Vancouver BC, Canada V6T 1Z3

The number and the diversity of secondary metabolites are subjects that have intrigued scientists for many years. I mean diversity in many senses: there is the diversity of chemical molecules, and the diversity of sources of these metabolites. One can argue that every living organism on earth either makes secondary metabolites, or, at the very least, participates in some form of secondary metabolism. It is important also to recognize the diversity of the potential functions of these molecules.

I would suggest that at least ten biological functions of secondary metabolites can be proposed (Table 1). I am not implying that these are the only functions, or that known secondary metabolites necessarily have these particular functions. The point of this list is to illustrate the wide diversity of functions that have been proposed for secondary metabolites, and the fact that we will, I hope, probably have additional functions suggested at this meeting. Some secondary metabolites are likely to have more than one biological role.

An interesting point about secondary metabolism is that whereas primary metabolism (intermediate metabolism) is *linear*, in the sense that its products stay with the organism and it is responsible for guaranteeing that an organism has sufficient nutrients and all the means it needs to produce the next generation,

TABLE 1 Some suggested biological functions for secondary metabolites

1. Competitive weapons against other bacteria, fungi, plants, amoebae, insects, etc. (Self-protection/exclusion)
2. Metal-transporting agents
3. Involved in plant–microbe symbiosis
4. Nematode–microbe symbiosis
5. Insect–microbe symbiosis
6. Sexual hormones (pheromones)
7. Differentiation effectors, between and within cells
8. Excretion of unwanted products
9. Products of 'selfish' DNA
10. Reserve pool of new pathways

secondary metabolism is a kind of 'lateral thinking' of microorganisms, or of any multicellular organism that produces a secondary metabolite. It is responsible for interactions between the organism and its environment. A good example is a class of compounds that one doesn't often think of as secondary metabolites, but which are in fact representative of secondary metabolism. This concerns pathogenic organisms. It is interesting to note that in microbial pathogens such as *Listeria*, the enzymes and toxins required for pathogenicity (that is, the interactions of the organisms with their mammalian host cells) are produced in a phase when the organism is not growing. It is a late phase of of development of the particular organism. So one can think of many substances associated with pathogenicity as being representative of secondary metabolism. I don't want to extend this analogy too far, but want to emphasize the point that the concept of secondary metabolism and the production of secondary metabolites seems to be concerned with what is going on outside the producing organism, rather than events going on inside. I think this is an important distinction.

There is much controversy about secondary metabolites in Nature, and in particular the question of what a secondary metabolite may actually do for the organism producing it. Many views have been expressed on this subject. Some people believe that we don't really know what secondary metabolites do, or at least that we can't establish what secondary metabolites may do for the organism concerned. Others take a different view. Dudley Williams, the co-proposer of this symposium, believes that secondary metabolites do play an important role in the life of the producing organism. This is something which clearly is open to discussion. I would like to have your suggestions on the functions of secondary metabolites in the organisms that produce them. I would like also to encourage you to discuss the whole gamut of possibilities with respect to the origins of secondary metabolites, and the production of secondary metabolites, in addition to their functions. We will never be able to comprehend the enormous diversity of these products, and the general considerations of what this kind of diversity means. This is what is interesting about this topic; it's why we are here, and why I am looking forward to a very interesting three days talking about this subject. Secondary metabolism has been ignored; there is nothing 'secondary' about its importance in biology!

Microbial secondary metabolism: a new theoretical frontier for academia, a new opportunity for industry*

Arnold L. Demain

Fermentation Microbiology Laboratory, Department of Biology, Massachusetts Institute of Technology, Cambridge, MA 02139, USA

Abstract. Microbial secondary metabolites are the low molecular mass products of secondary metabolism. They include antibiotics, pigments, toxins, effectors of ecological competition and symbiosis, pheromones, enzyme inhibitors, immunomodulating agents, receptor antagonists and agonists, pesticides, antitumour agents and growth promoters of animals and plants. They have a major effect on the health, nutrition and economics of our society. They have unusual structures and their formation is regulated by nutrients, growth rate, feedback control, enzyme inactivation and induction. Regulation is influenced by unique low molecular mass compounds, transfer RNA, σ factors and gene products formed during post-exponential development. The synthases of secondary metabolism are often coded by clustered genes on chromosomal DNA and infrequently on plasmid DNA. The pathways of secondary metabolism are still not understood to a great degree and thus provide a new frontier for basic investigations of enzymology, control and differentiation. Cloning and expression of genes in industrial microorganisms offer new opportunities for strain improvement and discovery. Microbial metabolites have already established themselves as coccidiostats, immunosuppressants, antihelminthic agents, herbicides and cholesterol-reducing drugs. Great potential exists for the discovery of antiviral, antiparasitic, antitumour and pharmacological compounds and new agricultural products. The future for natural products is bright indeed.

1992 Secondary metabolites: their function and evolution. Wiley, Chichester (Ciba Foundation Symposium 171) p 3–23

Secondary metabolites, also known as idiolites, are special compounds, often possessing chemical structures quite different from the primary metabolites (such as sugars, amino acids and organic acids) from which they are produced.

*Because of space limitations I have had to eliminate all citations prior to 1987, as well as citations to the work of my group. I apologize to all whose citations have been omitted and to my own students and associates: specific citations will be supplied upon request.

Idiolites from microorganisms are not essential for the growth of the producing culture but serve diverse survival functions in Nature. These special metabolites, in contrast to the general nature of primary metabolites, are produced only by some species of a genus, and by some strains of a species. Their unusual chemical structures include β-lactam rings, cyclic peptides containing 'unnatural' and non-protein amino acids, unusual sugars and nucleosides, unsaturated bonds of polyacetylenes and polyenes, and large macrolide rings. Idiolites are typically produced as slightly differing components of a particular chemical family, as a result of the low specificity of some enzymes of secondary metabolism. The main types of biosynthetic pathways involved are those forming peptides, polyketides, isoprenes, oligosaccharides, aromatic compounds and β-lactam rings. Knowledge of the pathways varies from cases in which the amino acid sequences of the enzymes and nucleotide sequences of the genes are known (for example, for penicillins and cephalosporins), to those in which even the enzymic steps are still unknown. Although most secondary metabolites are small (less than 1500 Da) and are produced by non-ribosomal systems, there does exist a family of ribosomally derived antibiotics of higher molecular weight (3000–4000 Da, 32–34 residues) known as lantobiotics (Bannerjee & Hansen 1988, Schnell et al 1988). These include nisin (produced by *Streptococcus lactis*), subtilin (*Bacillus subtilis*) and epidermin (*Staphylococcus epidermidis*).

Regulation of secondary metabolism

The intensity of secondary metabolism can often be increased by the addition of limiting precursors. Examples are shown in Table 1. Secondary metabolism occurs best at submaximal growth rates after growth has slowed down. The distinction between the growth phase (trophophase) and production phase (idiophase) is sometimes very clear, but in many cases idiophase overlaps trophophase. The timing between the two phases can be manipulated—the two phases are often distinctly separated in a complex medium favouring rapid growth, but overlap partially or even completely in a chemically defined

TABLE 1 Increase in intensity of secondary metabolism resulting from the addition of limiting precursors

Group	*Species*	*Secondary metabolite*	*Precursor*
Unicellular bacteria	*Bacillus polymyxa*	Colistin	Diamino-butyric acid
	Bacillus brevis	Gramicidin S	L- or D-Phenylalanine
Filamentous bacteria	*Streptomyces clavuligerus*	Cephamycin C	Lysine
Fungi	*Penicillium chrysogenum*	Penicillin G	Phenylacetic acid

medium supporting slower growth. A secondary metabolite is not 'secondary' because it is produced after growth, but because it is not involved in the growth of the producing culture. Thus, elimination of the production of a secondary metabolite by mutation will not stop or slow down growth; indeed, it may increase the growth rate.

The factors controlling the onset of secondary metabolism are complex and not well understood. Growth rate is important, but we do not know the mechanism(s) involved. Deficiencies in certain nutritional factors are also important, but again we are ignorant of the basic mechanisms.

The delay often seen before the onset of secondary metabolism was probably established by evolutionary pressures. Many secondary metabolites have antibiotic activity and could kill the producing culture if produced too early. Of course, the resistance of antibiotic producers to their own metabolites is well known (Cundliffe 1989 and this volume: 1992). Antibiotic-producing species possess suicide-avoiding mechanisms which are often inducible, but in some cases are constitutive. In the case of inducible resistance, death could result if the antibiotic is produced too early and induction is slow. Delay in secondary metabolite production until the starvation phase makes sense if the product is being used as a competitive weapon or endogenously as an effector of differentiation. In nutritionally rich habitats such as the intestines of mammals, where enteric bacteria thrive, secondary metabolite production is not as important as in soil and water, where nutrients limit microbial growth. Thus, secondary metabolites tend not to be produced by enteric bacteria such as *Escherichia coli* but by soil and water inhabitants such as bacilli, actinomycetes and fungi. Nutrient deficiency in Nature often induces morphological and chemical differentiation—that is, sporulation and secondary metabolism, respectively; both are beneficial for survival in the wild. Thus the regulation of the two types of differentiation is often related.

Most secondary metabolites are formed via enzymic pathways. The enzymes occur as individual proteins, free or complexed, or as parts of large multi-functional polypeptides carrying out a multitude of enzymic steps, as in polyketide synthases and peptide synthetases. The genes encoding the enzymes of secondary metabolism are usually chromosomal, but a few have been shown to be plasmid-borne, such as methylenomycin A of *Streptomyces coelicolor*. Whether chromosomal or plasmid-borne, the genes are usually clustered, especially in prokaryotes, but not necessarily as single operons. Expression of these genes is under strong control by nutrients, inducers, products, metals and growth rate. In most cases, regulation is at the level of transcription, as revealed by the absence of mRNA encoding idiolite synthases until growth rate has decreased.

Regulation by the carbon source

Glucose, usually an excellent carbon source for growth, interferes with the formation of many secondary metabolites. Polysaccharides (e.g. starch),

TABLE 2 Carbon sources interfering with secondary metabolism

Idiolite	*Interfering carbon source*	*Non-interfering carbon source*
Actinomycin	Glucose, glycerol	Galactose, fructose
Bacilysin	Glucosamine, starch, maltose, glycerol, ribose, xylose	Glucose
Benzodiazepine alkaloids	Glucose	Sorbitol, mannitol
Cephalosporin	Glucose, glycerol, maltose	Sucrose, galactose
Chlortetracycline	Glucose	Sucrose
Cycloserine	Glycerol	
Enniatin	Glucose	Lactose
Ergot alkaloids	Glucose	Polyols, organic acids
Erythromycin	Glucose, sucrose, glycerol, mannose, 2-deoxyglucose	Lactose, sorbose
Kanamycin	Glucose	
Oleandomycin	Glucose	Sucrose
Penicillin	Glucose, fructose, galactose, sucrose	Lactose
Peptide K-582	Glycerol	Glucose, sucrose, fructose, sorbitol
Puromycin	Glucose	
Rebecamycin	Sugars	Trisaccharides, polysaccharides
Tetracycline	Glucose	
Tylosin	Glucose, 2-deoxyglucose	Fatty acids

oligosaccharides (e.g. lactose) and oils (e.g. soybean oil, methyloleate) are often preferable for fermentations where secondary metabolism is desired. Examples of interfering carbon sources are given in Table 2. It should be noted that in certain cases (e.g. bacilysin) glucose is not an interfering carbon source, but other carbon compounds are.

In many secondary metabolite pathways, the enzymes subject to control by the carbon source are known. One is phenoxazinone synthase, an enzyme of the actinomycin pathway in *Streptomyces antibioticus*. Repression by glucose is exerted at the level of transcription; specific mRNA is low in trophophase, high in idiophase, and much lower in a glucose than in a galactose medium.

Regulation by the nitrogen source

Many secondary metabolic pathways are negatively affected by nitrogen sources favourable for growth—for example, ammonium salts. As a result, complex fermentation media often include a protein source (such as soybean

TABLE 3 Nitrogen sources interfering with secondary metabolism

Idiolite	*Interfering nitrogen source*	*Non-interfering nitrogen source*
Actinomycin	L-Glutamate, L-alanine, L-phenylalanine, D-valine	L-Isoleucine
Aflatoxin	Nitrate	NH_4^+
Alternariol	Nitrate, L-glutamate, urea	
Bikaverin	Glycine	
Candicidin	L-Tryptophan, L-tyrosine, L-phenylalanine, *p*-amino-benzoate	
Cephalosporin	NH_4^+, L-lysine	L-Asparagine, L-arginine
Chloramphenicol	NH_4^+	D-Serine, L-proline, DL-phenylalanine, DL-leucine, L-isoleucine
Erythromycin	NH_4^+	
Leucomycin	NH_4^+	Uric acid
Macbecin	L-Tryptophan, *p*-amino-benzoate, anthranilate	
Penicillin	NH_4^+, L-lysine	L-Glutamate
Rifamycin	NH_4^+, L-tryptophan, *p*-amino-benzoate	Nitrate, L-phenylalanine
Streptomycin	NH_4^+	Proline
Streptothricin	NH_4^+	DL-Aspartate, L-glutamate, DL-alanine, glycine
Tetracycline	NH_4^+	
Trihydroxytoluene	NH_4^+	
Tylosin	NH_4^+	Valine, L-isoleucine, L-leucine, L-threonine

meal) and defined media a slowly assimilated amino acid (such as proline) as the nitrogen source to encourage high production of secondary metabolites. Processes subject to regulation by the nitrogen source are shown in Table 3. Little information is available on the mechanisms underlying the negative effects of NH_4^+ and certain amino acids. In the production of tylosin, the sensitive enzyme appears to be valine dehydrogenase, which is repressed and inhibited by NH_4^+. Because valine is the best source of the acetate, propionate and butyrate precursors supplying the carbon atoms of the macrolide ring system, protylonolide, interference in valine degradation suppresses tylosin synthesis. In *Cephalosporium acremonium* (syn. *Acremonium chrysogenum*; *A. stricta*), at least two enzymes of the cephalosporin biosynthetic pathway, ACV synthetase and expandase, are repressed.

TABLE 4 Secondary metabolite processes susceptible to interference by inorganic phosphate

Bikaverin	Colistin	Nourseothricin
Butirosin	Ergot alkaloids	Streptomycin
Candicidin	Gentamicin	Tetracycline
Cephalosporin	Levorin	Tylosin
Clavulanic acid	Nanaomycin	Vancomycin

Regulation by the phosphorus source

Regulation by phosphorus sources (Table 4) includes both specific and general controls. A rather specific negative effect of inorganic phosphate arises from its ability to inhibit and/or repress phosphatases. Because biosynthetic intermediates of certain idiolite pathways (e.g. aminoglycoside antibiotics) are phosphorylated whereas the ultimate product is not, phosphatases are required in biosynthesis. For example, streptomycin biosynthesis by *Streptomyces griseus* includes at least three phosphate-cleavage steps and the process is very sensitive to phosphate concentration.

Phosphate also has a more general effect than the inhibition or repression of biosynthetic phosphatases; it appears to interfere in many secondary metabolic pathways not known to have phosphorylated intermediates. Such fermentations have to be conducted at levels of free phosphate (usually below 10 μM) which are sub-optimal for growth.

Induction of secondary metabolite synthases

In a number of secondary metabolite pathways, primary metabolites induce synthases and thus increase production of the final product. These include the induction of dimethylallyltryptophan synthetase by tryptophan in ergot alkaloid biosynthesis, leucine induction of bacitracin synthetase, and methionine induction of ACV synthetase, cyclase and expandase in the cephalosporin pathway of *C. acremonium*.

Of great importance in actinomycete fermentations is the inducing effect by endogenous metabolites which are not primary metabolites—for example, A-factor (2*S*-isocapryloyl-3*R*-hydroxymethyl-γ-butyrolactone) and related γ-butyrolactones. A-factor induces both morphological and chemical differentiation in *S. griseus* and *S. bikiniensis*, bringing on the formation of aerial mycelia, conidia, streptomycin synthases and streptomycin. A-factor induces at least ten proteins in *S. griseus* at the transcriptional level. One of these is streptomycin 6-phosphotransferase, an enzyme which functions both in streptomycin biosynthesis and in resistance. In an A-factor-deficient mutant, there is a failure of transcription of the entire streptomycin gene cluster (Mansouri et al 1989).

TABLE 5 Feedback regulation in secondary metabolism

Idiolite	*Enzyme*	*Mechanism*
Bacitracin	Bacitracin synthetase	Inhibition
Chloramphenicol	Arylamine synthetase	Repression
Cycloheximide	Unknown	
Ergot alkaloids	Dimethylallyltryptophan synthetase	Inhibition
Erythromycin	S-Adenosylmethionine: erythromycin C *O*-methyltransferase	Inhibition
Gramicidin S	Gramicidin S synthetases	Inhibition
Indolmycin	Initial enzyme	Inhibition
Kanamycin	Acetyltransferase	Repression
Lankacidin	Unknown	
Mycophenolic acid	*O*-Methyltransferase	Inhibition
Puromycin	*O*-Methyltransferase	Inhibition
Rubradirin	Unknown	
Tetracycline	Anhydrotetracycline oxygenase	Inhibition
Tylosin	S-Adenosylmethionine: macrocin *O*-methyltransferase	Inhibition

Many other actinomycetes produce A-factor, or related factors which differ in the length of the side chain. In those strains which do not produce streptomycin, the γ-butyrolactones induce the formation of the particular antibiotics that are produced, as well as morphological differentiation. A group of another five γ-butyrolactones, termed virginiae butanolides, are produced by the virginiamycin producer, *Streptomyces virginiae* (Kondo et al 1989), and by other streptomycetes.

Feedback regulation

Many secondary metabolites inhibit or repress their own biosynthetic enzymes (Table 5).

Control by growth rate

Growth rate control appears to be important in secondary metabolism and may be the overriding factor in the cases where nutrient limitation is needed for the production of secondary metabolites. For example, glucose or glycerol plus NH_4^+ support rapid growth but no bacitracin production by *Bacillus licheniformis* during trophophase. However, other more slowly used carbon sources (such as pyruvate, citrate or lactate) in the presence of NH_4^+, or other more slowly assimilated nitrogen sources (such as nitrate, alanine or glutamate) in the presence of glucose, allow bacitracin to be produced during trophophase. More importantly, the degree of bacitracin production varies inversely with the growth rate.

In contrast, both a low growth rate *and* a particular type of nutrient deficiency are needed to support secondary metabolism in other cases. For example, *Streptomyces cattleya*, in chemostat culture, requires both a reduced growth rate and a phosphate deficiency to produce thienamycin. However, to produce cephamycin C, the same strain needs only a low growth rate, which can be brought about by deficiency of carbon, nitrogen or phosphate.

The production of microcin B17, a large (3255 Da) ribosomally derived peptide antibiotic, by *E. coli* is derepressed when growth slows and thus it is produced temporally, in the stationary phase. Its single structural gene is plasmid-borne, carried on a 3.5 kb DNA cluster which also contains three biosynthetic genes (involved in processing the final translation product) and three resistance genes. Temporal production involves transcriptional control; that is, slow growth rates bring on β-galactosidase production in genetic fusions between the B17 structural gene and *lacZ* (Connell et al 1987). Even when the cells are growing exponentially, the level of production of β-galactosidase varies inversely with growth rate. Growth restriction could result from limitation of carbon, nitrogen or phosphate.

Enzyme decay

The production of secondary metabolites eventually stops, as a result of feedback regulation (see above) and decay of the synthase(s). For example, most of the synthases involved in the production of peptide antibiotics by bacilli disappear a few hours after they are formed. In the producer of gramicidin S, *Bacillus brevis*, the inactivation of the gramicidin S synthetases is oxygen dependent and independent of protease action. Decay is slowed by the presence of thiols, amino acid substrates and a utilizable energy source. The sites of inactivation appear to be the SH groups involved in binding the amino acids to the enzymes.

A new academic frontier

For many years, basic scientists (with the possible exception of organic chemists) were not interested in secondary metabolism. There were so many exciting discoveries to be made in the area of primary metabolism and its control that secondary metabolism was virtually ignored; study of this type of non-essential (luxury?) metabolism was left for industrial scientists and university pharmacognocists to dabble in. Today, the situation is different. The basic studies on *E. coli* and other microorganisms have elucidated virtually all of the primary metabolic pathways and most of the relevant regulatory mechanisms; many of the enzymes have been purified and the genes encoding them isolated, cloned and sequenced. Primary metabolism is no longer a frontier—it is a domesticated farm. The frontier is now secondary metabolism—a mysterious area of technology that has recently become science and poses many questions of considerable interest to academic scientists. What are the functions of idiolites in Nature? How are these molecules made? How are the pathways controlled? What are the origins of secondary metabolism genes? How did it happen that virtually the same genes, enzymes and pathways exist in organisms so different as the eukaryote *C. acremonium* and the prokaryote, *Flavobacterium* sp.? What are the origins of the resistance genes which producing organisms use to protect themselves from suicide? And are these the same genes as those found in clinically resistant bacteria? Fortunately, molecular biology has produced tools with which to answer many of these questions. It is clear that basic mechanisms controlling secondary metabolism are now of great interest to many academic (and industrial) laboratories throughout the world. In the following sections, I shall describe some of the academic progress being made in the area of the regulation of secondary metabolite biosynthesis.

Mechanism of general phosphate control

Although little is known about the mechanism of general phosphate control of secondary metabolism, phosphate control of candicidin production by *S. griseus* is certainly exerted at the level of transcription. The most advanced data on phosphate repression come from the *p*-amino-benzoic acid (PABA) synthase involved in candicidin production in *S. griseus*. Close to the 5′ end of the structural gene (*pabS*) is a 114 bp sequence and a promoter whose expression is controlled by the level of phosphate in the medium. The phosphate control sequence shows 78% identity to the ‘phosphate box’ of *pho* regulation in *E. coli* (Liras et al 1990).

The mechanism of γ-butyrolactone regulation of secondary metabolism in actinomycetes

Genes encoding proteins involved in the synthesis of A-factor are present on a plasmid in streptomycin producers such as *S. griseus* and *S. bikiniensis* but are chromosomally borne in *S. coelicolor*. In the latter organism, the structural gene *afsA* appears to be regulated by genes *afsB* and *afsR*; *afsB* independently controls synthesis of A-factor and the antibiotics actinorhodin, undecylprodigiosin, methylenomycin A and a calcium-dependent antibiotic (CDA). It encodes a potential DNA-binding protein which acts positively at the transcriptional level on *afsA* and regulatory genes controlling each antibiotic biosynthesis pathway in *S. coelicolor* (Horinouchi et al 1989). The *afsB* gene product is a 993 residue protein of M_r 105 600 containing two ATP-binding sequences and two DNA-binding sequences with a helix–turn–helix motif in its N-terminus.

In *S. griseus*, an A-factor-binding protein has been isolated which is the receptor of A-factor and a repressor of differentiation (Miyake et al 1989). When A-factor binds to this receptor, it is presumably removed from DNA, allowing the formation of aerial mycelia, conidia and streptomycin synthases. In *S. virginiae*, a butanolide C-binding protein (presumably the receptor of butanolide C and the repressor of differentiation in that organism) has been isolated (Kim et al 1989). In *S. coelicolor*, secondary metabolism and morphological differentiation appear to be independent of A-factor (Horinouchi & Beppu 1990). Thus the role of A-factor is unclear in *S. coelicolor*.

In *S. coelicolor*, gene *bldA* is required for morphological differentiation and chemical differentiation (i.e. the production of antibiotics, regulation and resistance). Differentiation shows signs of being controlled by a cascade-type mechanism (Hopwood 1988). At a relatively high level is *bldA*, which could be a master gene controlling the formation of aerial mycelia, conidia and secondary metabolites. It encodes a tRNA molecule which is involved in the translation of a rare leucine codon, UUA (TTA in the DNA). TTA is evidently absent in vegetative genes but present in some genes involved in differentiation, i.e. in genes encoding the regulation of antibiotic biosynthesis and resistance.

At the next lower level is presumably a gene containing a TTA codon which controls *afsB*. Once the *afsB* gene product has been made, it would turn on transcription of each of the specific antibiotic regulatory genes; it would also lead to A-factor synthesis.

With respect to the morphological differentiation cascade, below *bldA* would be *bldC*, which controls morphogenesis in *S. coelicolor* but not chemical differentiation. Under *bldC*'s control would be the *whi* loci which regulate conidia formation but not that of aerial mycelia or antibiotics. These loci include *whiG* for coiling, *whiA*, *whiB* and *whiI* for septation, *whiD* and *whiF* for wall-thickening and the rounding-off of spore compartments and *whiE* for spore pigmentation. *whiE* contains at least seven genes (Davis & Chater 1990). Other mutations (*absA*, *absB*) have been found in *S. coelicolor* which eliminate the formation of all four antibiotics without affecting morphological differentiation (Adamidis et al 1990). Although some information about the action of some of these genes is emerging—for example, *afsR* seems to specify a phosphorylation-dependent regulator (Stein & Cohen 1989, Horinouchi et al 1989)—it is not clear how their products cooperate to ensure efficient control of secondary metabolism (see Chater 1992: this volume).

Mechanism of growth rate (or temporal) control of secondary metabolism in bacilli

One set of important sporulation genes is *spoIIJ*, *spoOA*, and *spoOF*. Their products are part of a two-component signal transduction system (Antoniewski et al 1990, Perego et al 1989). Protein SpoIIJ appears to be the sensor kinase which phosphorylates proteins SpoOA and SpoOF. Protein SpoOA binds to a specific region of the promoter region of the *abrB* gene, downstream from the transcriptional start site, and represses its transcription (Strauch et al 1990). Protein AbrB represses sporulation, so protein SpoOA acts by negative control to turn on sporulation.

For sporulation to occur, a deficiency in the intracellular content of guanine nucleotides and/or a deficiency of the carbon, nitrogen or phosphorus source are required. Guanine nucleotide deficiency, which can be induced by the antimetabolite decoyinine, induces the expression of *spoOH*, a key early positively acting sporulation gene (Dubnau et al 1987).

An involvement of some sporulation genes in secondary metabolism is clear from the work of Marahiel et al (1987) on the cloning of the tyrocidin biosynthetic genes. Gene *tycA* encodes tyrocidin synthetase A in *B. brevis*. It was shown by *lac* fusions that expression of *tycA* in *B. subtilis* is temporally controlled at the level of transcription. Required for its expression are sporulation genes *spoOA*, *spoOB* and *spoOE*, but not *spoOC*, *spoOF*, *spoOH* or *spoOJ*. The *abrB* gene product represses expression of *tycA* by binding to the *tycA* promoter (Robertson et al 1989), and thus protein SpoOA prevents this repression of antibiotic synthesis by repressing transcription of the *abrB* gene (Perego et al 1988). These important studies show that control of morphological differentiation and control of chemical differentiation are not identical, but share certain common genes and their protein products.

Opportunities for industry

The pharmaceutical industry has been the home of commercial secondary metabolism for over 40 years. However, about 10 years ago, a number of companies became discouraged with the expense of finding new antibiotics and began to leave the field of natural product discovery. This was ill-timed, because it is now widely appreciated that secondary metabolites possess many activities unrelated to the killing of microorganisms and some have already been developed into major pharmaceutical compounds. Another opportunity for industry is the development of improved producing strains by the use of recombinant DNA technology.

New applications of secondary metabolites

The selective action that microbial secondary metabolites exert on pathogenic bacteria and fungi and on rapidly growing cells in general was responsible for ushering in the antibiotic era, and for over 40 years we have benefited from the remarkable properties of the 'wonder drugs', the antibiotics. Their success rate was so impressive that these molecules are the major ones used for antibacterial, antifungal and antitumour chemotherapy. As a result of these early successes, the pharmaceutical industry screened secondary metabolites for years almost exclusively for antimicrobial and antitumour activities. This biased and narrow view severely limited the potential application of microbial metabolites. Fortunately, this situation has recently changed, and we have now entered a new era in which microbial metabolites are being applied to diseases previously treated only by synthetic compounds.

For years, the major drugs for non-infectious diseases such as hypotensive and anti-inflammatory agents were strictly synthetic products. Similarly, major therapeutic agents (for example, coccidiostats and antihelminthics) used against parasitic diseases in animals arose solely from the screening of chemically synthesized compounds followed by molecular modification. Today, however, fermentation products such as monensin and lasalocid dominate the coccidiostat market. The avermectins, another group of streptomycete products, have high activity against helminths and arthropods. Indeed, their activity appears to be an order of magnitude greater than that of the previously discovered synthetic antihelminthic agents. Another natural compound of great interest is mevinolin (lovastatin), a fungal product which acts as a cholesterol-lowering agent. Mevinolin is produced by *Aspergillus terreus* and in its hydroxyacid form (mevinolinic acid) is a potent competitive inhibitor of 3-hydroxy-3-methyl-glutaryl-coenzyme A reductase from mammalian liver. Other successes from microorganisms include animal growth promoters, insecticides, herbicides, plant growth regulators, immunosuppressants, uterocontractants and oestrogenic agents for animals. A number of groups have isolated microbial secondary metabolites possessing important pharmacological activities through screening with simple enzyme or receptor assays. The future will feature further applications in important areas of medicine and agriculture, such as new antiviral, antiparasitic and antitumour agents, pharmacological compounds and pesticides. Furthermore, many of the protein products of the new biotechnology industry will eventually be replaced by low molecular mass idiolites from microorganisms.

Genetic manipulation to increase secondary metabolism

The clustering of biosynthetic, resistance and regulatory genes of secondary metabolism has facilitated the application of recombinant DNA technology. Such clustering is almost the rule in actinomycetes but also occurs in bacilli and sometimes even in fungi. Gene clusters generally contain more than a single operon plus individual genes and they have been detected in microorganisms producing the following metabolites: methylenomycin A, actinorhodin, streptomycin, erythromycin, tylosin, carbomycin, tetracycline, valinomycin, bialaphos, chloramphenicol, penicillin and cephalosporin. Clustering makes it possible to transfer operons or groups of operons from one organism to another by recombinant DNA technology, thus facilitating an increase in dosage of a gene encoding a limiting enzyme of a pathway. The presence of a resistance gene in the cluster makes the selection of recombinants very easy. That gene dosage is important in strain improvement is supported by the observation that gene amplification has been detected in high producers of penicillin obtained by non-recombinant methods (Smith et al 1989).

The regulatory genes are of great importance because production may be limited by the presence or absence of such genes, rather than by the dosage of a structural gene. Thus, disruption of the negatively acting regulatory gene *mmy* of methylenomycin biosynthesis increased production 17-fold. The introduction of one extra copy of the positively acting regulatory gene *act*II raised the synthesis of actinorhodin 30- to 40-fold. Regulatory gene *redD* increased undecylprodigiosin synthesis 30-fold, and *strR* increased streptomycin synthesis five- to seven-fold. It is clear that the exploitation of regulatory genes will play a major part in strain improvement in future industrial practice.

References

Adamidis T, Riggle P, Champness W 1990 Mutations in a new *Streptomyces coelicolor* locus which globally block antibiotic biosynthesis but not sporulation. J Bacteriol 172:2962–2969

Antoniewski C, Savelli B, Stragier P 1990 The *spoIIJ* gene, which regulates early developmental steps in *Bacillus subtilis*, belongs to a class of environmentally responsive genes. J Bacteriol 172:86–93

Banerjee S, Hansen JN 1988 Structure and expression of a gene encoding the precursor of subtilin, a small protein antibiotic. J Biol Chem 263:9508–9514

Chater KF 1992 Genetic regulation of secondary metabolic pathways in *Streptomyces*. In: Secondary metabolites: their function and evolution. Wiley, Chichester (Ciba Found Symp 171) p 144–162

Connell N, Han Z, Moreno F, Kolter R 1987 An *E. coli* promoter induced by the cessation of growth. Mol Microbiol 1:195–201

Cundliffe E 1989 How antibiotic-producing organisms avoid suicide. Annu Rev Microbiol 43:207–233

Cundliffe E 1992 Self-protection mechanisms in antibiotic producers. In: Secondary metabolites: their function and evolution. Wiley, Chichester (Ciba Found Symp 171) p 199–214

Davis NK, Chater KF 1990 Spore colour in *Streptomyces coelicolor* A3(2) involves the developmentally regulated synthesis of a compound biosynthetically related to polyketide antibiotics. Mol Microbiol 4:1679–1691

Dubnau EJ, Cabane K, Smith I 1987 Regulation of *spoH*, an early sporulation gene in bacilli. J Bacteriol 169:1182–1191

Hopwood DA 1988 Towards an understanding of gene switching in *Streptomyces*, the basis of sporulation and antibiotic production. Proc R Soc Lond B Biol Sci 235:121–138

Horinouchi S, Beppu T 1990 Autoregulatory factors of secondary metabolism and morphogenesis in actinomycetes. Crit Rev Biotechnol 10:191–204

Horinouchi S, Malpartida F, Hopwood DA, Beppu T 1989 *afsB* stimulates transcription of the actinorhodin biosynthetic pathway in *Streptomyces coelicolor* A3(2) and *Streptomyces lividans*. Mol & Gen Genet 215:355–357

Kim HS, Nihira T, Tada H, Yanagimoto M, Yamada Y 1989 Identification of binding protein of virginiae butanolide C, an autoregulator in virginiamycin production from *Streptomyces virginiae*. J Antibiot (Tokyo) 42:769–778

Kondo K, Higuchi Y, Sakuda S, Nihira T, Yamada Y 1989 New virginiae butanolides from *Streptomyces virginiae*. J Antibiot (Tokyo) 42:1873–1876
Liras P, Asturias JA, Martin JF 1990 Phosphate sequences involved in transcriptional regulation of antibiotic biosynthesis. Trends Biotechnol 8:184–189
Mansouri K, Pissowotzki K, Distler J et al 1989 Genetics of streptomycin production. In: Hershberger CL, Queener SW, Hegeman G (eds) Genetics and molecular biology of industrial microorganisms. American Society for Microbiology, Washington, DC, p 61–67
Marahiel MA, Zuber P, Czekay G, Losick R 1987 Identification of the promoter for a peptide antibiotic biosynthesis gene from *Bacillus brevis* and its regulation in *Bacillus subtilis*. J Bacteriol 169:2215–2222
Miyake K, Horinouchi S, Yoshida M et al 1989 Detection and properties of A-factor—binding protein from *Streptomyces griseus*. J Bacteriol 171:4298–4302
Perego M, Spiegelman GB, Hoch JA 1988 Structure of the gene for the transition state regulator, *abrB*: regulator synthesis is controlled by the *spoOA* sporulation gene in *Bacillus subtilis*. Mol Microbiol 2:689–699
Perego M, Cole SP, Burbulys D, Trach K, Hoch JA 1989 Characterization of the gene for a protein kinase which phosphorylates the sporulation-regulatory proteins SpoOA and SpoOF of *Bacillus subtilis*. J Bacteriol 171:6187–1696
Robertson JB, Gocht M, Marahiel MA, Zuber P 1989 AbrB, a regulator of gene expression in *Bacillus*, interacts with the transcription initiation regions of a sporulation gene and an antibiotic biosynthetic gene. Proc Natl Acad Sci USA 86:8457–8461
Schnell N, Entian KD, Schneider U et al 1988 Prepeptide sequence of epidermin, a ribosomally synthesized antibiotic with four sulphide-rings. Nature (Lond) 333:276–278
Smith DJ, Bull JH, Edwards J, Turner G 1989 Amplification of the isopenicillin N synthetase gene in a strain of *Penicillium chrysogenum* producing high levels of penicillin. Mol & Gen Genet 216:492–497
Stein D, Cohen SN 1989 A cloned regulatory gene of *Streptomyces lividans* can suppress the pigment deficiency phenotype of different developmental mutants. J Bacteriol 171:2258–2261
Strauch M, Webb V, Spiegelman G, Hoch JA 1990 The SpoOA protein of *Bacillus subtilis* is a repressor of the *abrB* gene. Proc Natl Acad Sci USA 87:1801–1805

DISCUSSION

Bu'Lock: What has changed in recent times is on one hand the capabilities of the biochemists, which have extended enormously; and, because they have become more fundamental, they have provided us with a much deeper insight into the way biochemistry works. On the other hand, as the molecular biologists have moved into areas such as antibiotic production, we have begun to get explanations in the sorts of molecular terms that morphologists are now using to explain morphological development. The work of David Hopwood is a clear example of how everything that has been learned by the study of molecular biology, and its developing techniques, is beginning to enlighten us about secondary metabolism.

So far as the excitement of natural product studies is concerned, all that has happened in the recent period is that the structures that structural chemists have come up with have become even more crazy! If you consider the structures of some of the macrocyclic lactones from streptomycetes, or of the low molecular mass compounds from marine organisms, or of, say, the toad toxins, from the chemist's point of view these are very unusual indeed.

Cane: It has been much less explored than in microorganisms, but is there any evidence that the theme of the clustering of biosynthetic genes applies to higher *plant* natural products?

Hopwood: For the anthocyanin biosynthetic pathway, which has been studied in plants such as maize and *Antirrhinum*, there is no gene clustering, as far as I know. There is the typical eukaryotic situation where the genes for different biosynthetic steps are scattered over the chromosomes. The real surprise as far as eukaryotes are concerned is the penicillin pathway in the fungi, where there is extensive clustering, as we shall hear later from Geoff Turner. It may have a special explanation, whereby the genes could have come from a prokaryote by lateral transfer.

Turner: I wonder if the clustering is as much of a surprise as we used to think. The earlier view was that in prokaryotes, pathway genes are clustered, and in eukaryotes they are not, but it's not just the antibiotic biosynthesis genes which are clustered; for a number of other pathways in the fungi there are two, three or four genes grouped together; for example, the proline and the quinic acid catabolism clusters in *Aspergillus nidulans* (Hullet al 1989, Hawkins et al 1988).

Demain: The fact is that even though the biosynthetic genes are clustered, they are not necessarily composed of single operons. I don't know of any single operons, with the possible exception of the peptide antibiotics from *Bacillus*, where just two or three enzymes are involved. In most cases of secondary metabolic pathways, there are a number of transcripts. With regard to clustering in fungi, the three genes of penicillin synthesis are clustered in *Penicillium*. However, for the longer pathway (the cephalosporin pathway) in the fungus *Cephalosporium acremonium*, we are dealing with genes on different chromosomes which are participating in a biosynthetic pathway.

Davies: Is anything known about the genetics of the biosynthesis of secondary metabolites in plants, in terms of the actual structure of biosynthetic clusters?

Bu'Lock: The Mendelian genetics of features like flower colour are well known; beyond that, there's little in terms of molecular genetics.

Cane: There are examples in various plants of biosynthetic enzymes whose genes have been cloned, for example strictosidine synthase from *Rauvolfia serpentina* and berberine bridge enzyme from *Eschscholtzia californica* (Kutchan et al 1991). But if one wants to produce desirable plant products in culture, say, on the notion that one could take an entire cluster of genes and move it into a more effective environment, such as one where the culture is more easily fermented, one will run into the problem of having to find the genes one at a time or in small groups, rather than taking a large piece of DNA and cloning it into a suitable host.

Demain: Let's take taxol as an example. Right now, it would be very desirable to make it in a microorganism (because of its potential use in cancer), but the enzymes are not known, in the yew tree. Since neither the enzymes nor the pathway are known, the ability to use recombinant DNA technology to synthesize taxol in microorganisms is limited.

Bu'Lock: It's because the biochemistry and now the molecular biology have been so much better worked out in microorganisms that secondary metabolism has come to be thought of as a feature of microorganisms; whereas in 1950, say, it was thought of purely as a characteristic of higher plants. This change has come because the microbial biochemistry and molecular biology of microorganisms are now so much better known than those for the classic plant systems.

Davies: I have suggested that all organisms make secondary metabolites. Is this likely to be true?

Bu'Lock: The scope for secondary metabolism has become restricted in the higher animals. There are things like the bufenolide toxins and so on, from amphibians, but the more sophisticated and highly integrated the organism, the less scope there seems to be for excursions into secondary metabolism. Animals have on the one hand a high requirement for integration and, on the other, so many other ways in which to display and exploit their genetic diversity.

Demain: When you made that statement, Julian, did you mean that every strain of, say, *Escherichia coli* has to make a secondary metabolite, or that there are some strains of *E. coli* that make secondary metabolites? At what level are you talking? We don't generally think of *E. coli* as a producer of secondary metabolites, but it is, especially strains isolated from the infant intestine.

Davies: I think any *E. coli* strain could make a secondary metabolite; the synthetic functions are likely to be coded on plasmids. I think humans make secondary metabolites. This is perhaps an incorrect way of describing detoxification by the liver, but it's a form of secondary metabolism, in my view, in that it's not something which is necessary for the organism to grow.

Nisbet: Surely the problem with secondary metabolites is that we have become stuck in a rut of describing them by the process by which they are made, or attributing the term 'secondary metabolite' to something that's made through so-called secondary metabolism. We need to think more broadly about the functions of these molecules. If one is looking for small molecules that have important physiological functions for the host, for its predators, or for its symbionts or parasites, or for interactions with mating types, then I agree with Julian Davies that what we call secondary metabolites in microbes have parallels in all biological systems. One might suggest that the endorphins are secondary metabolites that have been highly conserved. We have examples of reptilian and amphibian secondary metabolites with very potent actions, as neurotoxins or as antibiotics.

We should also question whether antibiotics *are* 'antibiotics'. Are we describing them as that because when we use them and give overdoses, that's what they do to bacteria? There is much evidence that these compounds are more potent in relation to other functions.

Demain: Yes. That is shown by some of the new microbial products such as the coccidiostats, which certainly are antibiotics; cyclosporin, used as an immunosuppressant, is also an antibiotic. They are not used in agriculture and medicine as antibiotics, but they have such activity.

Davies: The problem is that we are much too anthropomorphic about this!

Vining: Do we know of an archaebacterium that makes a secondary metabolite? I wonder whether a harsh environment, such as these bacteria live in, selects against extravagant kinds of metabolic pathways?

Demain: It might select *for* the production of a secondary metabolite!

Bu'Lock: In fact, the archaebacteria have not been a rewarding source of secondary metabolites. We have looked for them, as have others. But what has made the archaebacteria exciting has been the extraordinary variations in their primary metabolites. We became very excited about the unusual make-up of the archaebacterial cell membrane, for example. The same is true for archaebacterial cell proteins. I don't think the circumstances in which the archaebacteria live (e.g. in marginally nourished media with very poorly yielding energy metabolism) have given them very much scope for the development of secondary metabolism. In any case, there are suggestions that the surviving archaebacteria are relict species confined to the marginal habitats by the wider proliferation of eubacteria.

Cavalier-Smith: Most secondary metabolites in microorganisms are directed against other species. Since many archaebacteria live in habitats where they are the only species, because other organisms can't survive there, there would be less selection for such interspecific competition.

Chater: The notion of an organism in an environment where it's not in competition with any other organism has something false about it, because the organism is in competition with itself; evolutionary pressure will operate among siblings, as it were. This might mean that the secondary metabolites that would develop in such circumstances and would influence the interaction between members of the same species would have to be of a rather specialized kind; because if for example there was inhibition from one organism to another, it would have to be directed in a way which was not autoinhibitory.

Cavalier-Smith: That is why I suggest that it would be more difficult to evolve secondary metabolites; that's a restriction. I agree there would be competition between individuals of the same species.

Bu'Lock: In fact, there is quite a lot of autoinhibition in the extremophiles, both archaebacteria and eubacteria. This is one of their main characteristics; it's also one of the difficulties of working on them, because you can only get up to a certain culture density.

Chater: What is understood about these autoinhibitors?

Bu'Lock: Nothing!

Chater: But are they chemically extractable and purifiable?

Bu'Lock: If you are growing something at 85 °C and pH 1.0, the practical chances of extracting anything significant are quite low.

Wächtershäuser: How does the number of secondary metabolites extracted from aerobic microorganisms relate to the number extracted from anaerobes?

Demain: There are probably no *antibiotics* known to be produced by anaerobic microorganisms, but the screens have been very limited. This is an interesting question.

The statement about extreme environments not being conducive to antibiotic or secondary metabolism may have something to it. But, in those cases, organisms may use mechanisms for their survival and their competition other than those we traditionally look for. For example, yeasts, especially strains of *Saccharomyces cerevisiae*, make almost no antibiotics, but they make alcohol, a fairly potent compound! Perhaps yeasts have evolved to the point where they don't need these compounds that we call antibiotics, since they can inhibit other forms of life by alcohol production. We would not consider ethanol as a *secondary* metabolite because its production is part of the yeast growth process, but yeasts certainly use it as such.

Zähner: The anaerobic bacteria are very weak producers of secondary metabolites. We have tested some strains of spore-forming anaerobic bacteria for the production of secondary metabolites, without any success. But we have isolated an autoregulator, an autoinhibitor of spore germination, from *Streptomyces viridochromogenes*; it is a 2-hydroxy-pyrone.

Davies: We say that *antibiotics* are not produced by anaerobes, but what is known about secondary metabolites being made by anaerobes?

Demain: Professor Baldwin and I worked together at one time on β-lactam biosynthesis (Yoshida et al 1978) and he has subsequently examined some of the enzymes involved (Baldwin et al 1991). Most of them turn out to be oxygenases. Is there anything about making such compounds which limits an organism that's an anaerobe? Of course, they wouldn't make the oxygenases, but is there something chemical in Nature that restricts the production of these compounds, whether they be antibiotics or not, to oxygen-requiring reactions?

Baldwin: From what we know of the detailed chemical mechanism of the construction of penicillin and the cephalosporins, the whole basis of the system is oxygen and iron; a combination of iron and oxygen provides the chemical potential for the things that have to be done. I would be surprised if anyone were to find penicillins and cephalosporins in any true anaerobic organism.

I have a more general problem with this discussion, however, because it has never been clear to me what actually is secondary metabolism, as against primary. Is there a clear distinction, so that we can classify some groups of compounds as secondary metabolites and others as primary? Maybe there is a merging? Penicillin is certainly derived from amino acids, so could be regarded as a primary metabolite. And alkaloids are mostly synthesized from amino acids, too.

Demain: I think the simplest criterion is that if you eliminate a so-called secondary metabolic pathway by mutation, the organism will continue to grow. If you knock out genes encoding glycolytic pathway enzymes, the organism is probably not going to grow.

Cane: In the archaebacteria, some interesting metabolites are constituents of the membrane lipids. Perhaps we should talk about *interesting* metabolism and not secondary metabolism! This means that we can talk about ethylene, which is chemically simple but is made in an interesting way. Moreover, cholesterol, which has been the province of the organic chemist for some 60 years, is an essential cell membrane constituent, and has been worked on by natural products chemists for a long time. In fact, its study is considered to be one of the paradigms for the study of biosynthesis. It is important to think about some universality of mechanisms, and about the ways in which information is generated and stored, rather than to focus narrowly on an organism's mechanisms of defence.

Bu'Lock: The point Professor Baldwin made is important, and it comes back to one of the reasons why non-essential metabolism is not a marked feature of the anaerobic bacteria and the extremophiles. These groups have both colonized niches which do not give them much pay-off. Consider the number of times a methanogen must go round the metabolic cycle before generating a significant amount of ATP. These organisms are energy starved. I often see references to the 'energy burden' imposed by secondary metabolism, but where oxygen-driven enzyme systems are part of the synthetic mechanism, they are getting a free ride, because there is enormous chemical potential to be got out of the O_2 molecule. And the anaerobes have denied themselves that advantage. The extremophiles, certainly the extreme thermophiles, have largely done this, because of the low solubility of oxygen in water at high temperatures.

Cavalier-Smith: I don't think anaerobes are necessarily energy starved, particularly the phototrophic bacteria, which just lie in the sun, and surely get lots of energy?

Bu'Lock: They are getting a free ride too; I agree!

Cavalier-Smith: They can live in very dense populations, and you might expect them to produce secondary metabolites.

Demain: Jack Baldwin asked us what we use as a definition. We are not confusing natural products with secondary metabolites. Of course, natural products are made up of primary metabolites, amino acids and what have you, as well as secondary metabolites.

Davies: So a secondary metabolite is a natural product, but a natural product isn't necessarily a secondary metabolite.

Demain: Yes; that's what I am saying.

Orgel: Wouldn't it be more practical to list what you believe to be primary metabolites and say that anything else is a secondary metabolite?

Bu'Lock: This used to be done for you; the headings in the subject index of the *Journal of Biological Chemistry* were the primary metabolites! It has become more complicated, because those of us who are interested in secondary metabolites started asking questions which the molecular biologists have just got around to being able to answer.

Cane: Where do you put vitamins?

Demain: They are primary metabolites because organisms require them for growth. Most forms of life make them; if they don't, the vitamins have to be provided in the diet.

Wächtershäuser: I would say that any demarcation has a grey zone; we cannot avoid this. We have to classify, yet all classifications have grey zones, and we have to accept that.

Williams: The definition that best avoids a grey zone is the one that Arny Demain advocates, that a secondary metabolite is a substance whereby, if you remove the genes for its production, the organism can continue to survive. In other words, it has no role in the *internal* economy of the producer. So, on the basis of this definition, chlorophyll and haemoglobin, say, are primary metabolites rather than secondary metabolites.

You can still have situations which some regard as grey areas. Consider lanosterol, biosynthesized by sheep, and present in copious quantities in the wool; clearly, if you removed the genes for its production, since it is an intermediate in the biosynthesis of cholesterol, the sheep would die. On the other hand, lanosterol has a role that would normally be considered to be also that of a secondary metabolite; it is excreted by the organism, and gives a selective advantage by ensuring that water doesn't remain on the back of the producer. Nevertheless, within the above definition, lanosterol is a primary metabolite; but it is also performing a function that we usually associate with secondary metabolites.

Orgel: Are there any organic compounds elaborated by organisms which are so to speak in limbo; they don't make them either as primary or as secondary metabolites?

Demain: This is the 'playground' of Professor Zähner (1979), who says that there have been many such compounds, involved in evolution, but, once in a while, one of them is incorporated as a useful substance, and an organism suddenly depends on them for survival in the environment.

Zähner: Yes, that is our view. We published this hypothesis in 1982 and 1983 (see Zähner et al 1982, 1983).

Orgel: Is there agreement that certain compounds are definitely neither primary nor secondary metabolites?

Bu'Lock: Yes. I think the function of the carotenoids in photosynthesizing cells is pretty well defined in terms of photosynthetic mechanisms, but not the carotenoid that you get in *Dunaliella*, which is way over the top. The function of citric acid in the Krebs cycle and in carbon flow is entirely well defined. But the function of the accumulation of citric acid that you can get from growing the right strain of *Aspergillus niger* in the correct medium and in the right conditions is something else. No organism needs that much.

Demain: I don't think you can use over-production as part of the definition of secondary metabolism, because many primary metabolites, such as glutamic acid and riboflavin, are over-produced by industrial strains.

Bu'Lock: Over-production is Vanék's definition of what we are calling secondary metabolism (Krumphanzl et al 1982), and the concept owes much to Foster's ideas about 'overflow' metabolism and to practical experience in industrial microbiology.

References

Baldwin JE, Bradley M, Abbott SD, Adlington RM 1991 New penicillins from isopenicillin N synthase. Tetrahedron 47:5309–5328

Hawkins AR, Lamb HK, Smith M, Keyte JW, Roberts CF 1988 Molecular organisation of the quinic acid utilization (QUT) gene cluster in *Aspergillus nidulans*. Mol & Gen Genet 214:224–231

Hull EP, Green PM, Arst HN Jr, Scazzochio C 1989 Cloning and physical characterization of the L-proline catabolism gene cluster of *Aspergillus nidulans*. Mol Microbiol 3:553–559

Krumphanzl V, Sikyta B, Vanék Z (eds) 1982 Overproduction of microbial products. Academic Press, London & New York

Kutchan TM, Dittrich H, Bracher D, Zenk MH 1991 Enzymology and molecular biology of alkaloid biosynthesis. Tetrahedron 47:5945–5954

Yoshida M, Konomi T, Kohsaka M et al 1978 Cell-free ring expansion of penicillin N to deacetoxycephalosporin C by *Cephalosporium acremonium* CW-19 and its mutants. Proc Natl Acad Sci USA 75:6253–6257

Zähner H 1979 What are secondary metabolites? Folia Microbiol 24:435–443

Zähner H, Drautz H, Weber W 1982 Novel approaches to metabolite screening. In: Bu'Lock JD, Nisbet LJ, Winstanley DJ (eds) Bioactive microbial products, vol 1: Search and discovery. Academic Press, London & New York, p 51–70

Zähner H, Anke H, Anke T 1983 Evolution and secondary pathways. In: Bennett JW, Ciegler A (eds) Secondary metabolism and differentiation in fungi. Marcel Dekker, New York, p 153–171

Evolution of secondary metabolite production: potential roles for antibiotics as prebiotic effectors of catalytic RNA reactions

J. Davies*, U. von Ahsen**, H. Wank** and R. Schroeder**

*Department of Microbiology, University of British Columbia, 300—6174 University Blvd, Vancouver, BC, Canada V6T 1Z3 and **Institute for Microbiology and Genetics, University of Vienna, Althanstrasse 14, A-1090 Vienna, Austria

Abstract. It has been proposed that organic molecules related to known secondary metabolites have existed since the beginning of biochemical evolution and were present in primordial soups. Under primitive earth conditions certain of these molecules may have played roles as effectors in prebiotic reactions, especially those involving catalytic RNA (ribozymes). We demonstrate that a number of antibiotic-related secondary metabolites bind to group I introns and either inhibit splicing reactions or promote the formation of intron oligomers. This is consistent with the functional co-evolution of catalytic RNA and secondary metabolites as antibiotic inhibitors of translation, and supports the notion of an evolutionary relationship between group I introns and ribosomal RNA.

1992 Secondary metabolites: their function and evolution. Wiley, Chichester (Ciba Foundation Symposium 171) p 24–44

Investigations of the origins of life, especially with respect to the evolution of the mechanisms of replication, transcription and translation, have generally focused on the formation of macromolecular polymers from the components of 'primordial soup' mixtures or of extraterrestrial objects. Analyses of these mixtures tend to emphasize the presence of the protein amino acids or nucleic acid bases, ignoring the large number of other organic molecules present. Such studies have, we believe, underestimated the potential roles of other biologically active molecules which were likely to have been present in great variety, and even quantity, during biochemical evolution. Living organisms produce tens of thousands of so-called secondary metabolites: these are usually the products of the modification of intermediary metabolism (condensations, polymerizations, methylations, etc.) during late phases of the growth of the producing organism. Speculation on possible biological roles of secondary metabolites has been rife (see Introduction, p 1), but only in a few cases have plausible functions been identified. Among secondary metabolites a small number have been shown to have antibiotic activity; this property has been extraordinarily valuable in combatting and controlling a wide variety of infectious diseases. Consequently, secondary metabolites identified as antibiotics have been considered to play such a role in Nature—for example, by providing a selective growth advantage for a living organism to compete against other organisms in particular environments. Thus there is a widely held notion that all secondary metabolites with antibiotic activity play this role in Nature; in fact, this property was found only as a result of screening by pharmaceutical companies under laboratory conditions. Although positive antagonism has been demonstrated in a number of instances, it is unclear if this interpretation can be generally supported.

It has been proposed that molecules related to secondary metabolites (the 'forgotten' components of primordial soups) played important roles in biochemical evolution as modulators or effectors, enhancing or controlling the biological activities of primitive macromolecules (Davies 1990). For example, the recognized antibiotic inhibitors of ribosome function may have been effectors of a primitive translation system or of some other reaction involving RNA catalysis. We shall show that there are properties of some secondary metabolite inhibitors and their interactions with ribosomes which are consistent with this proposal.

However, to begin with, is it possible that molecules related to the organic compounds that we define as antibiotics were present during early stages of biochemical evolution? It is difficult to obtain evidence for this because there are no fossil records at the molecular level. Support for the existence of amino acids and nucleic acid bases in the primitive earth atmosphere has come from their identification as components of 'primordial soup' reactions (which simulate the reaction components believed to have been present on the primitive earth) and in chemical analyses of the constituents of meteorites (Miller 1987). Do these same reactions produce compounds related to known secondary metabolites? While most analyses have concentrated on the detection of protein amino acids, a number of other amino acids have been detected in simulated primeval earth reactions (Miller 1987) and, interestingly, many are known to be components of biologically active secondary metabolites, some of which have antibiotic activity (Table 1).

Our interest has been the evolution of protein synthesis, in particular the generation of the ribosome, and the possible role of secondary metabolites as mediators in this process. As Orgel (1987) has pointed out, this is a paradigm of the chicken-and-egg (or protein-and-nucleic acid) dilemma. The involvement of secondary metabolites allows one to make the following proposal for the evolution of a mechanism for ribosome-catalysed peptide bond formation in which low molecular mass effectors, and subsequently polypeptides, were used to modulate the activity and specificity of the ribozyme (the catalytic RNA core of the primordial ribosome structure).

TABLE 1 Amino acid components[a] of secondary metabolites found in primordial sources[b]

Amino acid	*Product*
Alanine	Gramicidin
β-Alanine	Destruxin
α-Amino-*n*-butyric acid	Lysergic acid
α-Amino-isobutyric acid	Alamethicin
β-Amino-*n*-butyric acid	Brevistin
γ-Amino-butyric acid	Subtilin
Diamino-butyric acid	Polymixin
α,β-Diamino-propionic acid	Edeine, viomycin
Glutamic acid	Bacitracin
Glycine	Nisin
α-Hydroxy-γ-amino-butyric acid	Butirosin
Isoserine	Edeine
***N*-Methylalanine**	Verticillin, tentoxine
Norvaline	Cyclosporin
Pipecolic acid	Actinomycin, ostreogrycin
Proline	Tyrocidin
Sarcosine	Cyclosporin, actinomycin
Valine	Gramicidin

[a]Non-protein amino acids are indicated in **bold** type.
[b]Detected in electric discharge reactions or in meteorites.

1. 'Primordial soup' components, or organic molecules from extraterrestrial objects, contain amino acids that could form simple oligopeptides by random chemical condensations.
2. RNA is produced, and an early form of ribozyme provides the catalytic function necessary for more effective peptide bond formation.
3. Low molecular mass components of the 'primordial soup', or their condensation products, bind to RNA and promote conformational changes that reduce activation energy or increase the specificity of peptide bond formation. (Alternatively, these molecules could have provided the means of initiating or terminating peptide chains.) These effectors can be considered as functional forerunners of ribosomal proteins. The use of stable, low molecular mass effectors in primordial reactions would be one way of overcoming the problem of the extreme conditions of the primeval environment (Pace 1991).
4. As larger polypeptides are produced, certain of them remain associated with the ribosomal ribozyme, which permits more effective catalysis of

peptide bond formation (ribosomal proteins are believed to be essentially involved in stabilization of the structure of the ribosome: Noller & Woese 1981). The RNA-binding peptides are assumed to be more efficient than the low molecular size effectors and displace the latter, generating a pro-ribosome.

5. The ribosome evolves to become a macromolecular RNA–protein complex which has improved capabilities for translation and for the production of larger polypeptides. The low molecular weight effectors, although displaced, still retain specific binding sites within the ribozyme. These sites are determined by RNA sequence and conformation.
6. The re-binding of these small effectors to the rRNA may interfere with the function of ribosomal proteins in maintaining active rRNA conformations, and thus the effectors become inhibitors of translation on the ribosome. The inhibitors bind not to protein but to RNA (ribozyme) in the macromolecular structures. Other chemical constituents of the 'soup' might bind to the primitive rRNA and exert different effects—for instance, stimulation of various steps of translation, or improvement of the accuracy of the process.

A number of different antibiotics which can inhibit translation, such as the aminoglycosides, interact with ribosomes by directly binding to specific RNA conformations, as demonstrated by chemical protection studies of ribosomal RNA or by analyses of resistance mechanisms in their producing microorganisms (reviewed by Noller 1991). We suggested (Davies 1990) that these secondary metabolites were one such class of RNA-catalysis effectors involved in translation and in other ribozyme-catalysed reactions at this early stage of biochemical evolution. While these molecules have pleiotropic (inhibitory and stimulatory) effects on cell metabolism, it is nonetheless generally considered that their primary mode of action is to affect translation at the ribosome level (Cundliffe 1981). One test of the hypothesis that certain antibiotics (or structurally similar molecules) played roles as effectors in prebiotic catalytic RNA reactions would be to demonstrate that translation inhibitors acted on the self-splicing of group I introns. In the first experiments, streptomycin and several analogues were shown to inhibit splicing at relatively high concentrations; on further investigation this inhibition was shown to be due to the presence of guanidino groups in these molecules (von Ahsen & Schroeder 1991). Guanine is required as the promoter of the self-splicing of group I introns, and competitive inhibition of guanine binding by guanosine analogues or a molecule containing a guanidino group, such as arginine, leads to a reduction in the formation of the products of the first step of splicing (Cech 1990). Kinetic studies of inhibition by streptomycin showed that it too was a competitive inhibitor of guanine's binding to its intron site (von Ahsen & Schroeder 1991).

In contrast, a number of aminoglycosides of the 2-deoxystreptamine (DOS) class (Fig. 1), such as gentamicin, tobramycin or neomycin, specifically inhibited

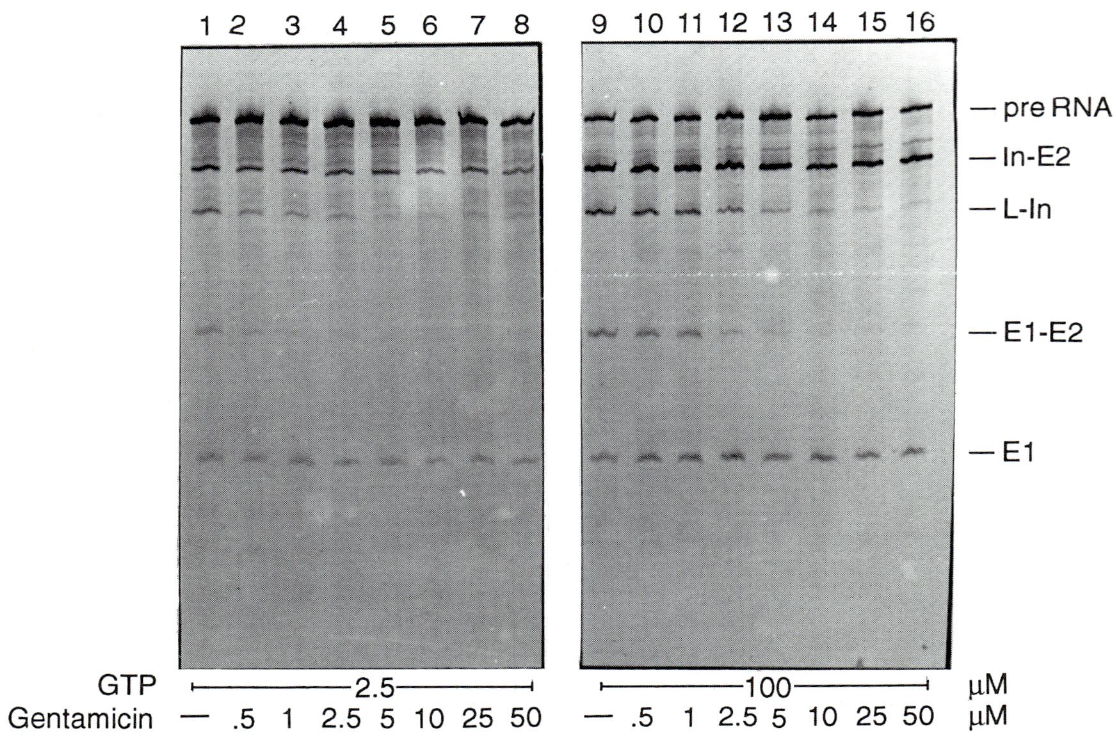

FIG. 1. The effect of the aminoglycoside gentamicin on the self-splicing of bacteriophage T4 *td* intron. Incubation was for 10 min at 37 °C at the antibiotic concentrations (μM) indicated. The reaction products were separated by electrophoresis and detected by autoradiography. For details, see von Ahsen et al (1991). preRNA, unspliced precursor RNA; In-E2, *td* intron–exon2 fragment; L-In, linear intron; E1–E2, spliced exon product; E1, exon 1. (Reprinted by permission from *Nature*, volume 353, p 368–370. Copyright © 1991 Macmillan Magazines Ltd.)

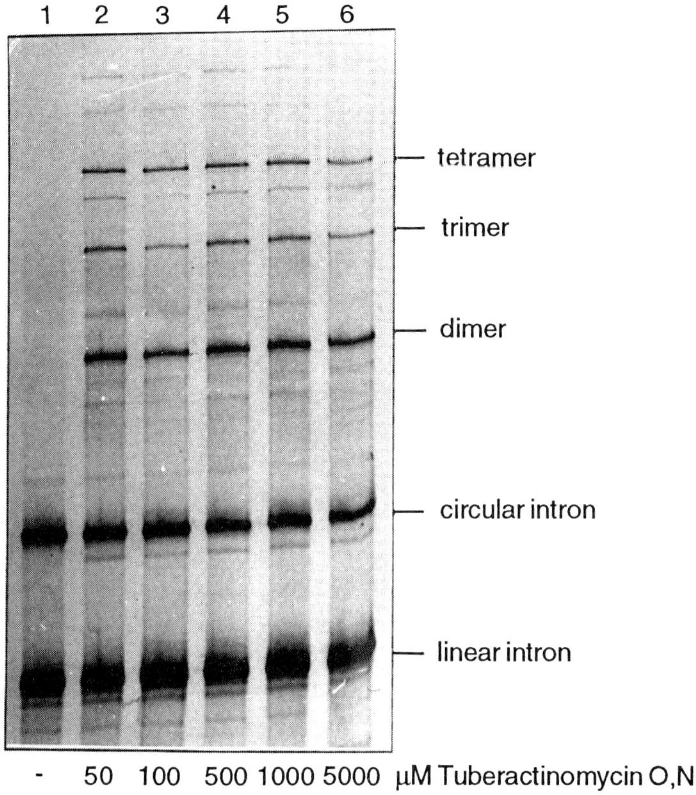

FIG. 2. Oligomerization of the phage T4 *td* intron in the presence of the peptide antibiotic tuberactinomycin. Linear intron from bacteriophage *td* was isolated by electrophoresis after *in vitro* transcription and splicing. The isolated intron RNA was incubated in Tris-$MgCl_2$-spermidine buffer for 10 min at 37° C with increasing concentrations of tuberactinomycin O,N. Reaction products were separated by electrophoresis on 5% polyacrylamide/7M urea and detected by autoradiography.

the second step of splicing in a reaction that was not competitive with guanine binding (von Ahsen et al 1991). Kinetic analyses confirmed this result; the DOS aminoglycosides must bind to the intron RNA at a site different from the guanine site (von Ahsen et al 1992). Having demonstrated the activity of the translation-inhibiting aminoglycosides as mediators of RNA-splicing reactions, we went on to ask if this was specific to this class of translation inhibitors. In the course of our tests we examined a number of other secondary metabolites, including such potent translation inhibitors as the macrolides or chloramphenicol: most had no detectable effect on either group I or group II splicing reactions (von Ahsen et al 1991). The fact that the inhibitors known to act on the large ribosomal subunit RNA were not active in inhibiting splicing reactions would be consistent with the existence of a functional and structural relationship between group I introns and 16S RNA. On the basis of the common activity of the aminoglycosides towards the two different RNAs, it has recently been proposed that the group I introns are evolutionarily related to the 16S RNA ribozyme (Schroeder 1992).

In further studies with other secondary metabolites, an interesting and unexpected result was obtained with the translation inhibitors belonging to the capreomycin/tuberactinomycin/viomycin group. These are closely related cyclic pentapeptides, consisting of four L-amino acids (two serine residues) and one unsaturated amino acid. These compounds have long been known to be highly potent inhibitors of translation at the ribosome level; they interact with both ribosomal subunits, although their sites of interaction within rRNA have not been identified (Cundliffe 1981). When tested for its effects on intron splicing *in vitro*, viomycin was found to be an inhibitor of the reaction, albeit less active than an aminoglycoside such as tobramycin. By contrast, at lower concentrations these peptide antibiotics, in particular tuberactinomycin, promote the formation of intron oligomers (H. Wank, U. von Ahsen, J. Davies & R. Schroeder, unpublished observations) (Fig. 2). The interaction of these small (primitive?) peptides with the ribozyme results in a reaction which is formally similar to RNA ligation. The peptide antibiotics may thus be considered as effectors of ligation systems and, like the secondary metabolites that influence splicing reactions, could have played roles in the biochemical evolution of nucleic acid structures. The joining of short RNA fragments, each with different roles or functions, to form multifunctional RNA molecules (for example, group I introns and ribosomal RNAs) might have occurred in the presence of secondary metabolites of the tuberactinomycin type formed under 'primordial soup' conditions.

One conclusion of these findings is that although investigations of antibiotic action are biochemically and intellectually interesting, the mode of action of pharmacological significance (e.g. how an antibiotic brings about bacteriostatic or bactericidal action in a test-tube) need not have any bearing on the true biological function of these compounds in Nature. Most, if not all, antibiotic inhibitors have pleiotropic effects; which of their effects are biologically significant, and in what circumstances? The mechanisms responsible for therapeutic and toxic effects might be quite distinct. For example, the aminoglycosides have well-recognized toxic side effects. Might these be due to interference by the antibiotic at the level of splicing reactions in mammalian cells?

A related question concerns the nature of the interactions between ribozymes and small molecules. We believe the latter to be effectors of ribozyme activity and, under some circumstances, substrates. RNA–protein interactions are vital to many aspects of pro- and eukaryotic cell function, being involved in the formation of macromolecular structures, gene regulation, transport of proteins and nucleic acids across membranes, and RNA processing, for example. The interaction of intron RNA with peptide antibiotics, as exemplified by our studies with tuberactinomycin, may be an evolutionary forerunner of some of these reactions and may thus provide a simple model for their study. The three-dimensional structures of a number of peptide antibiotics are known and could provide the basis for further studies of nucleic acid interactions. And what of their binding sites within ribozymes? The group I introns are mostly inserted within conserved regions of rRNA genes and can adopt complex conformations that extend over 400 bases or more in length (Cech 1990); the guanosine and arginine binding sites have been identified and site-directed mutagenesis studies have confirmed the overall structure of intron RNA. It will be important to isolate and study antibiotic-resistant mutations of the self-splicing introns, which will require the establishment of assays for antibiotic activity against catalytic RNA *in vivo*.

Finally, with respect to the biological roles of secondary metabolites, we submit that their functions and evolutionary significance should be re-evaluated. Antibiotic-type structures (or close analogues thereof), if present at early stages of biochemical evolution, could have played roles in the modulation of primitive biosynthetic reactions—in particular, for the cases we have examined, RNA-catalysed reactions. Like ribosomal proteins, secondary metabolites are 'facilitating' molecules and are not likely to be 'function-defining'; the existence of antibiotic-dependent mutants of bacteria (for example, streptomycin-resistant mutants) shows that antibiotics are effective in such a role. The biological functions of secondary metabolites in present-day reactions are likely to be some manifestation of their evolutionary function. The aminoglycosides and certain cyclic peptides are good models for investigation and additional examples are likely to be found when other secondary metabolites are examined under the appropriate conditions.

The evolution of the biosynthesis of secondary metabolites remains an enigma. We can imagine, as with amino acids and similar molecules present in primordial soups, that the biosynthetic pathways evolved because the end product was required for or conferred an advantage in some aspect of early cell physiology. A large number of functions for secondary metabolites are considered possible (see Table 1, p 1) and could have provided the necessary selection pressure; it is assumed that the biosynthetic pathways leading to secondary metabolites evolved later than pathways leading to their biosynthetic precursors: amino acids, sugars, and nucleic acid bases. We suggest that some secondary metabolites were present under prebiotic conditions on earth and, indeed, were important components of early 'biochemical' reactions; do simple oligopeptides related to secondary metabolites (tuberactinomycin-like, for example) exist in simulated 'soup' reactions? We have a lot of work to do in order to understand the evolutionary and modern history of secondary metabolism. Its end products are not simply substances provided for pharmaceutical screening; some may be ancient molecules that played key roles in the evolution of cell chemistry.

Acknowledgements

We wish to thank Austrian Fonds AF zur Förderung der Wissenschaft und Forschung for support and are extremely grateful to Dr H. Fukukawa for providing tuberactinomycin.

References

Cech TR 1990 Self-splicing of group I introns. Annu Rev Biochem 59:543–568
Davies J 1990 What are antibiotics? Archaic functions for modern activities. Mol Microbiol 4:1227–1232
Cundliffe E 1981 Antibiotic inhibitors of ribosome function. In: Gale EF, Cundliffe E, Reynolds PE, Richmond MH, Waring MJ (eds) The molecular basis of antibiotic action, 2nd edn. Wiley, Chichester, p 402–457
Miller SL 1987 Which organic compounds could have occurred on the prebiotic earth? Cold Spring Harbor Symp Quant Biol 52:17–27
Noller HF 1991 Ribosomal RNA and translation. Annu Rev Biochem 60:191–227
Noller HF, Woese CR 1981 Secondary structure of 16S ribosomal RNA. Science (Wash DC) 212:403–411
Orgel LE 1987 Evolution of the genetic apparatus: a review. Cold Spring Harbor Symp Quant Biol 52:9–16
Pace NR 1991 Origin of life—facing up to the physical setting. Cell 65:531–533
Schroeder R 1992 From ribozyme to ribosome: an antibiotic-mediated evolution. J Mol Evol, submitted
von Ahsen U, Schroeder R 1991 Streptomycin inhibits splicing of group I introns by competition with the guanosine substrate. Nucleic Acids Res 19:2261–2265
von Ahsen U, Davies J, Schroeder R 1991 Antibiotic inhibition of group I ribozyme function. Nature (Lond) 353:368–370
von Ahsen U, Davies J, Schroeder R 1992 Non-competitive inhibition of group I intron RNA self-splicing by aminoglycoside antibiotics. J Mol Biol 226:935–942

DISCUSSION

Turner: Thinking of your model for the evolution of the ribosome, do you see a parallel with the evolution of the spliceosome? You chose group I introns to study. What effects might these aminoglycosides have on the splicing of normal eukaryotic introns?

Davies: We chose to study group I introns first because they were easy to work with. We have also worked on group II introns, and some of the antibiotic inhibitors of ribosome function, such as tetracycline, do affect the splicing of group II introns. The aminoglycosides do not seem to work on group II introns; they influence the self-splicing of group I introns only.

Demain: What is the difference between the two types of introns?

Davies: The difference is in the way in which the reaction takes place in the final product. In the group II pathway of splicing, the intron that is cut out is in a lariat rather than linear form.

Turner: It is also a question of whether a protein is needed or not for splicing?

Davies: Proteins are probably needed for splicing in all cases. Even though group I introns are self-splicing *in vitro*, inside the cell they probably require accessory proteins. So both groups of introns require proteins for splicing reactions in the cell.

Turner: I was thinking of your model of a primitive ribosome, since some of the introns will splice out *in vitro* without any proteins, and your model for an 'early' RNA involved in the primitive ribosome was *not* associated with proteins; you had a low molecular mass compound and then later added in proteins, as a parallel with the development of splicing reactions.

Davies: There seems little doubt that there is a relationship between the origin of ribozymes and the formation of ribosomes. Harry Noller (personal communication) has extracted RNA from the 50S subunit of the ribosomes of a thermophilic bacterium and shown that this 23S RNA molecule catalyses the formation of peptide bonds. He demonstrated inhibition of this reaction by several antibiotics, known to inhibit peptide bond formation. It would be interesting to ask if one can get ribozymes to make peptide bonds.

Orgel: There are now experiments on that. Tom Cech has shown that a ribozyme can facilitate a reaction related to the synthesis of a peptide bond, namely the hydrolysis of an amino acid ester of an oligonucleotide. It's a relatively small effect, about a five-fold acceleration, over the uncatalysed reaction. He thinks that it will be possible to use an RNA molecule derived from a type I ribozyme to facilitate the formation of peptide bonds (Piccirilli et al 1992).

Davies: That's very nice; so now there is a relationship between the ribozyme, ribosomal RNA and antibiotic inhibitors for these functions.

Cundliffe: I was interested in your suggestion that as the ribosome evolved, the acquisition of proteins has displaced low molecular weight effectors. We have some (unpublished) observations on the binding of the antibiotic, thiostrepton (produced by *Streptomyces azureus*), to 23S ribosomal RNA. Thiostrepton binds to isolated 23S RNA in free solution about as tightly as most antibiotics bind to intact ribosomes. The dissociation constant for the binding, measured by equilibrium dialysis, is about 10^{-7} M. We can isolate from 23S RNA, which is about 3000 nucleotides long, an oligonucleotide 60 nucleotides long (2% of the RNA). This oligonucleotide binds thiostrepton about as well as does the intact RNA molecule. Within that oligonucleotide is the site of methylation that confers resistance to thiostrepton.

The other point we have shown is that thiostrepton does not bind to yeast ribosomes. We have taken from yeast the DNA that codes for the crucial piece of 23S-like ribosomal RNA and inserted it into an *E. coli* ribosomal RNA gene. What you then get, assembled inside the bacterial cell, are functional ribosomes that synthesize protein. They have the *E. coli* sequence for 23S RNA, with a 60mer or so sequence derived from yeast in the middle. The point is that although such ribosomes do not bind thiostrepton, the isolated RNA binds thiostrepton perfectly well, just like *E. coli* RNA. What has happened in this eukaryotic system (i.e. in yeast) is that the proteins have evolved to displace the antibiotic, which now can't even find the binding site on this mature ribosome, whereas the drug can still find a binding site in the prokaryotic ribosome. Although you didn't use this language, I see it as going part of the way towards substantiating your hypothesis that proteins have displaced these small effectors.

Davies: Thiostrepton binds very well to RNA, yet we have not found it to inhibit any splicing reaction. The antibiotic was tested in a number of splicing reactions. Clearly, we haven't done the right experiment.

Cavalier-Smith: One could interpret this in the opposite way, that the antibiotic evolved to damage protein synthesis by binding to 23S rRNA, and then some of the ribosomal proteins evolved to protect the ribosome from these antibiotics. That would be the simplest explanation of these facts.

Orgel: It would be an example of protein 'footprinting', to protect a region of the nucleic acid by binding of a protein to it.

Davies: That seems a very complicated way of going about the evolution of ribosomal proteins.

Cavalier-Smith: It's very simple; if antagonists evolve, you need protection against them; you would expect that to happen.

Davies: You are suggesting that there is a ribozyme, pure RNA, that is inhibited by a variety of low molecular weight compounds, and then proteins evolve and protect it. Don't think that the proteins evolved to enhance the activity of ribosomal RNA?

Cavalier-Smith: I imagine that ribosomal proteins have a variety of functions, but some of those functions could be to protect against antibiotics.

Davies: But the ribosomal proteins don't protect against antibiotics.

Chater: They do protect against *some* antibiotics, and those are the ones you no longer see as being antibiotics!

Davies: That may be true, but no ribosomal protein has been shown to bind to an antibiotic.

Wächtershäuser: Proteins are the products of ribosomes, and the ribosome has in its structure its own products; this means that those products have a feedback effect; they are autocatalytic for their own machinery. This theme recurs in many areas of metabolism. The simplest explanation is therefore that the ribosome evolved by using its own products to improve its function. Then what Professor Davies is saying is that there was a stage in evolution where protein synthesis hadn't reached its present level of sophistication; perhaps not all the amino acids were in place. Then, more complicated secondary products were made (from, say, peptides) that were effectors for the ribosome, and when the translation machinery was perfected and more complicated proteins could be made, these effectors were replaced by proteins. But your argument is of course totally independent of any theory on the origin of life.

Davies: Yes. I don't care (as it were) where the RNA molecules came from in the first place. I am interested in the evolutionary stages which led to the role of the RNA and the ribosome in protein synthesis.

Piepersberg: Do you know what happens if you take random RNA or DNA molecules and try to see whether they have ligand-binding functions for a random organic molecule such as the aminoglycosides you used, like certain dyes in the experiments of Ellington & Szostak (1990, 1992), where you find very similar secondary or tertiary structures of molecules unrelated in their primary structure? Wouldn't this say that you have just picked out at random aminoglycoside-binding RNA tertiary structures which are similar to many others? If in turn you take many different ligands, and find binding of one or the other to the same RNA molecule, then this is just a random effect, rather than that the self-splicing RNA molecules are somehow primordial aminoglycoside-binding ribozymes which have a specific similar function. Also, if you pick out RNA molecules of similar tertiary structure, would you say that these could all have a similar ribozyme or other catalytic function?

Davies: Not necessarily. I think a very specific structure is needed.

Piepersberg: However, this is the opposite way to that in the experiment that you did. You took a series of ligands, and picked out one which could bind to that intron-splicing structure. If you had taken the *same* ligand for random RNA molecules, you probably would have found ones similar in tertiary structure to which that ligand would have bound. And what you consequently would say is that all these must be self-splicing ribozyme-like structures.

Davies: I would have to demonstrate that they have this activity. In fact, that experiment is being done, to ask whether one can, by selection, find RNA molecules that will bind to the low molecular weight inhibitors, and to ask whether these molecules have any function.

Cane: I would like more information on the inhibition experiments. You have shown that increasing amounts of antibiotic prevent one or the other portion of the splicing reaction. Do you know anything about the type of inhibition? That is, do you know whether the antibiotic interacts with the same local sequence that guanosine is interacting with, or is it a non-competitive inhibition which can be overriden? This would speak to the issue of whether it's a specific interaction or a general interaction with the polymer, which might have a serious enough effect on its conformation. Do you know anything about the multiplicity of binding of inhibitor to target?

Davies: These are all the questions to which I would like to know the answers! In terms of the 2-deoxystreptamine-based compounds, and their inhibition of the splicing of group I introns, we cannot demonstrate competitive inhibition with any other molecule, including guanosine. Some of the 2-DOS molecules are good inhibitors and others work very poorly. In terms of their ability to bind to RNA in ribosomes, they are all much the same, but, at the ribozyme level, some are clearly reacting better than others. We don't know where they bind at the moment, and we would like to try to do binding experiments between antibiotics and introns by covalent attachment, to see where they attach, and whether the antibiotics bind at one or many sites. I can't tell you anything about the stoichiometry; I would hope it's 1:1.

Vining: The idea of low molecule mass effectors improving the efficiency of protein synthesis is very attractive, as is your inference that secondary metabolism is an ancient activity. However, there is a problem in suggesting that the effector role of these substances was a 'temporary' one. Once proteins began to displace the low molecular mass effectors from the ribosome, there was no longer a need for them. When the selective advantage they conferred disappeared, why did they persist? Why would biosynthetic pathways evolve for substances that had become obsolete?

Davies: I think these molecules had other functions.

Vining: Do you suggest that they had other, adventitious functions, that conferred long-term advantages and ensured selection?

Davies: Yes. And these other functions, I would argue, evolved in some cases by different mechanisms, unrelated to the mechanism by which these molecules first evolved.

Chater: Aren't we avoiding the central problem that the selection, and anything which operates in terms of the biosynthetic pathways, is post-ribosomal; it all involves protein? So if secondary metabolites, or, rather, compounds that are chemically the same as them, occurred in the primordial soup and facilitated the development of ribosomes, that doesn't bear on the evolution of pathways for the biosynthesis of anti-ribosomal antibiotics that we now have?

Davies: They are completely independent; it's just a question of whether the low molecular weight compounds existed. The evolution of the biosynthetic pathways could have been much later.

Cavalier-Smith: If you are saying that the proteins that now make antibiotics evolved much later, and independently of the earlier evolutionary stages, you are just dealing with a coincidence; that is, proteins later evolved that by chance happen to make secondary metabolites that were similar to some of those present a lot earlier. This is just coincidence from the point of view of the origin of the proteins making the metabolites, and therefore implies no genetic or evolutionary continuity.

Wächtershäuser: Coincidence is the working of evolution! You can have either opportunism, or foresight; these are the two types of evolutionary theory. You have either what amounts to foresight, or what amounts to opportunism. That's all there is.

Bu'Lock: Francis Bacon spoke of idols in thought—the distortions which originate externally but which can determine a person's thinking. Some of us have a reverence for the Creator which ends up in the view that 'everything is purposeful'. But this is neoLamarckism at its worst; it is Platonic; it is fundamentally a religious approach, which drives us to keep asking the question of 'what is it all for?'.

Contrary to this is the Hegelian approach that everything is possible. There is also the lure of the market place, which means that the only things which are interesting are those we can sell! This is responsible for the confusion between the idea of secondary metabolites and of antibiotics.

There is also the mystical antiquarian approach, in this case the retreat into unknowable protobiology. In the protobiotic situation, the 'primordial soup', the key developments were those to do with the replication of the system as it had developed so far, maybe taking it a bit further. At that stage, the mechanisms of cell formation and cell growth by replication were evolving. Everybody accepts that the invention of the self-replicatory system is protobiology.

From the degree of similarity between all known terrestrial biotic systems (and I include the extraordinary ones like the archaebacteria) it is clear that this sort of system can only have evolved once. The archaebacteria indeed have very peculiar bases in their transfer RNA and so on, but basically they are using the same biochemistry as you and I. There is no room here for the multiple re-entry kind of targetting that is associated with secondary metabolism.

What you have done, Professor Davies, in the work you described, has illuminated a very interesting area of secondary metabolite action. And I accept that it is action on a system which has a very early evolutionary role. But I don't believe that because some metabolites interfere with something very primordial, they had to be there in order for that primordial system to develop. Other secondary metabolites interfere with some very sophisticated and advanced systems. So I don't think you can solve the problems by going back into an experiment that we really know nothing about, and cannot conduct.

Chater: If you can show effects on intron splicing *in vitro*, one would expect that those antibiotics should have significant activity *in vivo* against eukaryotic cells. Is that observed?

Davies: I agree with you that this inhibition should take place; in fact, a number of aminoglycosides are toxic to eukaryotic cells. We have additional experiments in progress.

Turner: To go back to my earlier point about the type of intron you studied, how widespread are group I introns? Mitochondria don't have introns at all in higher eukaryotes, for example.

Davies: No, but I think the experiments should be done in bacteria or fungi to begin with, and nobody has yet looked for splicing inhibition in higher eukaryotes.

Cavalier-Smith: Have you even tried it with *Tetrahymena* group I introns? It might not even be true for all group I introns.

Davies: We know there are many antibiotic compounds that block the growth of *Tetrahymena*; we can assume that they work at the level of protein synthesis, because the resistant mutants are altered in RNA sequence, and some of these, in fact, are in regions close to the intron–exon boundary. Although one can show effects at the ribosome level, it is not possible to eliminate the possibility that inhibition takes place at the level of intron splicing. The experiments still have to be done.

Demain: What about the specificity of the *in vitro* effects? How many compounds have been tested which do *not* normally act by reaction with RNA?

Davies: If you consider inhibitors like chloramphenicol, erythromycin or tetracycline, which also react with RNA, these molecules do not inhibit the self-splicing of group I introns, whereas the aminoglycosides do. We have not randomly tested a large number molecules which are not known to be inhibitors of ribosome function. Novobiocin and other inhibitors of DNA synthesis have been tested, but they did not inhibit splicing of group I introns.

Baldwin: Have you tried very simple things, like guanidines? All of these compounds have a common structural motif.

Davies: Guanidine is a weak inhibitor. The polyamines (such as spermidine) do not work in the group I self-splicing experiments that we have carried out.

Orgel: Arginine is a very powerful inhibitor?

Davies: Yes, but there is a specific binding site within the group I intron structure for arginine binding.

Baldwin: So this is not a very specific process?

Davies: The inhibition of self-splicing by aminoglycosides occurs by a mechanism that is different from the inhibition by arginine or by analogues of guanosine. However, with streptomycin, there is competition at the guanine/arginine site. The 2-deoxystreptamine compounds presumably act at a different site within the intron, because inhibition is not reversed by the addition of guanosine.

Demain: How about 2-deoxystreptamine itself?

Davies: It doesn't work.

Orgel: I noticed a considerable jump between the main part of your paper, the experiments, and the conclusions. The talk referred to the action of a small group of molecules already known to interact with polynucleotides, but your conclusions were very broad and global. Perhaps you should restrict the conclusions to those antibiotics which are known to interact with RNA in this way? I wouldn't think that strychnine, for example, should be considered on quite the same basis as an antibiotic inhibitor.

Secondly, in many of these evolutionary problems, you have to ask how much can be attributed to coincidence and how much to similarity by descent. You could ask whether the fact that an antibiotic works both on ribozymes and on ribosomes is because ribosomes and ribozymes are related by descent, or because the ribosomal RNA and the ribozyme have some common structural feature. Obviously, I don't know the answer, but it is an answerable question and one that could be solved within a short time: you make a column, attach your antibiotic to it, and select out a number of RNAs that bind.

Suppose that it turns out to be very difficult to make any RNA that binds to one of the aminoglycosides. Then you have to say that it probably isn't a coincidence, because there are very few structures that bind. One might then think that these antibiotics were primitive. The step in the argument that puzzles me is how you proceed from the idea that some antibiotics are primitive to the idea that they were effectors in 'the RNA world' (Joyce 1989).

Davies: The short answer is that I don't really know! However, I am glad you pointed out that I was being very global in my conclusions, because I think you *can* indeed apply this argument to other known inhibitors; perhaps not for strychnine, but for other inhibitors similar types of argument apply—although not for all known secondary metabolites, of course.

Orgel: What you are saying seems to me to be interesting and important; there are many different sorts of molecules which people have chosen to call secondary metabolites, but the group that interacts with RNA has a special interest, if you believe in an 'RNA world'.

Davies: You don't have to believe in an RNA world, for my hypothesis. All you need to say is that in order to make a ribosome, you must have an RNA already. This doesn't have anything to do with the concept of an RNA world.

Orgel: You have to believe in ribozymes.

Davies: You can believe in ribozymes without believing in an 'RNA world'.

Cavalier-Smith: If you don't believe in 'an RNA world', your hypothesis makes no sense, because it says that the origin of antibiotics was pretranslational, affecting RNA; such a pretranslational world with RNA functions *is* the basic definition of an RNA world.

Davies: No; I am saying that RNA is the basis for protein synthesis.

Cavalier-Smith: If you are not now claiming that antibiotics go back to before the evolution of translation, this is a very different hypothesis from what you earlier apparently suggested. So you seem to have two versions, one that accepts the RNA world and says that antibiotic functions preceded the evolution of translation, and another version that says that they evolved after translation.

Davies: I am saying that the hypothesis is independent of whether or not there was an RNA world.

Wächtershäuser: In the 'RNA world', there is a smoke screen thrown up. Formerly, there was a belief in a primitive soup that contained all kinds of things. All workers in the field believed in a soup of nucleotides, and they believe that the first organism was a living RNA molecule that was an extreme food specialist, feeding on activated nucleotides, and reproducing itself. When Gilbert introduced the term 'RNA world' in 1986, he defined it as an organism feeding on a nucleotide soup. Now the workers in the field have become embarrassed, and say that the term simply means that at some point in the evolution of cells or organisms, RNA became involved, and then the RNA 'world' started, but there is no reason to use that term for such an event, because there could be numerous worlds: protein, or sugar, or lipid worlds.

Orgel: There was a period in which most people believed that the first form of genetic organization was based on RNA replication with a preformed supply of nucleotides. But in 1986, Shapiro emphasized the difficulty of making nucleotides. There has been a shift of opinion, and many people now believe that while there was an 'RNA world', other forms of biological organization preceded it. But this all seems irrelevant to the present discussion, which refers to the period when the 'RNA world' was already working. How it got started, while interesting, is not relevant to the present hypothesis.

You haven't yet explained why you believe that there was ever need for 'effector' molecules. Surely, these antibiotic molecules could have been invented to do one of the things they do now, for example to prevent other organisms from splicing the producer's RNA; why don't you believe that?

Davies: I'm not saying I don't believe it; I am saying that low molecular weight molecules (such as peptide antibiotics) were likely to be present in primordial soups, and they could bind to RNA; this would lead to conformations which result in the stimulation of catalytic RNA reactions. I can give you many examples of low molecular weight (antibiotic) molecules stimulating protein biosynthesis. And if there are examples of these molecules stimulating peptide bond formation now, I would argue that their role at the very beginning of biochemical evolution, whenever that was, was also to stimulate catalytic RNA reactions, not to inhibit them. It was only after the ribosome was formed that the effectors acted as inhibitors.

Orgel: This seems to be a complicated hypothesis. A far simpler one is that the aminoglycosides have always done what they do now.

Williams: The antibiotics that you used, like gentamicin, have a lot of asymmetric centres in them, and are clearly today the products of reactions catalysed by many proteins and therefore, at an earlier stage, also ribosomes. Let us go back to their possible role as prebiotic effectors. As you point out, in some cases the amino acid constituents were possibly present in a prebiotic soup, but you have the problem not only of putting the molecules together spontaneously, but also of making them with the appropriate stereochemistry. Therefore, on a statistical basis one would say *a priori* that their concentration would be very small. Even if you give some kind of thermodynamic or kinetic preference to the desired stereochemistry, it is still perhaps a problem for your hypothesis.

Davies: It is a problem, but I have no sense of the concentrations of molecules that existed in prebiotic conditions. The concentrations of organic molecules believed to have been on the earth 3.7 billion years ago have been estimated, but I have no sense of what the concentration of a particular amino acid would have been, at this time.

Bu'Lock: Besides this problem of concentrations, there is also the problem of the relative concentrations. In a soup which contains all the amino acids, there are a lot of dipeptides, 26^2 already, so you have to divide the total available dipeptide concentration by that value; and by more if you include the unnatural amino acids.

Another problem is that the structure, chemistry and functionality of nucleic acids in general are dominated by binding interactions between the components, with base-pairing leading to ordered and meaningful structures. This is a process which is intrinsically liable to be disrupted by some component which is analogous to but not part of the normal systems. As soon as you begin to look at molecules that are liable to interact with RNA, like the various arginine derivatives, you start to arrive, with more or less specificity, at something which is intrinsically likely to interfere with RNA function, and you can find some very specific examples within that area. So I think that RNA binding of one sort or another is probably tremendously important as the target site for natural products which confer properties that can be selected for. But I don't think the fact that a large number of the natural products do interact with RNA is surprising, because interactions with RNA are very likely to be interfered with by small molecules.

Davies: I don't believe there are a large number of molecules that interact with RNA!

Orgel: Your hypothesis has one further interesting implication. The aminoglycosides are, to an organic chemist, far too complicated molecules to have arisen by the spontaneous association of ammonia, formaldehyde and whatever else was in the primitive soup. To make such complex molecules you need an organized biological system of some kind. If your hypothesis is correct, it says something interesting about the metabolism of the organisms at that point, namely that they were capable of making aminoglycosides, and therefore there was already an established mechanism of specific carbohydrate synthesis.

Cane: They certainly knew how to make ribose!

Orgel: That is my point; if they knew how to make ribose, they had to know how to make carbohydrates, and therefore to make aminoglycosides; if aminoglycosides were there then, it says a lot about the way the 'RNA world' was at the time.

Davies: Has anybody looked in primordial soups for a variety of saccharides?

Orgel: Well, let me give you an idea of the complexity of what you get in an unorganized system which hasn't gone through an evolutionary development. If you take a favourite molecule, formaldehyde, and polymerize it, you get about 30 known sugars and a large number of molecules you can't identify. Among the products, ribose is represented maybe at the level of 2%. If you looked at the disaccharides, they would be made up from all these monosaccharides. I think you could count on not finding more than 0.05% of any one of them. By the time you get to a trisaccharide with an amine here and a guinidine there, 10^{-8} would be an optimistic view of the yield you could get. So if these aminoglycosides were there early on, they weren't made as random structures out of the prebiotic soup; they were made as a consequence of the evolution of complex catalysts in the soup.

Davies: One doesn't have to say that the antibiotics that I have studied were the same molecules doing the job millions of years ago. You need only to say that they are related, perhaps evolutionary precursors, according to my hypothesis.

Orgel: But wouldn't the same arguments apply to any molecule of equivalent complexity?

Davies: I don't know.

Wächtershäuser: The way I understand your theory is that it is totally independent of the notion of the prebiotic broth. The prebiotic broth could well be a misconception; but your theory could still hold water. Life then arose *without* a prebiotic broth and made all its constituents by 'inventions'; somewhere along the line, at an early stage, it invented things like your effectors. Therefore, we should forgo references to a prebiotic broth, because they are not necessary to your theory.

Leadlay: I am very interested in the chemistry of the tetramerization of the intron. If I understood correctly, there is no energy input in the form of GTP, so what drives it? Is it a simple transesterification?

Davies: Yes.

Leadlay: Is it accurate?

Davies: We don't know yet. We can identify polymers, and we know that they are head to tail.

Leadlay: And is there a dimeric intermediate? Can you dimerize the dimers?

Davies: We haven't tried.

Orgel: Would it be fair to say that the peptide converts the ribozyme reactions from intramolecular to intermolecular reactions? Normally, the ribozyme attacks itself, but now the peptide holds two RNA molecules together so that one can attack the other?

Davies: That's correct, but I don't know anything more about the reactions. These experiments are very recent and we know very little about what is going on.

Baldwin: The proposal that you will make peptide bonds in the presence of the RNA in aqueous solution is very difficult to get to grips with from a chemist's point of view, because the ribosome itself has to have some provision of free energy, to drive the formation of the polyamide system in water; ATP is consumed. If you don't, in your theory of the evolution of the ribosome, provide for the energetics required for this, you are missing out on the most important thing. You can imagine in your picture something that begins to look structurally like a ribosome, but you are not dealing with the problem of how you actually make amide bonds in water grow into a reasonably large protein. Shouldn't energetics be at the root of any proposal?

Davies: Yes, and the energetics is in the RNA; it's in the transesterification reaction.

Baldwin: I am referring to the peptide bond formation.

Davies: I think that takes place in the same way.

Baldwin: But that's an equilibrium process; we will never get anywhere.

Davies: It doesn't have to be an equilibrium process, does it?

Baldwin: The equilibrium constant in water for the formation of an amide bond is not far short of one.

Davies: I am not sure that the transesterification is necessarily taking place in water.

Williams: If you don't put in ATP or some energy source, it is a problem in equilibrium thermodynamics and we have to say that polypeptides will not spontaneously accumulate to a large extent over the constituent amino acids in aqueous solution.

Wächtershäuser: The assumption of pre-existing RNA presupposes ATP and a complex metabolism. So, when you talk about RNA and effectors for RNA functions, you presuppose the evolution of a metabolism that has produced all that already.

Orgel: I think the natural hypothesis is that there was some simple form of activation which put the amino acids together.

Baldwin: The problem is that all these discussions take place on that assumption; that's added on at the end, to take care of ΔG, but it might be better to start from that and deal with the energetics first.

Davies: What does Tom Cech do to get peptide bonds forming on catalytic RNA?

Orgel: He puts the energy in to begin with, by forming the esters of the amino acids which are equivalent to the terminus of a loaded tRNA. But I don't see this as an objection. If you assume your energy source for making RNA, why shouldn't you use the same one for making protein? Or even a different source?

Baldwin: The energy source itself, if we are talking about chemical energy to drive a reaction in a certain direction, is not just, say, heat energy; it's a specific form of chemical energy, which means that that type of energy source determines what can be done with it. You cannot just say 'I am going to have a dehydrating agent to make anything I want'. You have to specify a dehydrating agent that will make phosphate bonds, and couple them; some will make anhydrides and some, amides. They don't all do the same. This should be very carefully analysed. You cannot just add 'energy' and then think you can deal with all the problems.

Wächtershäuser: Basically, two types of energy sources have been postulated: condensation energy sources, like ATP or polyphosphates; and redox energy sources. All living organisms use primarily redox energy sources, even the photosynthesizers; they boost electrons. As a primary energy source, condensation energy sources are unknown in the living world. So the essential question is: how is a redox energy source converted into a condensation energy source?

Williams: I don't think this is a problem. The source of non-equilibrium thermodynamics is initially capturing photons from the sun (if we exclude thermal vents). As you have just said, you thereby create a redox system, and you can use, gradually, the lowering of the energy of the electron, which was raised through the capturing of the photon, to synthesize ATP. Once you have ATP, you have your condensing source.

Wächtershäuser: How is that done?

Williams: That is done because along the pathway of photosynthesis, as the electron loses its energy, and as energy goes into smaller quanta, you can drive any reaction going from non-equilibrium towards an equilibrium state, by having a negative ΔG. I don't think there's a problem in principle in getting from one reaction type to the other.

References

Ellington DE, Szostak JW 1990 *In vitro* selection of RNA molecules that bind specific ligands. Nature (Lond) 346:818–822

Ellington DE, Szostak JW 1992 Selection *in vitro* of single-stranded DNA molecules that fold into specific ligand-binding structures. Nature (Lond) 355:850–852

Gilbert W 1986 The RNA world. Nature (Lond) 319:618

Joyce GF 1989 RNA evolution and the origins of life. Nature (Lond) 338:217–224

Piccirilli JA, McConnell TS, Zaug AJ, Noller HF, Cech TR 1992 Aminoacyl esterase activity of the *Tetrahymena* ribozyme. Science (Wash DC) 256:1420–1424

Shapiro R 1986 Origins. A skeptic's guide to the creation of life on earth. Summit Books, New York

Why are secondary metabolites biosynthesized? Sophistication in the inhibition of cell wall biosynthesis by vancomycin group antibiotics

Dudley H. Williams and Rachel A. Maplestone

University Chemical Laboratory, Lensfield Road, Cambridge CB2 1EW, UK

Abstract. The evidence that secondary metabolites serve sophisticated roles in the survival strategy of the producer is briefly reviewed. This evidence stems from the common involvement of tens of kilobases of DNA in the programming of their synthesis, of up to several tens of discrete enzymic conversions in their biosynthesis, and of the existence of sophisticated mechanisms in the producers for resistance against their physiological effects. It also stems from a study of the molecular basis for these physiological effects. The molecular basis for the antibacterial action of the vancomycin group antibiotics is presented, and demonstrates that essentially every portion of these molecules appears to be finely honed to promote efficient antibacterial action.

1992 Secondary metabolites: their function and evolution. Wiley, Chichester (Ciba Foundation Symposium 171) p 45–63

A paper from our laboratory has recently argued that secondary metabolites (defined as substances appearing to have no role in the internal economy of the organisms producing them) improve the evolutionary fitness of the producer specifically by effecting a physiological response in another organism (Williams et al 1989). These arguments support the view previously taken by Katz & Demain (1977) and Demain (1980, 1989). Haslam (1986) had summarized the six main hypotheses put forward (by others) to account for secondary metabolite production; on the basis of the arguments which we presented, and in line with Occam's razor (the principle that hypotheses should not be allowed to multiply unnecessarily), we rejected all the hypotheses other than the one indicated above.

A secondary metabolite could improve the evolutionary fitness of the producer either by having an adverse effect on a competitor, or by helping to engage a second organism in an activity which is advantageous to the producer. In either case, natural selection would be expected to work to elicit the desired physiological response at the minimum necessary cost to the producer. A common feature of such a minimum cost may be anticipated to be a highly specific molecular interaction between the secondary metabolite and a 'receptor' site in the receiving organism (albeit that in the case of an antagonistic interaction, the 'receptor' site in the second organism may be a 'reluctant receptor').Thus, we argued that secondary metabolites may frequently be found to bind to target sites in a second organism with a molecular complementarity reminiscent of that found in enzyme–substrate interactions (Williams et al 1989). This viewpoint seems eminently reasonable when it is remembered that even phenotypic features of organisms are shaped by long-standing interactions between two species (for example, between bees and flowers, hummingbirds and flowers, and butterflies and flowers; Harborne 1988). The argument for molecular complementarity was supported by detailing the specific molecular complementarity existing between the cell wall precursor mucopeptides terminating in di-*N*-acetyl-L-Lys-D-Ala-D-Ala (or *N*-acetyl-D-Ala-D-Ala) and antibiotics of the vancomycin group, and also by pointing to the remarkable complementarity between actinomycin D and the self-complementary tetranucleoside triphosphate d(A-G-C-T) (Reid et al 1983).

More recently, we have reinforced the viewpoint that secondary metabolites have evolved to serve a crucial survival function for the producer by using also the evidence of the great complexity of the genetic apparatus normally involved in their production (Maplestone et al 1992). The evidence, from the work of others and calling upon and following from the pioneering work of Hopwood (1988), is that secondary metabolite production can require coding by tens of kilobases of DNA. Thus, the biosynthesis of a complex secondary metabolite may involve several tens of discrete enzyme-catalysed reactions, with the accumulated molecular masses of the enzymes involved being well in excess of one million daltons. For example, the biosynthesis of the aglycone core of erythromycin requires six consecutive cycles of condensation and reduction, believed to involve a total of 28 discrete enzyme-catalysed reactions (Cortes et al 1990, Donadio et al 1991, Bevitt et al 1992). The final production of erythromycin requires numerous additional steps, including formation of the macrocyclic lactone, hydroxylations, and the biosynthesis of neutral and amino sugars and attachment of these sugars to the macrolide—in total, a formidably long and complex pathway.

For antibiotics, the importance of, and the selection pressures for, their production are supported by the clustering of the genes which encode the proteins on the biosynthetic pathway. Additionally, Davies (1986) has proposed that antibiotic resistance determinants orginate in the organisms which produce the antibiotics—thus, the producer protects itself against the potentially lethal effects of its own product. For example, the actinomycete *Saccharopolyspora erythraea* (formerly *Streptomyces erythraeus*), which produces the macrolide erythromycin mentioned above, must protect itself against the potentially lethal inhibition of protein synthesis that erythromycin can cause. In other susceptible bacteria, this inhibition of protein synthesis is achieved by binding of the antibiotic to their ribosomes. *S. erythraea* possesses a gene which codes for a protein (ribosomal RNA methylase) which is responsible for dimethylation of a single adenine base in its own ribosomal RNA (Dhillon et al 1989). This modification blocks the inhibition of protein synthesis by erythromycin in the producer. Further sophistication is indicated by the clustering of the genes for resistance, biosynthesis, and the regulation of biosynthesis. Plausible, if speculative, routes to gene clustering can be envisaged (see, for example, Stone & Williams 1992). Gene duplication followed by mutation can account for two steps which are chemically similar but operate with somewhat different substrates (such as in polyketide production), whereas fusion of two hitherto distinct pathways can obviously occur by the transposition and bringing together of the two genes involved. If the genes had been sited initially at distant positions on the chromosome, the production of the secondary metabolite would be favoured if the genes were moved closer together, because this would facilitate mechanisms for their simultaneous expression, and would increase the probability of them being passed on as a unit to subsequent generations or, via a plasmid, to other species. Such a clustering of genes would be selected for only if the secondary metabolite conveyed a selective advantage on the organism.

Here, we summarize some of the sophistication involved in the molecular basis of interaction of vancomycin group antibiotics with the above-mentioned cell wall mucopeptide precursors. This increasingly evident sophistication gives further support to the assertion that the structures of these antibiotics have been honed under selection pressures to bind to peptides which are part of cell walls. Also, we note, such is the simplicity and complementarity of the interactions involved, that they can be used to deduce the intrinsic binding energies associated with interactions between specific functional groups, which are useful in the understanding of biological molecular recognition (Gerhard et al 1992).

The vancomycin family of antibiotics

The vancomycin group of antibiotics is a family of glycopeptides, all members of which possess closely related chemical structures. These structures are characterized by a heptapeptide backbone which is conformationally restricted by cross-linking of the side chains of amino acid residues 2 and 4, 4 and 6, and 5 and 7. In some members, for example ristocetin A (**1**), the side chains of residues 1 and 3 are also cross-linked, giving rise to a structure in which the amide NH and CO groups of the amide backbone are all specifically restricted such that they reside on one or other side of the antibiotic. This conformation is reproduced in Fig. 1, which shows an exploded view of the complex that the antibiotic forms with the cell wall mucopeptide precursor analogue *N*-Ac-D-Ala-D-Ala; hydrogen bonds formed within the complex are

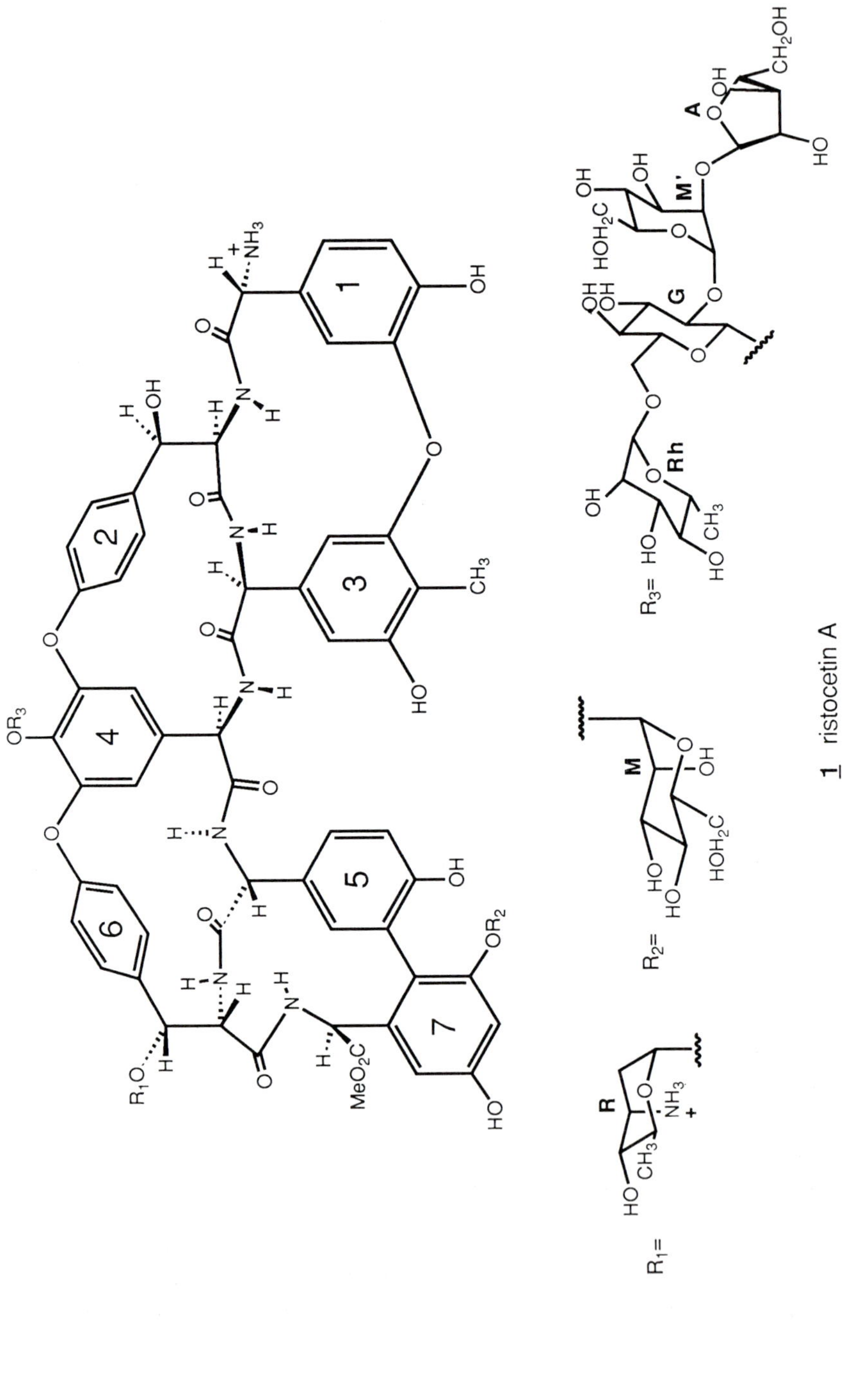

1 ristocetin A

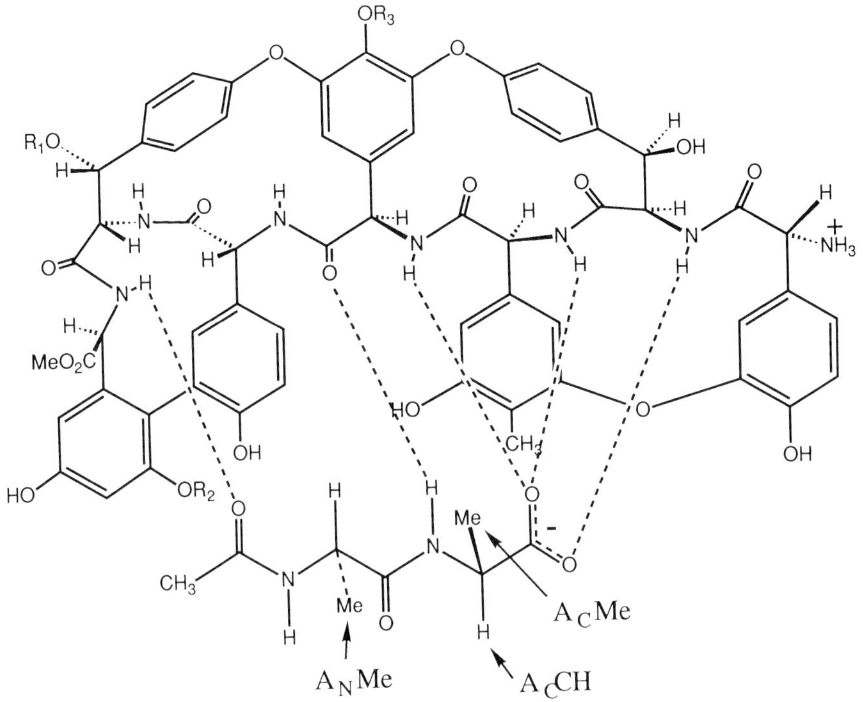

FIG. 1. Schematic representation of the complex formed between ristocetin A and a bacterial cell wall peptide model, *N*-Ac-D-alanyl-D-alanine. The broken lines indicate the positions of intermolecular hydrogen bonds.

indicated by dotted lines (Barna & Williams 1984). The binding constant for this interaction is approximately $10^5\,M^{-1}$ (Perkins 1982, Williamson et al 1984, Rodriguez-Tebar et al 1986).

Adaptations in the operation of the carboxylate anion-binding pocket

The amide backbone of ristocetin A is not in the form of a β-strand, in which adjacent C=O and NH groups alternate up and down, despite the fact that successive amino acid side chains alternate on opposite sides of the backbone. This absence of a β-strand backbone is at least in part due to the fact that the absolute configurations of the amino acids (from the N- to the C-terminus) are *RRSRRSS*. In particular, the combination of the cross-linking of residues 1 and 3 and 2 and 4 and the *RRSR* stereochemistries for the first four amino acids has been shown (Williams & Waltho 1988) to lead to the formation of a carboxylate anion-binding pocket in which all three amide NH groups of residues 2,3 and 4 are forced to point in the same direction (Fig. 1). Additionally, when

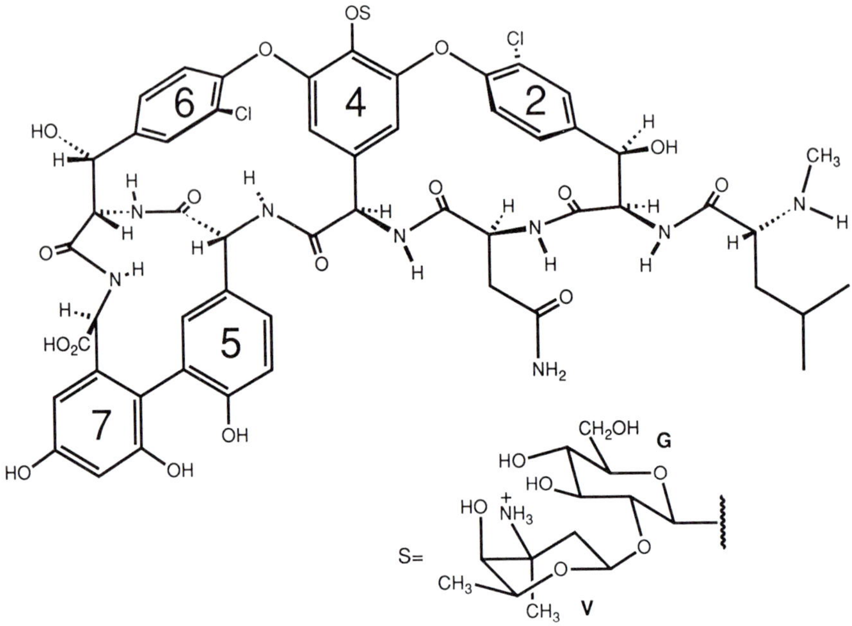

2 vancomycin

this carboxylate anion-binding pocket is not occupied by such an anion, the side chain of amino acid residue 1—leucine in the case of vancomycin (**2**) and a substituted benzene ring in the case of ristocetin A (**1**)—does not form a 'folded-in' hydrophobic wall of the pocket. Rather, in vancomycin (**2**), the leucine side chain rotates relatively freely and more rapidly than the amide backbone (Waltho et al 1988). In ristocetin A, the Nuclear Overhauser Effects (NOEs) from the α-CH proton of residue 1 (x_1, Fig. 2a) to protons 1f and 1b are both small. Because the intensity of an NOE is an inverse function of the sixth power of the distance between two protons, this result establishes that neither 1f nor 1b approaches its minimum possible distance from x_1, and that the two protons lie at a similar distance from x_1. Thus, the vector from C_α to proton x_1 must be roughly orthogonal to the plane of the substituted benzene ring which constitutes the side chain of residue 1 (Fig. 2a). That is, the pocket consisting of the three backbone NH groups of residues 2, 3 and 4 has a hydrophobic wall to its upper side (made up from the substituted benzene ring which is part of the side chain of residue 2, see Fig. 1), but the potential hydrophobic wall, that is, the side chain of residue 1, of the lower side of this pocket is 'down' when the pocket is not occupied by a carboxylate anion (Fig. 2a).

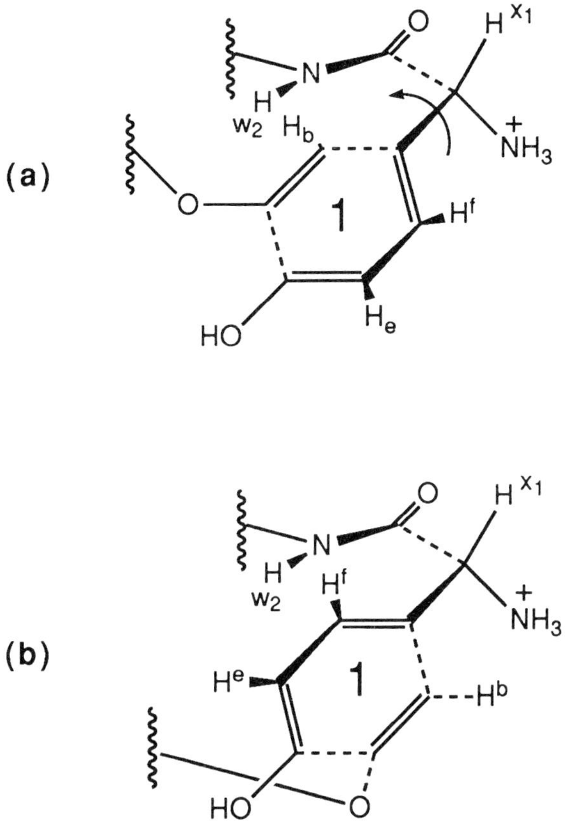

FIG. 2. The orientation of ring 1 of ristocetin A, (a) in free solution and (b) in the presence of di-*N*-Ac-L-Lys-D-Ala-D-Ala. Note that in (a) the protons 1f and 1b are both relatively distant from x_1, whereas in (b) protons 1f and x_1 are adjacent.

Upon formation of a complex between ristocetin and di-*N*-Ac-L-Lys-D-Ala-D-Ala, the NOE between x_1 and 1f increases vastly in build-up rate and intensity, whereas that between x_1 and 1b remains small. It must be concluded that the $C-H_f$ and $C-H_x$ vectors are now approximately eclipsed, and that the introduction of the carboxylate anion into the pocket has caused the erection of the lower hydrophobic wall (side chain of residue 1, Fig. 2b; Waltho & Williams 1989). Thus, when the cell wall analogue is *bound* to ristocetin A, its CO_2^- group is bound to three antibiotic NHs and it has hydrophobic walls on *both* sides. The story is analogous for vancomycin; although there are no NOEs from the leucine side chain to the central core of the antibiotic when the antibiotic is free (Waltho et al 1988), when di-*N*-Ac-L-Lys-D-Ala-D-Ala binds, the leucine side chain folds into the main part of the structure, as evidenced by the intramolecular NOEs $AsnCH_2$–LeuMe and AsnNH–LeuMe.

FIG. 3. A space-filling model of the complex formed between vancomycin and di-*N*-Ac-L-Lys-D-Ala-D-Ala, showing the spatial proximity of the non-C-terminal alanine methyl (A) and the leucine methyl (B). The carbon-bound protons of the tripeptide have been hatched for clarity.

Additionally, there is an NOE from the methyl group of the non-C-terminal alanine of the tripeptide cell wall analogue to LeuMe of the antibiotic (Williams et al 1983). Again, a second hydrophobic wall is 'raised' on binding of the cell wall analogue, and the observed NOEs are consistent with the conformation shown in Fig. 3.

Adaptation in the role of the sugars

Ristocetin A contains a tetrasaccharide attached to the side chain of residue 4, an amino sugar attached to the benzylic hydroxyl group of residue 6, and mannose attached to the side chain of residue 7 (**1**). Vancomycin contains a disaccharide attached to residue 4 (**2**). The sugars increase their antibiotics' solubilities in aqueous media, but it is becoming increasingly clear that both the positions and the structures of the sugars have probably been selected to promote efficient antibiotic action in subtle ways.

The conformation of the disaccharide in vancomycin is such that the C-6 methyl group of the amino sugar vancosamine is positioned as a 'plug' at the top end of the substituted benzene ring which constitutes the side chain of residue 4. Thus, when the cell wall analogue di-*N*-Ac-L-Lys-D-Ala-D-Ala binds to the antibiotic, the methyl group of the C-terminal alanine fits neatly into the pocket formed by the side chains of residues 2, 4 and 6, and the C-6 methyl group of vancosamine. This snug fit is established by appropriate NOEs (Rajamoorthi et al 1988). In this way, strong complexation to *completed* cell walls (at which stage the antibiotic cannot exert its killing action, and which complexation would sequester antibiotic, lessening its killing action on *growing* cell walls) is avoided. For example, Gram-positive bacteria with Type I cell walls (such as *Bacillus* species) contain within their completed walls the sequence shown in Fig. 4a, with a large side chain attached to the C-terminal amino acid. This large side chain cannot be accommodated in a pocket possessing a snug fit for a methyl group. It is therefore suggested that vancomycin evolved when its producer was in competition with organisms possessing Type I (or similar) cell walls (Williams & Waltho 1988). Lacking the disaccharide that is present in vancomycin, ristocetin A would not possess this selectivity, and indeed it has been demonstrated that ristocetin A binds strongly to cell wall analogues with large side chains at the C-terminal amino acid, whereas vancomycin does not (Nieto & Perkins 1971, Williams & Waltho 1988).

One respect in which ristocetin A differs from vancomycin is in its possession of mannose on residue 7 (**1**). The anomeric proton of this mannose gives an NOE to the methyl group of the non-C-terminal alanine of a cell wall analogue (Williams et al 1983). In the antibiotic–analogue complex, a binding pocket for this methyl group is provided by hydrophobic portions of the side chains of residues 1, 5 and 7 and mannose. The presence of mannose ensures that groups larger than methyl can only be tolerated with difficulty at this site. Interestingly, Gram-positive bacteria with Type III cell walls (e.g. *Micrococcus* species) contain in their completed cell walls the peptide sequence shown in Fig. 4b. Ristocetin A, because of its mannose residue, is unlikely to bind strongly to such structures, and so avoids sequestration in a non-lethal interaction. This conclusion is supported by an approximately 100-fold reduction in binding constant of ristocetin A to tripeptide cell wall analogues in which the central residue has a large side chain (such as glutamic acid) (Nieto & Perkins 1971, Williams & Waltho 1988). Vancomycin, lacking such selectivity at this site, binds to such analogues with little change in binding constant.

(a)

(b)

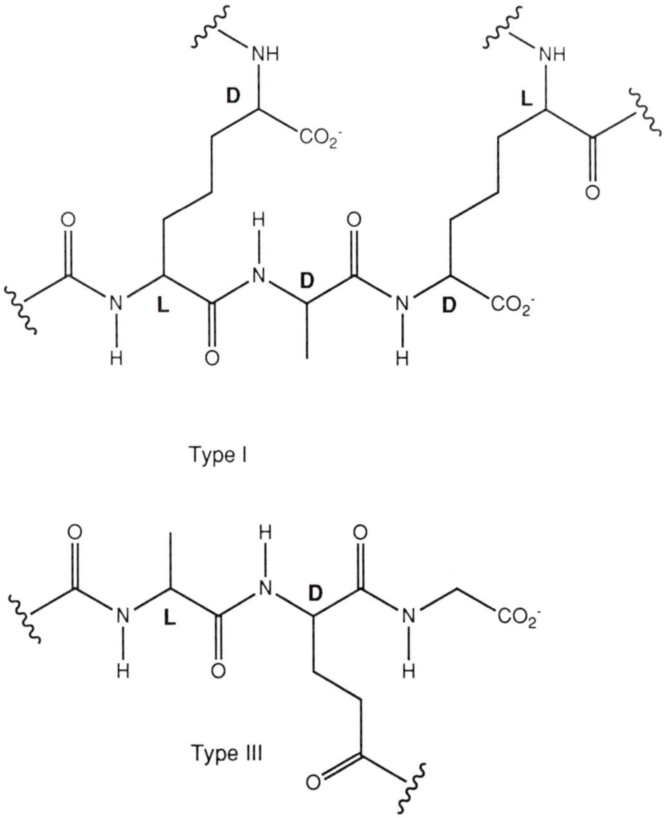

FIG. 4. Examples of peptides found within mature cell walls of (a) Type I bacteria and (b) Type III bacteria (see Ghuysen 1980) that have the correct stereochemistry for binding to vancomycin group antibiotics.

In summary, the selection pressure which has led to the incorporation of this particular disaccharide into vancomycin might have been competition with bacteria possessing Type I cell walls, whereas that leading to the incorporation of mannose on residue 7 of ristocetin A might have been competition with bacteria possessing Type III cell walls. Further evidence to support this view is provided by the finding that the lethal concentration of antibiotic for the relevant bacteria is reduced by the presence of the appropriate sugars in the antibiotic molecule (Williams & Waltho 1988).

We have previously reported that ristocetin A dimerizes with an association constant of about 2000 M^{-1}, and have deduced structural details of the dimer (Fig. 5; Waltho & Williams 1989). In our most recent work, it has become evident that the sugars of ristocetin A, vancomycin (probably), and the related antibiotic eremomycin promote dimerization (J. Mackay, U. Gerhard, R. A. Maplestone

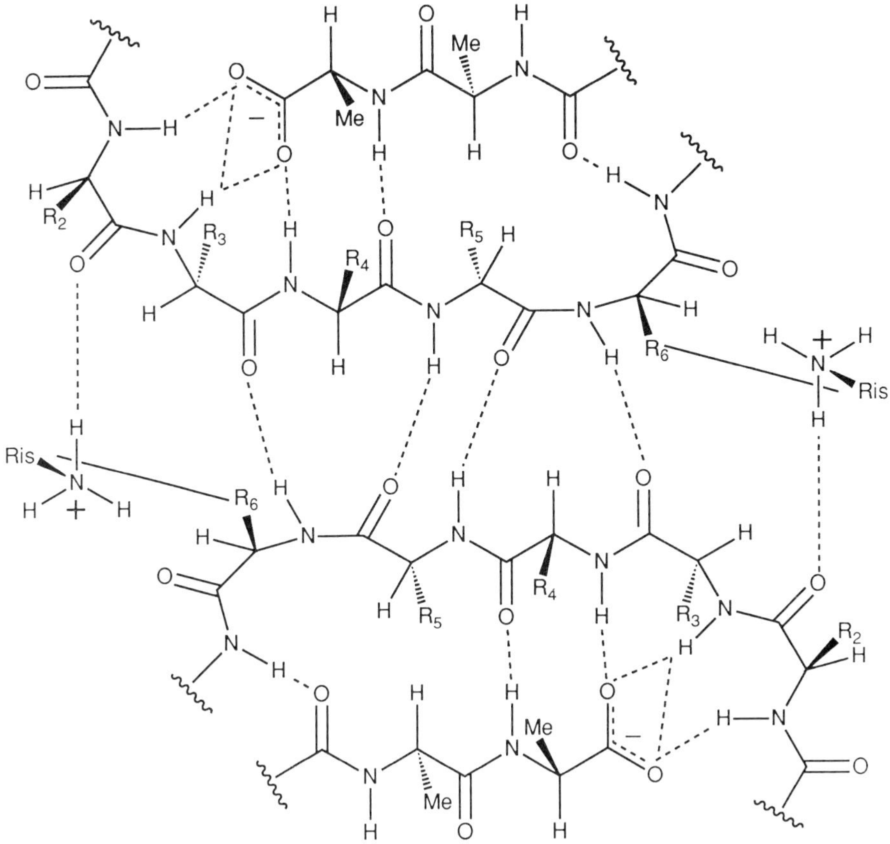

FIG. 5. Hydrogen-bonding network of the dimer formed between two molecules of ristocetin A when bound to a tripeptide. The broken lines indicate the positions of intermolecular hydrogen bonds.

& D. H. Williams, unpublished work 1992). In particular, eremomycin (**3**, Good et al 1990) dimerizes in aqueous solution with a dimerization constant $>10^5\,M^{-1}$, and this is reduced on removal of the disaccharide epivancosaminyl-glucose from the side chain of residue 4 to about $10^4\,M^{-1}$. A model of the eremomycin dimer, based on the structural details previously deduced for the ristocetin dimer, and on chemical shift changes and NOEs observed on eremomycin dimerization, shows an astonishing complementarity between the two halves of the head-to-tail dimer. In particular, this complementarity results in no small measure from the positions and structures of the eremomycin sugars (including, in addition to the disaccharide mentioned above, epivancosamine attached to the benzylic hydroxyl group of residue 6). These observations suggest that there has been a selection pressure for the evolution of a structure able

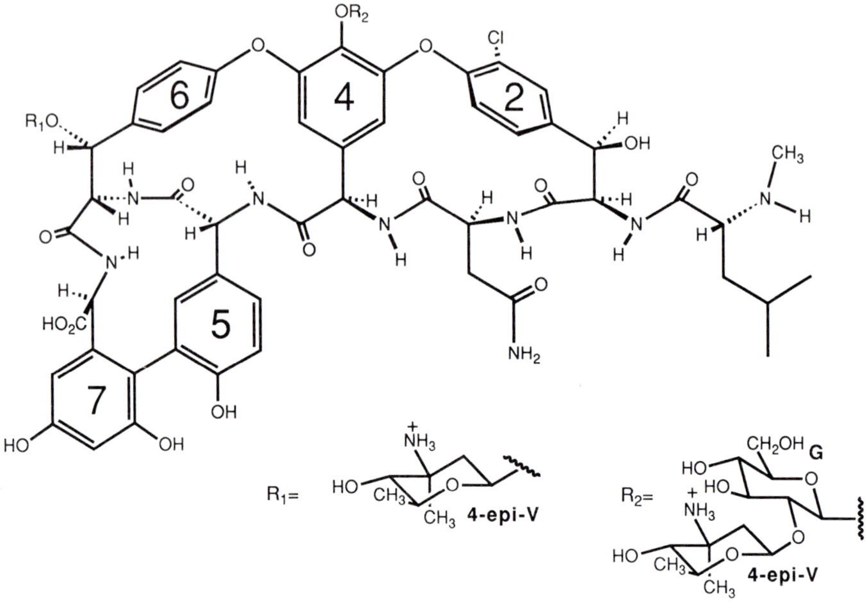

3 eremomycin

to dimerize. In this connection, it is interesting that although vancomycin binds peptide cell wall analogues somewhat more strongly than does eremomycin, eremomycin is a somewhat more efficient antibiotic than vancomycin in inhibiting bacterial cell growth (Good et al 1990). This observation suggests that dimerization (which occurs more efficiently in eremomycin than in vancomycin, the latter having a dimerization constant of only about 800 M^{-1}) may be useful in the action of the antibiotics. It is possible that it is a dimer which binds to cell wall precursors in the growing bacterial cell wall.

In this connection, two points are of interest. First, because the head-to-tail dimer has a twofold axis with respect to its potential to bind simultaneously to two cell wall mucopeptide precursors, it may be that the growing cell wall structure has a twofold axis; such an axis is not evident in currently accepted structures for completed cell walls (Ghuysen 1980), but could conceivably arise between successive layers of peptidoglycan in the multilayered cell wall structure. Second, it is striking that cell wall fragments can be induced to aggregate even by the weakly dimerizing vancomycin (Dr M. N. Gwynn, personal communication). This behaviour could be rationalized by an antibiotic dimer (AA), able to provide two binding pockets within one structure, associating with cell wall fragments (BB′) containing two or more peptide fragments to which the antibiotics can bind.

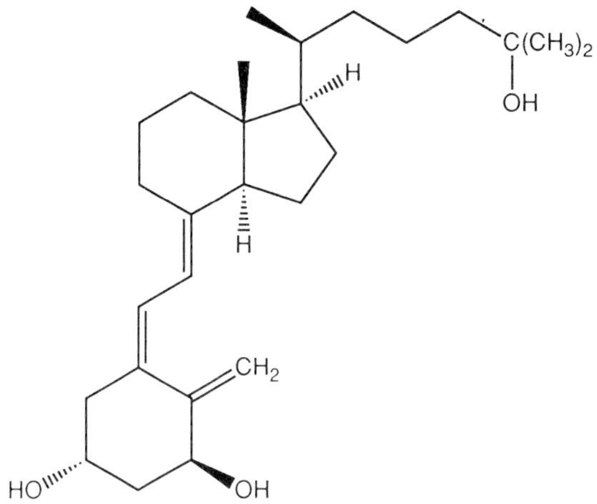

4 1α, 25-dihydroxycholecalciferol

Through such interactions, extended oligomeric aggregates (–AABB′AABB′–) can in principle be formed. The detailed molecular basis for this dimerization, and its possible physiological relevance, are currently under investigation.

Conclusions

The evidence now available on the complex structures and organizations of genes, and on the associated sophisticated biosynthetic pathways which programme and lead to the production of secondary metabolites, can leave no reasonable doubt as to the importance of the end products in the producer's fight for survival. This view is supported by the antagonistic, or sometimes cooperative, physiological effects known to be exerted by many of these products. An analysis of the molecular basis of action of one group of antibiotics, the vancomycin group, has revealed an amazing level of sophistication in their mode of action. These specializations are clearly not due to 'accidental fits', and lead to the conclusion that the interactions of secondary metabolites with their target sites probably involve beautiful complementarities rivalling those of enzyme–substrate interactions (Williams et al 1989).

In the area of secondary metabolite studies, many fascinating problems remain to be studied. Important among these are experiments to support, or refute, the possibility that there has been gene transfer between species even more widely separated within the evolutionary tree than those species between which gene transfer has already been established. In this respect, the production of 1α, 25-dihydroxycholecalciferol (**4**; dihydroxy-vitamin D_3) both by mammals (including humans), in which it is the hormone promoting Ca^{2+} absorption from the intestine, and by the plant *Solanum malacoxylon* Senoltner (as a glycoside), is of particular interest (Williams et al 1989). Cattle eating the plant develop *enteque seco*, a wasting disease typified by abnormally high concentrations of calcium and potassium in the blood. Adverse effects of high calcium concentrations in humans can also be induced by high levels of the hormone. The potential advantage to the plant of deterring heavy grazing is obvious, but the origins of the genes which code for the proteins affecting the 1α- and 25-hydroxylations invite curiosity. Even evidence supporting the less revolutionary idea of independent evolution of these hydroxylases in mammal and plant would attest further to the power of molecular mimicry in interacting species which are distant in evolutionary terms.

Acknowledgements

We thank the Science and Engineering Research Council, the Upjohn Company (Kalamazoo) and SmithKline Beecham for financial support. We also gratefully acknowledge the financial support received in some of the later phases of this work from Eli Lilly, Roussel Uclaf, Pfizer and Schering.

References

Barna JCJ, Williams DH 1984 The structure and mode of action of glycopeptide antibiotics of the vancomycin group. Annu Rev Microbiol 38:339–357

Bevitt DJ, Cortes J, Haydock SF, Leadlay PF 1992 6-Deoxyerythronolide B synthase 2 from *Saccharopolyspora erythraea*: cloning of the structural gene, sequence analysis and inferred domain structure of the multifunctional enzyme. Eur J Biochem 204:39–50

Cortes J, Haydock SF, Roberts GA, Bevitt DJ, Leadlay PF 1990 An unusually large multifunctional polypeptide in the erythromycin-producing polyketide synthase of *Saccharopolyspora erythraea*. Nature (Lond) 348:176–178

Davies JE 1986 Origins and evolution of antibiotic resistance. Alliance for the Prudent use of Antibiotics Newsletter (Fall):7–8

Demain AL 1980 Do antibiotics function in nature? Search (Syd) 11:148–151

Demain AL 1989 Functions of secondary metabolites. In: Hershberger CL, Queener SW, Hegeman G (eds) Genetics and molecular biology of industrial microorganisms. American Society for Microbiology, Washington, DC, p 1–11

Dhillon N, Hale RS, Cortes J, Leadlay PF 1989 Molecular characterisation of the gene from *Saccharopolyspora erythraea* which is involved in erythromycin biosynthesis. Mol Microbiol 3:1405–1414

Donadio S, Staver MJ, McAlpine JB, Swanson SJ, Katz L 1991 Modular organisation of genes required for complex polyketide biosynthesis. Science (Wash DC) 252:675–679

Gerhard U, Searle MS, Williams DH 1992 The free energy change of restricting a bond rotation in the binding of peptide analogues to vancomycin group antibiotics. Bioorg Med Chem Lett, in press

Ghuysen J-M 1980 Antibiotics and peptidoglycan metabolism: In: Sammes P (ed) Topics in antibiotic chemistry. Ellis Horwood, Chichester, vol 5:15–117

Good VM, Gwynn MN, Knowles DCJ 1990 MM 45289, a potent glycopeptide antibiotic which interacts weakly with diacetyl-L-lysyl-D-alanyl-D-alanine. J Antibiot (Tokyo) 43:550–555

Harborne JB 1988 Introduction to ecological biochemistry, 3rd edn. Academic Press, London
Haslam E 1986 Secondary metabolism—fact and fiction. Nat Prod Rep 3:217–249
Hopwood DA 1988 Towards an understanding of gene switching in *Streptomyces*, the basis of sporulation and antibiotic production. Proc R Soc Lond B Biol Sci 235:121–138
Katz E, Demain AL 1977 The peptide antibiotics of *Bacillus*: chemistry, biogenesis and possible functions. Bacteriol Rev 41:449–474
Maplestone RA, Stone MJ, Williams DH 1992 The evolutionary role of secondary metabolites. Gene (Amst) 115:151–157
Nieto M, Perkins HR 1971 The specificity of combination between ristocetins and peptides related to bacterial cell wall mucopeptide precursors. Biochem J 124:845–852
Perkins HR 1982 Vancomycin and related antibiotics. Pharmacol & Ther 16:181–197
Rajamoorthi K, Harris CM, Harris TM, Waltho JP, Skelton NJ, Williams DH 1988 Function of the amino sugar and N-terminal amino acid of the antibiotic vancomycin in its complexation with cell wall peptides. J Am Chem Soc 110:2946–2953
Reid DG, Salisbury SA, Williams DH 1983 Proton NOE effect study of an actinomycin D complex with a self-complementary tetranucleoside triphosphate. Biochemistry 22:1377–1385
Rodriguez-Tebar A, Vasquez D, Velazquez JPL, Laynez J, Wadso I 1986 Thermochemistry of the interaction between peptides and vancomycin and ristocetin. J Antibiot (Tokyo) 39:1578–1583
Stone MJ, Williams DH 1992 On the evolution of functional secondary metabolites (natural products). Mol Microbiol 6:29–34
Waltho JP, Williams DH 1989 Aspects of molecular recognition: solvent exclusion and dimerization of the antibiotic ristocetin when bound to a model bacterial cell wall precursor. J Am Chem Soc 111:2475–2480
Waltho JP, Williams DH, Stone DJM, Skelton NJ 1988 Intramolecular determinants of conformation and mobility within the antibiotic vancomycin. J Am Chem Soc 110:5638–5643
Williams DH, Waltho JP 1988 Molecular basis of the activity of antibiotics of the vancomycin group. Biochem Pharmacol 37:133–141
Williams DH, Williamson MP, Butcher DW, Hammond SJ 1983 Detailed binding sites of the antibiotics vancomycin and ristocetin A: the determination of intermolecular distances in antibiotic–substrate complexes by the use of time-dependent NOE. J Am Chem Soc 105:1332–1340
Williams DH, Stone MJ, Hauck PR, Rahman SK 1989 Why are secondary metabolites biosynthesized? J Nat Prod (Lloydia) 52:1189–1208
Williamson MP, Williams DH, Hammond SJ 1984 Interactions of vancomycin and ristocetin with peptides as models for protein binding. Tetrahedron 40:569–577

DISCUSSION

Davies: There has been a substantial increase in vancomycin resistance, and I believe that one mechanism of resistance is an alteration in the cell wall peptide.

Williams: Yes. This is work done in Chris Walsh's lab. The resistant species may have D-hydroxybutyrate at the C-terminal position instead of D-alanine (Bugg et al 1991). How such resistance has come about in such a relatively short space of time is remarkable, unless there was already this variant there, or unless the appropriate genes have been 'highjacked' from elsewhere.

Davies: Have you looked at this structure with reference to your model, to examine how this prevents binding of the antibiotic to the cell wall?

Williams: We looked at the binding of *N*-acetyl-D-Ala-D-lactate into the binding site of ristocetin A. Instead of having an NH group, as you would in D-Ala, to bind to a carbonyl amide, you have an oxygen. The interaction in our measurements reduced the binding constant by *more* than a factor of 10^2. An NH, as we know in proteins, will hydrogen bond to a carbonyl; in contrast, with an oxygen up against a carbonyl, you would expect a repulsion; so this is in agreement with the reduced binding constant that we observed.

Cane: Has it been demonstrated *in vitro* that the ristocetin dimer can be dissociated by the addition of D-Ala-D-Ala?

Williams: No; I am not suggesting that it can be dissociated. In fact we have recently shown that the dimerization of ristocetin A is promoted by *N*-Ac-D-Ala-D-Ala. We take an NMR solution of ristocetin A where we know we have dimer partially formed; and we then add more antibiotic (to which *N*-Ac-D-Ala-D-Ala is already bound in a 1:1 complex) at the same concentration. From the change in the NMR spectrum we can tell that the presence of the piece of cell wall actually *promotes* the dimerization process; so there's a hint here of cooperativity. This you would expect, because what happens is as follows. In the dimer, the amino group of the amino sugar attached to residue 6 hydrogen bonds to the backbone carbonyl group of residue 2; so you have a protonated amino group hydrogen bonded to the CO. This increases the polarization of the amide bond and so increases the binding of the carboxylate anion into the binding pocket. In effect, we now have a salt bridge mediated through an amide functionality. This is the hypothesis; it's supported by the fact that dimerization increases when we add the ligand. So the amino sugar is positioned, hypothetically, very nicely to promote dimerization and to increase the binding of bacterial cell wall. So it's no coincidence, if that hypothesis is correct, that there is an amino sugar in precisely that position.

Cane: If your model for dimerization is correct, with a sufficient amount of D-Ala-D-Ala you should reverse that expansion of the inhibition zone, and then begin to compete out the second site.

Williams: That is found, and has been published, by Good et al (1990).

Piepersberg: Would you suggest that we have to look for well-defined receptors for all secondary metabolites?

Williams: Yes; it is my hypothesis that we have these very sophisticated antibiotic structures because there is, out there, some molecular target to which we expect them to bind. To me it's striking that in the few cases where we know the receptor, which is double-helical DNA, the secondary metabolites that bind to DNA display wonderful sophistication in the way they do it. People say that if you propose the existence of receptors for secondary metabolites, it is philosophy rather than science. One answer is that if you look through a collection of natural products for those that might bind to DNA, even before they have been shown to do so, you find molecules which have very suggestive adaptations to do precisely this. Of course, daunomycin was known to bind to DNA before we were coming up with these sort of ideas. Daunomycin is used in anticancer therapy. It has been shown by X-ray crystallography that when it intercalates, the amino sugar binds into the minor groove of DNA (Moore et al 1989). When you look through a collection of natural products you can ask if certain structures might be regarded as more sophisticated versions of daunomycin, and bind to DNA. We did this in the Upjohn collection of natural products, and we found distrisarubicin B. We now find that a sugar that binds into the minor groove has been extended by more sugar units. Moreover, there is a trisaccharide at the other side of the intercalating chromophore which would also continue to bind in the minor DNA groove, at the other side. Presumably it is no accident that since there is an approximate two-fold axis in DNA, we also found a two-fold axis in the distrisarubicin B molecule; so this extra hydroxyl is put into daunomycin. It isn't down, like the one attaching the first saccharide unit, which would destroy the two-fold axis, but it is up. Thus you achieve an approximate two-fold axis in the antibiotic, which matches that of DNA. The same is true of actinomycin D, which has essentially the two-fold symmetry that is found in DNA.

Cavalier-Smith: I think you are right that in a high proportion of cases you will be able to find specific receptors that the secondary compound is directed against. But there are some cases where this is unlikely; so one shouldn't over-generalize. For example, some plants produce resins whose function seem to be tangling up the legs or mouth parts of insects; this is a simple physical process that does not require a specific chemical receptor. And even some chemically orientated antagonists, like perhaps polyphenols, may attack a whole class of compounds in a very non-specific way; for example, attacking proteins and causing general problems, rather than being highly specific antagonists.

Williams: I agree with that; I am guilty perhaps of over-generalization here, and of using rather sloppy language—but, as you suggest, the legs or the mouth of an insect constitute a rather 'sloppy' receptor. The resins are there, nevertheless, for a function, and if you have something that in quite a sophisticated way can make a glutinous mass and at the same time adhere to the legs of insects, then that's sophistication, and it's harmful to something that may be a threat to the producer. So even this example fits in the general philosophy that I am putting forward.

Lehrer: I wonder about ristocetin. I believe that ristocetin also fits extraordinarily well to another molecule which is related to the human blood coagulation system, mediating the binding of von Willebrand's factor to platelet glycoprotein Ib (Gralnick et al 1984). This is the reason ristocetin is not used clinically. Although ristocetin 'fits' here extremely well, is that an accident, or by design?

Williams: I don't know. I have speculated that perhaps one reason that ristocetin does this at the molecular level is because the tetrasaccharide that is part of ristocetin A in some way simulates epithelial saccharides. When platelets hit that epithelial polysaccharide, in wounding, you get blood coagulation. Maybe there is a molecular similarity which allows the ristocetin polysaccharide to cause coagulation also; but I have no idea whether it is an evolved fit or an accidental one. Clearly, in many cases there will be accidental fits, whereas the ones that we have looked at in detail (vancomycin and ristocetin) are so sophisticated that in my opinion it's not accidental there. But since biology is sloppy in many places, I am sure we will also find accidental fits!

Haslam: As a more general point, thinking of where we chemists fit into this discussion group, I suspect that the mainstream consists of microbiologists, biochemists and geneticists, and that chemists are on both the left wing and the right wing; individually they are pretty radical, but they are also a highly conservative group!

Perhaps therefore I can put a point of view which is both radical and highly conservative. It derives from a point made by Hans Krebs in the 1950s, when considering the idea of purpose in biology. He reflected on teleological arguments and said that teleology is a lady with whom no self-respecting biologist can live without, but whom he is ashamed to be seen escorting in public! You have described a high degree of sophistication in a binding site, and have therefore, as it were, asked what is the *purpose* of this. But then you have said that the metabolic event is caused by the purpose, and I am not sure you can turn the argument around in that way. To put it more specifically: yesterday I had the great pleasure of meeting your charming wife; clearly, at some stage during evolution, you and she met; but I think it was an entirely random *meeting*, and the binding has gone on *since*. Could you respond to that point?

Williams: Yes! First of all, I'm not putting forward any proposal of 'purpose' in natural selection. This comes back to the point we were talking about earlier, when Dr Wächtershäuser implied that Darwinian selection is a matter of opportunism (p 36). I follow the philosophy of Ernst Mayr (1988) when he says that Darwinian selection doesn't set itself goals in the future; it is merely opportunism. But when opportunism in natural selection is expressed, then we do get high degrees of sophistication. When we see such highly sophisticated adaptations, I see a functional use for the end product, in the sense that if I look at a motor car, say, and I find it has all the functional parts that allow you to get from A to B, I assume that it has been made for a use. My viewpoint is the same for the antibiotics.

Rinehart: In relation to the availability of receptors for secondary metabolites, what about morphine and the existence of the enkephalin receptors? Does that impart any advantage to the opium poppy? Or is it sheer accident that morphine fits a receptor for small oligopeptides?

Williams: I don't know whether it is an accidental fit, whereby opium has this analgesic effect on vertebrates, or not. If it is an accident that morphine binds to vertebrate enkephalin receptors and has an effect on them of the type that enkephalin produces (that is, the ligand formed in the organism itself), I would still maintain the idea that morphine has such a sophisticated structure, and probably requires so much DNA and protein for the coding of its synthesis, that it must have, or once have had, some functional role that was advantageous to the poppy, even if it is not the role we have just discussed. (The role just discussed could, incidentally, have been as a deterrent to grazing animals.)

References

Bugg TD, Wright GD, Dutka-Malen S, Arthur M, Courvalin P, Walsh CT 1991 Molecular basis for vancomycin resistance in *Enterococcus faecium* BM4147: biosynthesis of a depsipeptide peptidoglycan precursor by vancomycin resistance proteins VanH and VanA. Biochemistry 30:10408–10415

Good VM, Gwynn MN, Knowles DJC 1990 MM 45289, a potent glycopeptide antibiotic which interacts weakly with diacetyl-L-lysyl-D-alanyl-D-alanine. J Antibiot (Tokyo) 43:550–555

Gralnick HR, Williams SB, Coller BS 1984 Fibrinogen competes with von Willebrand factor for binding to the glycoprotein IIb/IIIa complex when platelets are stimulated with thrombin. Blood 64:797–800

Mayr E 1988 Towards a new philosophy of biology. The Belknap Press of Harvard University, Cambridge, MA, p 43

Moore MH, Hunter WN, d'Estaintot BL, Kennard O 1989 DNA–drug interactions. The crystal structure of d(CGATCG) complexed with daunomycin. J Mol Biol 206:693–705

Origins of secondary metabolism

T. Cavalier-Smith

Canadian Institute for Advanced Research Evolutionary Biology Program, Department of Botany, The University of British Columbia, Vancouver, BC, Canada V6T 1Z4

Abstract. Secondary metabolites generally benefit their producers as poisons that protect them against competitors, predators or parasites. They are produced from universally present precursors (most often acetyl-CoA, amino acids or shikimate) by specific enzymes that probably arose by the duplication and divergence of genes originally coding for primary metabolism. Most secondary metabolites are restricted to single major taxa on the universal phylogenetic tree and so probably originated only once. But different secondary metabolic pathways have originated from different ancestral enzymes at radically different times in evolution. Secondary metabolites are most abundantly produced by microorganisms in crowded habitats and by plants, fungi and sessile animals like sponges, where chemical defence and attack rather than physical escape or fighting are at a premium. The first secondary metabolites were probably antibiotics produced in microbial mats over 3500 million years ago. These first ecosystems probably consisted entirely of eubacteria: archaebacteria and eukaryotes arose much later. As a phylogenetic context for considering the earliest origins of antibiotics I summarize a cladistic analysis of the explosive eubacterial primary diversification. This suggests that the most primitive surviving cells are the photosynthetic heliobacteria. Study of these and of the nearly as primitive chloroflexibacteria, spirochaetes and deinobacteria may provide the best evidence on the origins of secondary and primary metabolism.

1992 Secondary metabolites: their function and evolution. Wiley, Chichester (Ciba Foundation Symposium 171) p 64–87

Secondary metabolism is metabolism that is not essential for the growth and reproduction of the organism itself (Mann 1987). But it cannot be sharply distinguished from primary metabolism—the interlinked central metabolic network catalysed by several hundred enzymes that provides the raw materials for making the lipids and macromolecules from which organisms are made.

Many of the enzymes of primary metabolism are present in all living organisms and were undoubtedly therefore present in the most recent ancestor of all life. By contrast, the enzymes of secondary metabolism, and the genes coding for them, are restricted to a very small subset of all living organisms. Secondary metabolites are however produced from universally present precursors, most often acetyl-CoA, amino acids or shikimate. The genes coding for the enzymes that make them probably arose by duplication and divergence of genes originally involved in primary metabolism. It is clear that some of these duplications are very old, whereas others are much more recent, and that different secondary metabolites have originated in this way independently on many different occasions in many different branches of the evolutionary tree of life.

This means that secondary metabolism has had many origins and not a single one: that is to say, it is polyphyletic. Although secondary metabolites are often considered less essential than primary metabolites to their producers, this is a somewhat misleading view, because the genes coding for them could not have evolved and been conserved for many tens, hundreds or sometimes thousands of millions of years without being of considerable selective advantage to their hosts (Vining 1990). Secondary metabolites are most often beneficial to their producers as agents of the chemical warfare which is perpetually being waged against competitors, predators and parasites. This general selective force for the often haphazard and opportunistic modification of pre-existing enzymes, producing new and more effective poisons, must have been present even before the origin of the first cell. In this natural poison warfare between organisms, secondary metabolites are probably more important and diverse than macromolecular toxins because they are small molecules that can diffuse more easily into cells. Protein toxins, such as the lethal colicins of bacteria or the proteins of animal venoms, act primarily on the plasma membrane of cells, whereas secondary metabolites not only can do this, but also can attack DNA, transcription, ribosomes, metabolic enzymes, microtubules (e.g. colchicine, vinblastine), microfilaments (e.g. phalloidin) and endomembranes (e.g. monensin).

Because secondary metabolism has had so many separate origins, with many completely different genes being ancestral to the newly recruited enzymes, and because the particular selective forces that have favoured the evolution of poisonous small molecules will differ in detail according to the organisms (both producers and targets) involved and also to the ecological theatre in which their antagonisms are played out, I cannot in a brief paper discuss 'the origin [*sic*] of secondary metabolism' as I was originally invited to do. I shall therefore concentrate on three things: (1) discussing the origins of secondary metabolism in the earliest living organisms; (2) providing an overall phylogeny of living organisms, so as to place some of the origins of important secondary metabolic pathways in a proper phylogenetic context; and (3) criticizing the view (Davies 1990) that secondary metabolism originally had a positive physiological role in precellular biochemical evolution. Even though secondary metabolism as a whole is polyphyletic, it is clear that this is not true for the majority of specific metabolic pathways. Most of these are clearly restricted to one single branch (clade) of the universal phylogenetic tree and therefore evolved once only and were then inherited by all (or many) descendants of that original ancestor. Table 1 summarizes the presently known systematic distribution of the major classes of secondary metabolites in the eight kingdoms of life and their subkingdoms, following the classification of Cavalier-Smith (1991a).

TABLE 1 The systematic distribution of major classes of secondary metabolites

	Derived from acetyl-CoA					*Derived from amino acid pathways*					*Mixed pathways*		
	Poly-acetylenes	*Prosta-glandins*	*Cyclic polyketides /polyphenols*	*Isoprenoids*	*Derived from shikimate*	*β-lactams*	*Peptides*	*Alkaloids*	*Others*	*Derived from purines*	*Flavonoids*	*Other heterosides*	*Others*
Eukaryotes:													
ANIMALIA													
Radiata[1]	+	+	+	+					+	+		+	
Bilateria[2]	+	+		+									
FUNGI[3]	+		+	+	+	+	+	+	+	+		+	+
PLANTAE													
Viridiplantae[4]	+		+	+	+		+	+	+	+	+		+
Biliphyta[5]				+					+				
CHROMISTA													
Cryptista[6]													
Chromobiota[7]			+	+					+				
PROTOZOA													
Dictyozoa[8]				+					+	+			
Percolozoa[9]													
ARCHEZOA[10]													
Bacteria:													
ARCHAEBACTERIA				+									
EUBACTERIA													
Posibacteria			+		+	+	+	+	+			+	
Negibacteria				+		+							

1. Sponges, Placozoa, Cnidaria and Ctenophora. 2. All other animals. 3. Includes lichens with over 500 secondary metabolites (including 100 unique). 4. Green plants: green algae, bryophytes and vascular plants. 5. Red algae and Glaucophyta. 6. Cryptomonads and *Goniomonas*. 7. Heterokonts (e.g. brown algae, diatoms and oomycetes) and haptomonads. 8. Protozoa with Golgi dictyosomes (17 phyla: ciliates, sporozoans, most flagellates and most sarcodines). 9. Protozoa lacking dictyosomes e.g. *Naegleria*, *Percolomonas*. 10. Primitively amitochondrial eukaryotes: phyla Archamoebae, Metamonada, Microsporidia.

References: Luckner 1990, Mann 1987, Blunden & Gordon 1986, Ragan & Glombitza 1986. There must be several lacunae in this table resulting from incomplete knowledge (both on my part and generally).

Though Table 1 includes examples from higher organisms, the rest of this chapter concentrates on the origins of poison warfare in the bacteria that made up the world's first crowded ecosystems: the microbial mats that would have covered the bottoms of shallow water and moist soils on land, over 3500 million years ago. Though we still know very little about the chemical ecology of microbial mats (Cohen & Rosenberg 1989), we are beginning to learn more about the phylogeny of the organisms that make them.

The universal tree of life

The history of life can be divided almost equally into two parts: the age of bacteria, 3500–1700 million years ago, and the age of eukaryotes, from 1700 million years ago to the present (Cavalier-Smith 1990a). For the first half of biotic history it was a purely prokaryotic world, with no organisms more complex than filamentous photosynthetic bacteria. Figure 1 summarizes the most significant steps in the history of the living world, which gave rise to the eight kingdoms. The evidence that eukaryotes and archaebacteria are sister groups, that were derived from a single common ancestor—a mutant stable L-form derived from a eubacterium by the loss of murein peptidoglycan—comes from a combination of molecular, ultrastructural and fossil evidence and is discussed in detail elsewhere (Cavalier-Smith 1987a, 1991a,b). Contrary to what has sometimes been thought, the most recent common ancestor of living organisms was certainly a well-developed bacterial cell (Cavalier-Smith 1987b, 1991a) and not a crude precellular progenote (Woese & Fox 1977). Although the balance of evidence and of reasoning about the directions of mega-evolutionary transitions in bacterial evolution both favour the view that the ancestral cell was a eubacterium (Cavalier-Smith 1987a, 1991a,b), the possibility that it was an archaebacterium (Woese & Wolfe 1985) instead cannot yet be totally ruled out.

From 16S rRNA-based phylogeny (Woese 1987), it is clear that about 12 major evolutionary lineages of eubacteria diverged from each other at essentially the same time, which I (Cavalier-Smith 1987b) and Pierson & Olson (1989) interpret as the explosive radiation of the eubacterial cell following the origin of photosynthesis and the first cell. Because of the sudden nature of this radiation it is not possible to use the rRNA tree to determine what sort of eubacterial cell is the most primitive. It is fundamentally important to determine whether the first eubacterial cell was photosynthetic or not, and whether it was bounded by only a single membrane as in eukaryotes and archaebacteria or instead by two bounding membranes. [Eubacteria with only a single membrane (the plasma membrane) are placed in the purely non-photosynthetic subkingdom, Posibacteria (Cavalier-Smith 1987a), which comprises three phyla or divisions: the true Gram-positive bacteria (Firmicutes), the mycoplasmas and spiroplasmas (Mollicutes), and the Togobacteria (Cavalier-Smith 1991a,b)—i.e. the recently

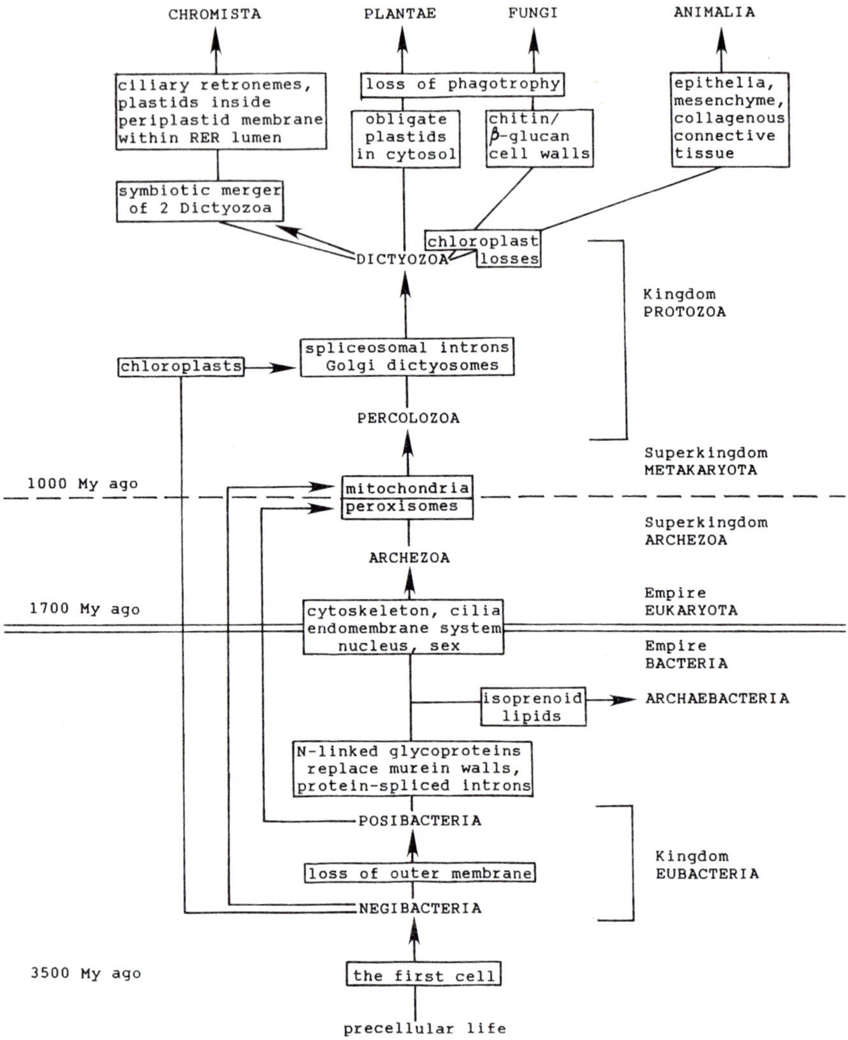

FIG. 1. The eight-kingdom phylogeny of life. The major events in the diversification of life are shown in the boxes. Two of the kingdoms, Eubacteria and Protozoa, have such fundamental distinctions between their subkingdoms that these also are shown. The history of life involves not only major branchings and phyletic transitions but also four major symbiotic events: the origin of mitochondria, peroxisomes (Cavalier-Smith 1990b), chloroplasts (Cavalier-Smith 1992), and the kingdom Chromista (Cavalier-Smith 1989). Note that, contrary to what is often asserted, symbiosis was not involved in the origin of eukaryotes. It was instead the origin of metakaryotes that involved the simultaneous conversion of two different eubacteria into respiratory organelles (mitochondria and peroxisomes) within the cytoplasm of an archezoan eukaryote (probably a tetraciliate retortamonad: Cavalier-Smith 1991a).

described Thermotogales (Huber et al 1989).] All other eubacteria have a second (outer) membrane outside the plasma membrane and are placed in the subkingdom Negibacteria. Negibacteria are so diverse that they are here divided into 10 divisions, of which five are entirely or partly photosynthetic (Table 2). This indicates that photosynthesis arose very early in eubacterial evolution and that much of early eubacterial diversification involved the origin of a great variety of photosynthetic pigment systems, presumably in part to exploit the full and changing spectrum of light that is successively altered as it filters through microbial mats (Cavalier-Smith 1987b, Pierson & Olson 1989).

The first cells: the origin of primary metabolism

The traditional view that the first cell had only a single membrane (Goldacre 1958, Hargraeves & Deamer 1978) implies that negibacteria evolved from posibacteria by adding an extra membrane: only Dawes (1981) has suggested how this might have occurred, but his model seems mechanistically implausible. It would be much easier to convert a negibacterium into a posibacterium by the loss of the outer membrane (Blobel 1980, Cavalier-Smith 1980), which happens from time to time in modern L-forms (Madoff 1986) as a result of the loss of murein, rather than by the hypertrophy of murein, which I (Cavalier-Smith 1980, 1987a,b) proposed as the mechanism of origin of posibacteria. Moreover, one can simultaneously account for the origin of the first cell and of the negibacterial outer membrane by the folding up of a postulated precellular entity, the 'inside-out' cell (Blobel 1980) or obcell (Cavalier-Smith 1987b), in which genes and ribosomes were initially located on the outside of a phospholipid vesicle in full contact with their molecular food in the prebiotic soup (Fig. 2). As I (Cavalier-Smith 1987b) and Pierson & Olson (1989, Olson & Pierson 1987a,b) have independently argued, it would have been much easier to have evolved a simple porphyrin-based photosynthetic system than the glycolytic fermentative one that has so often been assumed to have been the first bioenergetic system. According to these analyses the most likely ancestral cell would have been an anaerobic anoxygenic photoheterotroph. I suggest that, of all modern organisms, the recently discovered brownish Heliobacteria (Gest & Favinger 1983) most closely resemble the ancestral cell in structure and physiology. They are strictly anaerobic photoheterotrophs with the simplest known porphyrin-based photosynthetic reaction centre [containing bacteriochlorophyll g′ (Kobayashi et al 1991a,b), a quinone and an iron sulphur protein, just as in the primitive postulated reaction centre (Cavalier-Smith 1987b, Olson & Pierson 1987a,b, Pierson & Olson 1989)], which could be ancestral both to the Type 1 reaction centres of Chlorobiales and Cyanobacteria and to the Type 2 reaction centres of Chloroflexales, purple bacteria and Cyanobacteria. They are able to fix N_2 but, unlike other photosynthetic bacteria, heliobacteria cannot fix CO_2 or use inorganic

TABLE 2 The 15 phyla of bacteria and their 30 classes

Kingdom 1 EUBACTERIA*
(Bacteria with acyl ester membrane lipids)

Subkingdom 1 NEGIBACTERIA* Bacteria with a second (outer membrane) lipid bilayer outside the plasma membrane; murein usually present as a thin layer between the two membranes

Infrakingdom 1 Eobacteria*† Murein walls containing ornithine; lipopolysaccharide and diaminopimelic acid absent, no cytochrome *aa*3 or rubisco; small citrate synthase

1. Heliobacteria** Anaerobic photoheterotrophs with bacteriochlorophyll g and g′, lacking chlorosomes or *b* cytochromes
2. Chlorobacteria**† Aerobic gliding photoautotrophs with bacteriochlorophyll c and *b* cytochromes; flagella absent
   Class 1. Eochlorea**† lacking chlorosomes (e.g. *Heliothrix*)
   Class 2. Chloroflexibacteria**† having chlorosomes (e.g. *Chloroflexus*)
3. Hadobacteria† Heterotrophs
   Class 1. Deinobacteria Radiation resistant (e.g. *Deinococcus*)
   Class 2. Eothermea† Heat resistant (e.g. *Thermus*)

Infrakingdom 2 Endoflagellata Flagella in periplasmic space; lipopolysaccharide absent

4. Endoflagellata Spirochaetes and leptospiras
   Class 1. Spirochaetea† Spirochaetes: murein walls with ornithine
   Class 2. Leptospirea† Leptospiras: murein walls with diaminopimelic acid

Infrakingdom 3 Pimelobacteria† Murein walls with diaminopimelic acid; lipopolysaccharide present in outer leaflet of outer membrane

5. Sphingobacteria* Have sphingolipids, lack flagella: usually glide
   Class 1. Chlorobea**† Anaerobic phototrophs with chlorosomes: Chlorobiales**
   Class 2. Flavobacteria Aerobic heterotrophs (e.g. *Flavobacterium*, *Cytophaga*)
6. Selenobacteria† Heterotrophs lacking sphingolipids
   Class 1. Selenomonadea† Endospores absent (e.g. *Selenomonas*)
   Class 2. Sporomusea† Endospores present (e.g. *Sporomusa*)
7. Cyanobacteria** Oxygenic photosynthesizers with chlorophyll and phycobilisomes
   Class 1. Gloeobacteria**† Thylakoids absent (e.g. *Gloeobacter*)
   Class 2. Myxophycea** Thylakoids present: the classical blue-green algae; syn. Cyanophyceae
8. Proteobacteria* Includes purple bacteria and *Escherichia coli*: endospores sphingolipids and phycobilisomes absent
   Class 1. Alphabacteria*†
   Class 2. Betabacteria*†
   Class 3. Gammabacteria*†
   Class 4. Deltabacteria†
   } Respectively the α, β, γ, δ purple bacteria, as defined by Woese 1987

Infrakingdom 4 Planctobacteria Aerobic respirers lacking murein but possessing lipopolysacharide

9. Planctobacteria (e.g. *Planctomyces*)

Infrakingdom 5 Chlamydiae Murein and ATP synthesis absent: obligate intracellular parasites

10. Chlamydiae (e.g. *Chlamydia*)

(Continued)

TABLE 2 *(Continued)*

Subkingdom 2 POSIBACTERIA Non-photosynthetic bacteria with no outer membrane

11. Firmicutes Gram-positive bacteria: thick murein wall with teichoic acids
    Class 1. Clostridea† Endospores often present: DNA with low GC content (e.g. *Bacillus*, *Clostridium*)
    Class 2. Actinobacteria Exospores often present: DNA with high GC content (e.g. *Streptomyces*)
12. Mollicutes Murein absent: mycoplasmas and spiroplasmas
    Class 1. Mycoplasmea† Non-motile (e.g. *Mycoplasma*, *Ureaplasma*)
    Class 2. Spiroplasmea† Motile with corkscrew movement (e.g. *Spiroplasma*)
13. Togobacteria Thermotogales: obligately anaerobic thermophiles with very thin murein walls and an external non-lipid toga

Kingdom 2 ARCHAEBACTERIA
(Bacteria with isoprenoid ether membrane lipids)

14. Euryarchaeota Four classes: Halobacteria (halophiles), Methanobacteria (methanogens), Thermoplasmea† (no cell wall: *Thermoplasma*), Thermococcea† (walled thermophiles: Thermococcales)
15. Crenarchaeota 'Eocytes', Sulfobacteria: sulphur dependent
    Class 1. Sulfolobea† Aerobic or facultatively aerobic S oxidizers (e.g. Sulfolobales)
    Class 2. Thermoprotea† Anaerobic S reducers (e.g. Thermoproteales)

*Often photosynthetic. **Always photosynthetic. †New names proposed here.
This division into kingdoms, phyla and classes (based in eubacteria largely on the characters of the cell envelope and type of photosynthesis) is totally consistent with the rRNA tree, except that the latter (because of its bushy character) provides no evidence concerning the sub- and infrakingdom groupings proposed here. Clearly, these require more rigorous testing. The rRNA data suggest that the Firmibacteria were derived from Selenobacteria, and the Mollicutes from Clostridea.

reductants such as H_2 or H_2S or H_2O: like the first cells (Oparin 1938), they can only use organic compounds from their environment. Their antenna pigments are the simplest known, namely bacteriochlorophyll g and only one carotenoid (neurosporene, which is near the base of the carotenoid biosynthetic pathway): bacteriochlorophyll g could have isomerized to yield chlorophyll a (characteristic of cyanobacteria and chloroplasts), and by adding extra biosynthetic steps could have evolved into bacteriochlorophyll a, which characterizes the green and purple bacteria. The heliobacterial electron transport chain has only one cytochrome (C553) and lacks cytochrome *b* found in most others. Their membrane lipid composition is a simple mixture of phosphatidylethanolamine, and mono- and di-phosphatidylglycerol—as simple as any in eubacteria. Some (*Heliobacterium*) are gliders, others have flagella (*Heliobacillus*, *Heliospirillum*), but none has gas vacuoles or chlorosomes. Like several other putatively primitive phyla, their outer membrane lacks lipopolysaccharide, one of the most complex molecules in most negibacteria, and their peptidoglycan has ornithine instead of diaminopimelic acid in the peptide cross-links.

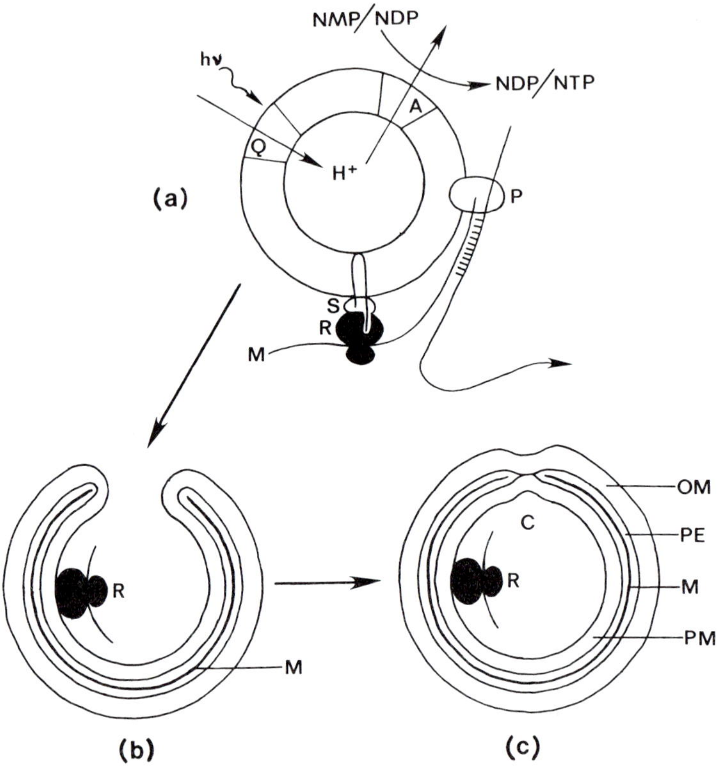

FIG. 2. Origin of the first cell and of primary metabolism. According to the 'obcell' theory (Cavalier-Smith 1987b), protein synthesis and co-translational protein insertion by a ribonucleoprotein signal recognition particle originated on the outer surface of a liposome-like obcell, or inside-out cell, using amino acids and nucleotides from the prebiotic soup, and the obcell evolved porphyrin-based proton pumping and chemiosmotic proton-driven phosphorylation before any metabolism (a). Later, the origin of peptidoglycan murein synthesized inside the obcell caused it to fold up (b) and the lips of the closing obcell to fuse together to form the first true cell (c). The interior of the obcell became the negibacterial periplasm, PE, and the newly enclosed portion of the exterior medium became the cytoplasm (C). Only then could soluble cytoplasmic biosynthetic pathways and primary metabolism have evolved, mainly using newly evolved enzymes, but in part using pre-existing 'obcellular' enzymes that were for the first time able to lose the signal peptides that formerly attached them to or translocated them across the obcell membrane to its interior; previously, deletion of signal peptides would have caused soluble proteins to be lost from the organism to the external medium. The two membranes derived from the 'gastrulation' of the obcell differentiated into the plasma membrane (PM) and outer membrane (OM) of the first negibacterium. DNA replication and DNA co-segregation with the murein division may have evolved before or during the conversion of obcell to cell (Cavalier-Smith 1987b). Q, light-driven Mg^{2+}/Fe^{2+} porphyrin/quinone-based proton pump; A, proton-driven P-type ATP/NTP synthetase; P, Membrane-bound RNA polymerase; R, ribosome; S, signal recognition particle; and M, coding (messenger) strand of RNA genome.

Bacterial diversification

Figure 3 is a cladistic analysis of character distribution in the eubacteria. It can be seen that treating the heliobacteria as the ancestral group gives a steady increase in both structural and metabolic complexity throughout the tree (with certain secondary simplifications, of which the loss of the outer membrane and the polyphyletic loss of photosynthesis are the most important). In my view, no other group than Heliobacteria could be chosen as the most primitive that would give as logical, and selectively and mechanistically comprehensible, a progression of steps from the simple to the complex. This tree is unlikely to be correct in every detail, but it will serve as a good working hypothesis for discussing metabolic evolution, and will, I hope, stimulate the search for additional cladistically useful conserved characters with which to test and refine it. Only two of the 13 eubacterial phyla have been well studied by molecular biologists (Proteobacteria e.g. *Escherichia coli*, and Posibacteria e.g. *Bacillus subtilis*): it is clear that these, together with the cyanobacteria and the parasitic chlamydias and mollicutes, are the most derived and advanced of all the eubacterial phyla. We need to know much more about the six putatively most primitive phyla, so as to establish more firmly the branching order and to test that proposed in Fig. 3, and thereby better reconstruct the metabolic capabilities of the ancestral cell.

All the differences between the eubacterial phyla probably arose in less than 100 million years, around 3.5×10^9 years ago. Pierson & Olson (1989) have lucidly summarized the driving forces for the diversification of the five photosynthetic phyla in the first microbial mats. All the non-photosynthetic bacteria probably evolved from them by loss of photosynthesis (Woese 1987, Cavalier-Smith 1987b, Pierson & Olson 1989). Their evolution can be divided into two phases: before and after the origin of oxygenic photosynthesis (and of aerobic respiration, which would have evolved very soon afterwards, probably in the first oxygenic photosynthesizers themselves: the common ancestors of cyanobacteria, proteobacteria and posibacteria).

The first anaerobic ecosystems

The first ecosystem after the origin of cells may have consisted purely of photoheterotrophic heliobacteria. Today, some are gliders and some flagellate swimmers: either method of motility could have originated either in the precellular obcell (Cavalier-Smith 1991b) or after it folded up to make the first cell. Benthic habitats (gliders) and planktonic ones (flagellates) were probably both colonized very early. It seems likely that antagonistic interactions were most strongly favoured in benthic habitats and that secondary metabolism would have initially co-evolved with primary metabolism in pure heliobacterial mats. Since there would then have been no primary producers (no CO_2 fixation), competition for organic resources remaining from the prebiotic soup would have been intense between these photoheterotrophs. Unfortunately, nothing is known about antibiotics in heliobacteria—or indeed any of the four most primitive phyla—but it seems highly probable that some of them would have evolved both antibiotics and cell-lytic agents very early on: antagonistic interactions of this sort occur in modern microbial mats (Marshall 1989) and deserve to be studied much more intensively.

I suggest that spirochaetes evolved from heliobacteria immediately after the

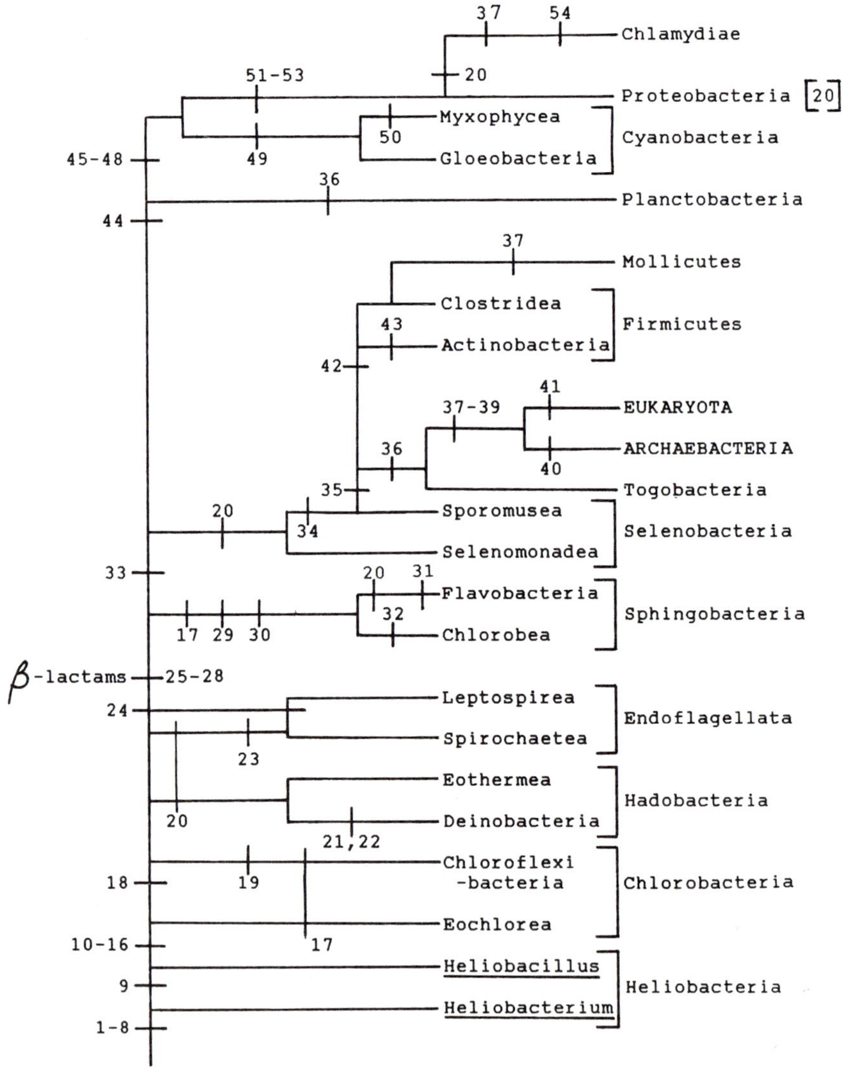

origin of flagella and that the first non-photosynthetic saprotrophic bacteria in the mats would have been the spirochaetes, whose unique mode of motility would allow them to corkscrew readily through the glutinous microbial mats of heliobacteria, in which they would also have been well protected from ultraviolet radiation. The highly radiation-resistant Deinococci may have been the first saprotrophs in the more exposed plankton or near the surface of the mats or of moist soil.

The long-term future of the ecosystem and its further increase in complexity depended on the origin of CO_2 fixation. This seems to have been first achieved by the green bacteria, probably independently in Chloroflexibacteria (by a still poorly characterized mechanism) and Chlorobiales (by the reductive tricarboxylic acid cycle). After this, another group of anaerobic chemoheterotrophs, the Flavobacteria, evolved, almost certainly from the Chlorobiales (Woese et al 1990); like both groups of green bacteria they are all non-flagellate, predominantly gliding bacteria, which include not only saprotrophs such as *Flavobacterium* and *Flexibacterium* but also the predatory *Cytophaga* species that actively lyse other cells. Since many of them make β-lactam antibiotics (Ono

FIG. 3. Cladistic analysis of the eubacteria. The position of 54 major innovations and losses in the history of the eubacteria are indicated by the numbered bars. 1, photosynthetic reaction centre Type 1; 2, separate plasma membrane and outer membrane; 3, neurosporene (a carotenoid); 4, mono- and diphosphatidylglycerol and phosphatidylcholine; 5, small type citrate synthase; 6, murein with ornithine instead of diaminopimelic acid; 7, gliding motility; 8, N_2 fixation; 9, flagella outside outer membrane; 10, Type 2 photosynthetic reaction centre; 11, bacteriochlorophyll a; 12, cytochrome *b*; 13, gas vacuoles; 14, lycopene, γ-carotene and more complex carotenoids; 15, phosphatidylinositol; 16, galactolipids; 17, loss of flagella; 18, bacteriochlorophyll c and chlorosomes; 19, CO_2 fixation (by unknown mechanism differing from reductive pentose phosphate and reductive tricarboxylic acid cycles); 20, loss of photosynthesis; 21, murein greatly thickened: Gram-positive staining; 22, high radiation resistance; 23, endoflagella (flagella relocated into periplasmic space); 24, ornithine in murein replaced by diaminopimelic acid; 25, β-lactam antibiotics; 26, oxygenic photosynthesis; 27, aerobic respiration: cytochromes aa_3 and *o* as terminal oxidases; 28, lipopolysaccharide in outer leaflet of outer membrane; 29, sphingolipids; 30, loss of Type 2 photosynthetic reaction centre and of oxygenic photosynthesis; 31, sulphonolipids; 32, CO_2 fixation by the reductive tricarboxylic acid cycle; 33, ribulose bis-phosphate carboxylase and carbon fixation by the reductive pentose phosphate cycle; 34, endospores with dipicolinic acid; 35, loss of outer membrane; 36, thinning of murein; 37, loss of murein; 38, N-linked glycoproteins; 39, protein-spliced introns; 40, isoprenoid ether lipids; 41, the eukaryote cell (23–29 new characters: Cavalier-Smith 1987a); 42, hypertrophy of murein; 43, exospores; 44, ubiquinone; 45, group I self-splicing introns in tRNA; 46, large form of citrate synthetase; 47, sulpholipids; 48, loss of chlorosomes; 49, phycobilisomes; 50, thylakoids; 51, chromatophores (invaginations of plasma membrane) bearing reaction centre Type 2; 52, bacteriochlorophyll b; 53, loss of photosynthetic reaction centre Type 1; 54, loss of ability to make ATP. The various taxa are defined in Table 2.

& Harada 1990), it is highly probable that β-lactam antibiotics originated in anaerobic microbial mats around 3500 million years ago, before the origin of oxygenic photosynthesis and respiration.

The origins and evolution of β-lactam antibiotics

According to Fig. 3, the target of the β-lactams, murein peptidoglycan, was present in the first cell; I suggested that murein evolved even earlier in the obcell and may have been instrumental in folding it up to form the first negibacterial cell (Cavalier-Smith 1987b). Therefore, β-lactams might be much older than indicated in Fig. 3. Many of the genes coding for their biosynthesis are so highly conservative that it should not be too difficult to use conserved primers to isolate them by the polymerase chain reaction (PCR). It is highly desirable to search for them in the four most primitive eubacterial phyla in order to test the hypothesis that many antibiotics may have originated during precellular evolution.

The strong homology between the fungal and the bacterial enzymes has led several authors (e.g. Kleinkauf & von Döhren 1990) to suggest that the fungi acquired them by lateral gene transfer from bacteria. This is certainly possible, because the bacterial genes are often clustered and the selective advantage of acquiring them could have been very great. However, this homology could equally well be accounted for by direct descent of the metabolic pathway from bacteria to fungi via the archezoan and protozoan ancestors of the fungi (Cavalier-Smith 1987a). I am confident that nobody has ever even looked for β-lactam antibiotics or the genes for the pathway, either in archezoa, or in the protozoa most likely to be ancestral to fungi, the choanoflagellates (Cavalier-Smith 1987c). Until this is done by modern PCR methods it will be unwise simply to assume by default that they are absent. To favour the lateral transfer over the direct descent hypothesis, one would not only have to show that the genes are uniformly absent from protozoa and archezoa, but also to show that the fungal genes resemble those of some specific group of bacteria substantially more closely than do the genes in different phyla of bacteria. Moreover, such evidence for lateral transfer would only be convincing if the bacteria with the most similar genes turned out to be a different group from those most likely to be ancestral to eukaryotes.

Are secondary metabolites relics of precellular evolution?

Davies (1990) proposed that antibiotics and other secondary metabolites originally served a positive function in the very reactions that they now inhibit in the days before the origin of coded protein synthesis, and are present today largely because of such primordial functions. This hypothesis has three major defects which lead me to reject it most strongly.

1. It fails to explain how the enzymes that now make the antibiotics originated. If the first antibiotics were indeed made prebiotically, there would have been no genetic continuity between their synthesis and the presence of the same antibiotic today: the hypothesis suffers from the same fundamental defect as does de Duve's (1991) hypothesis of the takeover of a pregenetic metabolism by a genetically based one. There is no known or readily conceivable way by which this could have been done. For this reason I am also sceptical of Lipmann's (1971) suggestion that the peptide antibiotic synthesis mechanism, involving a series of unique enzymes, might have preceded protein synthesis by ribosomes.
2. The phylogenetic evidence strongly indicates that the majority of secondary metabolites are of recent origin and are not primordial. A few, such as the β-lactams and peptide antibiotics, may well go back to the earliest bacteria and therefore have arisen very early. Since these mostly target peptidoglycan biosynthesis, membranes or ribosomes, and these structures may all have arisen in a precellular phase of evolution (the inside-out cell or obcell), they could have arisen even before the origin of the first cell. But even if this were true for some of these antibiotics, there is no reason to suppose that the majority of secondary metabolites have such an early origin. Moreover, even for those that evolved very early, there is an absolutely fundamental distinction between an origin soon after the evolution of translation mechanisms, which is quite plausible, and an early origin before the origin of translation, which is not, for the reasons given in the previous paragraph.
3. One cannot explain the persistence and diversification of genes for the enzymes of secondary metabolism, for hundreds or thousands of millions of years, by postulating an archaic function for the same metabolites before the origin of protein synthesis. The metabolites must have been continuously beneficial to the producing organism for the spread and survival of the genes coding for them to be favoured. Without such a continuous benefit, the genes would have been lost billions of years ago by degenerative mutations. The fact that antagonistic ecological roles have not been demonstrated for the majority of antibiotics in Nature is no reason to postulate instead vague prebiotic functions. (In no case is it suggested precisely how such functions would have contributed to the survival/reproductive success of the precellular entities.) The present situation largely reflects our general ignorance of the molecular population ecology of microorganisms, which is a much more difficult and grossly neglected field than the molecular physiology that still dominates the thinking of molecular biologists. For animals and plants, which are much easier to study, there is much evidence for a general antagonistic role for secondary metabolites (Luckner 1990).

The idea that antibiotics were the functional precursors of ribosomal proteins (Davies 1990) seems far-fetched even for antibiotics that now block protein synthesis, but it could hardly begin to be applied to those that attack membranes or peptidoglycan synthesis. Still less would it apply to secondary metabolites that attack eukaryote-specific proteins (e.g. colchicine, vinblastine and numerous other diverse plant products that target microtubules, or fungal metabolites like phalloidin or cytochalasin that affect actin filaments). The targets for these and numerous other secondary metabolites probably did not originate until 1.7×10^9 years after the origin of life, while the plant and fungal biosynthetic pathways that make them probably originated over 10^9 years after their targets. So these metabolites can have played no role in the origin of the eukaryote cytoskeleton. In my view, primary metabolism originated largely after the origin of the first cell [whether by folding of an obcell (Cavalier-Smith 1987b) or by encapsulation by a single membrane (Hargraeves & Deamer 1978, Morowitz et al 1988)] and received a further major boost with the origin of CO_2 fixation. This is because both the enzymes and the intermediates are water soluble and would have had to be enclosed by a lipid membrane.

Since this is also true of most secondary metabolism, I think that antibiotics made by several enzymic steps involving soluble intermediates would not have originated prior to the origin of the first closed cell. However, if an antibiotic could have been made by a single enzymic step by a membrane-bound enzyme—or if all intermediates were membrane associated—ready diffusion away of the product to attack other precellular entities would be an advantage, not a drawback. Whether such precellular entities could be sophisticated enough to avoid harm to themselves in any of the ways now achieved by bacteria (Cundliffe 1989) is not clear. But, if they could, one could, with the above provisos, envisage the evolution of at least some antibiotics before the origin of cells. Whether any modern antibiotic biosynthetic pathways go back as far as that can best be approached by seeing whether or not they are present in the most divergent eubacteria shown in Fig. 3. We shall not discover this merely by speculation or by confining our studies to advanced bacteria like *E. coli*, *Bacillus subtilis* or *Streptomyces*.

Envoi

Secondary metabolism does not make sense in terms of the physiological concept of function, whether contemporary or prebiotic. But the perpetual ecological battle for numerical supremacy provides a unifying selective force for the origins of secondary metabolites by the opportunistic recruitment of numerous bizarre chemicals as agents of chemical warfare, by the haphazard modification of genes for pre-existing metabolic enzymes.

Acknowledgements

I thank T. Chappell for typing, E. Chao for help with the figures, and the Natural Sciences and Engineering Research Council of Canada for an operating grant. The author is a fellow of the Canadian Institute for Advanced Research.

References

Blobel G 1980 Intracellular membrane topogenesis. Proc Natl Acad Sci USA 77: 1496–1500

Blunden G, Gordon SM 1986 Betaines and their sulphonio analogues in marine algae. Prog Phycol Res 4:39–80

Cavalier-Smith T 1980 Cell compartmentation and the origin of eukaryote membranous organelles. In: Schwemmler W, Schenk HEA (eds) Endocytobiology: endosymbiosis and cell biology, a synthesis of recent research. Walter de Gruyter, Berlin, p 893–916

Cavalier-Smith T 1987a The origin of eukaryote and archaebacterial cells. Ann NY Acad Sci 503:17–54

Cavalier-Smith T 1987b The origin of cells: a symbiosis between genes, catalysts, and membranes. Cold Spring Harbor Symp Quant Biol 52:805–824

Cavalier-Smith T 1987c The origin of fungi and pseudofungi. In: Rayner ADM, Brasier CM, Moore D (eds) Evolutionary biology of the fungi. Cambridge University Press, Cambridge (Symp Br Mycol Soc 13) p 339–353

Cavalier-Smith T 1989 The kingdom Chromista. In: Green JC, Leadbeater BSC, Diver WC (eds) The chromophyte algae: problems and perspectives. Oxford University Press, Oxford, p 379–405

Cavalier-Smith T 1990a Microorganism megaevolution: integrating the living and fossil evidence. Rev Micropaleontol 33:145–154

Cavalier-Smith T 1990b The symbiotic origin of peroxisomes. In: Nardon P, Gianinazzi-Pearson V, Grenier AM, Margulis L, Smith DC (eds) Endocytobiology IV. Institut National de la Recherche Agronomique, Paris, p 515–521

Cavalier-Smith T 1991a The evolution of cells. In: Osawa S, Honjo T (eds) Evolution of life. Springer-Verlag, Tokyo, p 271–304

Cavalier-Smith T 1991b Evolution of prokaryotic and eukaryotic cells. In: Bittar GE (ed) Fundamentals of medical cell biology. JAI Press, Greenwich, CT, vol 1:217–272

Cavalier-Smith T 1992 The origin, losses and gains of chloroplasts. In: Lewin RA (ed) Symbiogenesis: prochlorophytes and the origins of plastids. Chapman & Hall, New York, p 291–349

Cohen Y, Rosenberg E 1989 Microbial mats: physiological ecology of benthic microbial communities. American Society for Microbiology, Washington, DC

Cundliffe E 1989 How antibiotic producing organisms avoid suicide. Annu Rev Microbiol 43:207–234

Davies J 1990 What are antibiotics? Archaic functions for modern activities? Mol Microbiol 4:1227–1232

Dawes IW 1981 Sporulation in evolution. In: Carlile MJ, Collins JF, Moseley BEB (eds) Molecular and cellular aspects of microbial evolution. Cambridge University Press, Cambridge, p 85–130

de Duve C 1991 Blueprint for a cell: the nature and origin of life. Neil Patterson Publishers, Burlington, NC

Gest H, Favinger JL 1983 *Heliobacterium chlorum*, an anoxygenic brownish-green photosynthetic bacterium containing a new form of bacteriochlorophyll. Arch Microbiol 136:11–16

Goldacre RJ 1958 Surface films, their collapse on compression, the shapes and sizes of cells and the origins of life. In: Danielli JF, Parkhurst KGA, Riddifin AC (eds) Surface phenomena in chemistry and biology. Pergamon Press, New York, p 278–298

Hargraeves WR, Deamer DW 1978 Origin and early evolution of bilayer membranes. In: Deamer DW (ed) Light transducing membranes: structure, function and evolution. Academic Press, New York, p 23–59

Huber R, Woese CR, Langworthy TA, Fricke H, Stetter KO 1989 *Thermosipho africanus* gen. nov., represents a new genus of thermophilic eubacteria within the '*Thermotogales*'. Syst Appl Microbiol 12:32–37

Kleinkauf H, von Döhren H 1990 Bioactive peptides—recent advances and trends. In: Kleinkauf H, von Döhren H (eds) Biochemistry of peptide antibiotics. Walter de Gruyter, Berlin, p 1–31

Kobayashi M, Van de Meent EJ, Erkelens C, Amesz J, Ikegami I, Watanabe T 1991a Bacteriochlorophyll g epimer as a possible reaction centre component of heliobacteria. Biochim Biophys Acta 1057:89–96

Kobayashi M, Watanabe T, Ikegami I, Van de Meent EJ, Amesz J 1991b Enrichment of bacteriochlorophyll g in membranes of *Heliobacterium chlorum* by ether extraction. FEBS (Fed Eur Biochem Soc) Lett 284:129–131

Lipmann F 1971 Attempts to map a process evolution of peptide biosynthesis. Science (Wash DC) 173:875–884

Luckner M 1990 Secondary metabolism in microorganisms, plants and animals, 3rd edn. Springer-Verlag, Berlin

Madoff S (ed) 1986 The bacterial L-forms. Marcel Dekker, New York

Mann J 1987 Secondary metabolism, 2nd edn. Clarendon Press, Oxford

Marshall KC 1989 Cyanobacterial–heterotrophic bacterial interaction. In: Cohen Y, Rosenberg E (eds) Microbial mats: physiological ecology of benthic microbial communities. American Society for Microbiology, Washington, DC, p 239–245

Morowitz HJ, Heinz B, Deamer DW 1988 The chemical logic of a minimum protocell. Origins Life Evol Biopshere 18:281–287

Olson JM, Pierson BK 1987a Evolution of reaction centers in photosynthetic prokaryotes. Int Rev Cytol 108:209–248

Olson JM, Pierson BK 1987b Origin and evolution of photosynthetic reaction centres. Origins Life 17:419–430

Ono H, Harada S 1990 Discovery of new β-lactam and β-lactam-like antibiotics from bacteria. In: Kleinkauf H, von Döhren H (eds) Biochemistry of peptide antibiotics. Walter de Gruyter, Berlin, p 131–158

Oparin AI 1938 The origin of life. Macmillan, New York

Pierson BK, Olson JM 1989 Evolution of photosynthesis in anoxygenic photosynthetic procaryotes. In: Cohen Y, Rosenberg E (eds) Microbial mats: physiological ecology of benthic microbial communities. American Society for Microbiology, Washington, DC, p 402–427

Ragan MA, Glombitza K-W 1986 Phlorotannins, brown algal polyphenols. Prog Phycol Res 4:129–141

Vining LC 1990 Functions of secondary metabolites. Annu Rev Microbiol 44:395–427

Woese CR 1987 Bacterial evolution. Microbiol Rev 51:221–271

Woese CR, Fox GE 1977 The concept of cellular evolution. J Mol Evol 10:1–6

Woese CR, Wolfe RS 1985 Archaebacteria: the urkingdom. In: Woese CR, Wolfe RS (eds) The bacteria. Archaebacteria. Academic Press, New York, vol 8:561–564

Woese CR, Mandelco L, Yang D, Ghema R, Madigan MT 1990 The case for relationship of the Flavobacteria and their relatives to the green sulfur bacteria. Syst Appl Microbiol 13:258–262

DISCUSSION

Turner: I worry when I see diagrams in the literature of when various events happened, like the eukaryotes splitting from prokaryotes, because the dates can vary by a billion years either way. This makes it hard to get to grips with ideas about the lateral transfer of gene clusters, and the possible dates when they occurred, because the sequencing of one extra gene may add another billion years to the estimate. Presumably, somewhere the dates must go back to fossil records, to provide some timing? Many people are tracking back gene sequences, and you don't know how fast a gene is changing without independent data to relate change to. How reliable are microbial fossil records? Is it give or take half a billion years, or what?

Cavalier-Smith: It depends. The data are becoming more reliable, but considerable uncertainty remains. Micropalaeontologists have long regarded about 1400–1500 million years ago as the time when eukaryotes first appeared in the fossil record. More recently, this date has increased to about 1700 million years, a distinct change; it is based solely on a marked increase in cell size in the fossil record. A few people maintain that cell size doesn't tell you anything about the time of origin of eukaryotes, but in fact prokaryotes seem unable to grow beyond a certain size, whereas all the major phyla of eukaryotic protists (i.e. single-celled eukaryotes) have the capacity to make very big cells. Even some archezoa are as much as 1 mm in diameter. The basic eukaryotic structures of the cytoskeleton and nucleus are probably required for this larger size. Since all eukaryotes have the capacity to evolve large cells and there would always have been selective advantages for different species being numerous different sizes, the first really large cells would have almost certainly evolved during the initial adaptive radiation, very soon after the origin of eukaryotes (Cavalier-Smith 1991). So I think fossil cell size is a very important and reasonably accurate indicator of the approximate time of origin of archezoan eukaryotes.

A second objection raised to this is that protists such as amoebae are soft-bodied and don't leave fossils; they therefore could have existed much earlier than the fossil record indicates. But that is a fallacy, because every group of protists has some stage in its life history where cell walls are formed. Many amoebae form cysts and these fossilize. There is a far better fossil record for eukaryotic single-celled organisms than for any multicellular group: since about 700 million years ago you can identify some of these into major groups (e.g. foraminiferans, radiolarians and dinoflagellates). Before then, you can't, because they seem to have been very simple morphologically, which suggests that some major change happened about that time. I suggested (Cavalier-Smith 1990) that this was the origin of the Golgi dictyosome, which is involved in making and secreting the extracellular materials for the cell walls. But except for some multicellular cyanobacteria you can't assign a fossil bacterium to any major group; a fossilized 'bacterial' cell could even really be a small eukaryote. But you can say that for the first 1700 million years, there is no evidence whatever for eukaryotes.

Bu'Lock: You are very keen to place the eubacteria a long way from the fungi; but you are also clear that the β-lactam system got into the fungi by lineal descent. This seems contradictory, because in fact the β-lactams didn't get into the *kingdom* of the fungi, but only into two or three species. I find this a much more compelling argument for lateral transfer than for lineal descent. If it went by lineal descent, it presumes that in all other species of fungi and in all known species of higher plant, β-lactam synthetic capability was deleted. That doesn't seem to me economical.

Cavalier-Smith: I agree that it is not economical, if it is really true that β-lactams are not found in other fungi. People used to make similar suggestions about the occurrence of haemoglobin in plants; they thought it was present only in legumes. It has since been discovered in a much wider variety of plants, but is difficult to detect. We don't know whether the genes for β-lactam synthesis are present in the other fungi but the compounds are made at an undetectably low level.

Secondly, to demonstrate lateral transfer from bacteria, what you need to show is not just the presence of β-lactams in the two groups. You must demonstrate that the amino acid sequences of their biosynthetic enzymes in the fungi are closer to those of one particular bacterial group than are those of different bacteria to each other. You must also show that the ancestral bacterium was a different organism from the one that was ancestral to the eukaryotes. There's no indication that that is true. There seems to be a great deal of divergence between the eukaryotic and the streptomycete sequences—as much divergence as you might expect on the direct descent theory.

Davies: Has the presence of β-lactam *rings* been examined in various organisms? Is the distribution of molecules that have β-lactam rings known?

Bu'Lock: If you just look at these structural features, then you have to add *Bacillus* and of course *Streptomyces* to the list of microbial groups.

Turner: The problem is that most of the data on how many species of fungi make β-lactams haven't come through academic research, but from pharmaceutical companies screening huge numbers of organisms. Some discoveries were made relatively recently—for example, the discovery of β-lactam production in *Flavobacterium* (Singh et al 1984). The outcome depends on the sensitivity of the techniques. Cohen et al (1990) suggest that it would be interesting to hunt for antibiotic pathways with gene probes in a wide range of microorganisms.

Cavalier-Smith: The sequences are so conserved that you should be able to pick them out by PCR, even in an organism making them at a very low level. If that search were totally negative in the majority of fungi and also in archezoa and the relevant protozoa, the case for direct descent would indeed be weakened.

Bu'Lock: An interesting facet of secondary metabolism is that in so many cases the extent to which it actually proceeds is so small. When you are looking for antibiotics you can come up with an organism which makes a very small amount of something very interesting. You may spend the next 10 years trying to get it to produce more, *and it never will.* Many of these metabolites are produced only in very small amounts. The genes and the enzymes are there, but they are not very functional! I don't know the significance of this, but it's often overlooked.

Wächtershäuser: Professor Cavalier-Smith, you say that the double cell wall of Gram-negative bacteria is ancestral, and the single cell wall of Gram-positive organisms arose by loss of the outer wall. How do you account for the archaebacterial cell wall, which is found in the methanogens, and has been argued to have arisen separately in one branch of the methanogens? But recently a new methanogen was found, *Methanopyrus kandleri*, which branches off at a very different place, very close to the root of the archaebacteria, and it also has a cell wall.

Cavalier-Smith: I was talking about the outer *membrane*, not the cell wall. The outermost wall layer in *M. kandleri* (Kurr et al 1991) is probably a proteinaceous S-layer, as in other archaebacteria as well as eubacteria, not a second lipid bilayer. What I have described still stands, namely that the eukaryotes and archaebacteria share the presence of only a single membrane with the eubacteria that I call the Posibacteria. I regard these three groups as related to each other. There was only a *single* loss of the outer membrane in the common ancestor of all of them, I suggest. These organisms are clearly related to each other by many criteria (Cavalier-Smith 1987); the actinomycetes especially share features in common with eukaryotes. I think my cladogram gives a reasonable phylogeny. This point concerns just the lipids; I wasn't talking about the cell walls.

Vining: Is it possible that some of the secondary metabolic pathways considered to have been transferred laterally might have entered the eukaryotes by endosymbiosis? Just as chloroplasts provided photosynthetic capabilities that were positively selected for, so perhaps a β-lactam-producing bacterium might have given considerable advantages to a host fungus. The β-lactam biosynthetic genes might have been transferred in, not by some kind of genetic transformation, but rather in a whole organism?

Cavalier-Smith: It is perfectly possible and I would be surprised if it hadn't happened. I am not opposed to lateral transfer, either of DNA or by endosymbiosis. Both are events for which there is evidence, and they have to be considered. But I would like people not merely to postulate what might have happened, but to look to see what phylogenetic evidence there is as to whether it did happen. If a secondary metabolic pathway did come from a symbiont, you ought to be able to show a much closer relationship to some particular symbiont; for example, to cyanobacterial genes for a chloroplast ancestry, or to purple bacterial genes for a mitochondrial ancestry. In the absence of such evidence the simplest hypothesis is direct descent.

Davies: Dr Vining, you are talking about a form of symbiosis in which a whole organism is incorporated?

Vining: Yes; I am suggesting an analogy to the widely accepted origin of chloroplasts; an organism such as an ancestral form of actinomycete with the β-lactam pathway might have been the endosymbiont.

Davies: Has any secondary metabolic synthetic pathway in eukaryotes been found that is associated with organelles? This might support this idea?

Vining: I don't know of any mitochondrial or chloroplast biosynthetic pathway for a secondary metabolite, but I don't think it's essential to the idea for the organelle to have persisted.

Chater: I'd like to mention two potentially relevant aspects of actinomycetes. Dr Cavalier-Smith talked about the size of cells. There is a circumstance in which *Streptomyces* produces cells that may be 100 μm across: during protoplast regeneration, huge protoplasts are formed and then hyphae emerge from them.

Secondly, in sections of *Streptomyces* colonies one can sometimes see one hypha growing up inside another. When that happens, I presume you have two membranes with the right topology to fit with what you would like to see. One could view that as having happened either by invasion, or alternatively as mycelial branching from a cross-wall rather than from a lateral wall.

Cavalier-Smith: That's similar to the hypothesis put forward by Dawes (1981) for the origin of the outer membrane. He suggested that it came from the membrane surrounding the endospore, which is formed by invagination of the plasma membrane.

Chater: Then you would have the wrong topology of the two membranes; they would be facing each other.

Cavalier-Smith: The outer membrane has completely different proteins from the plasma membrane and would have to have changed radically, by any hypothesis, from how it first started, so I wouldn't worry too much about that: there is no evidence of conserved protein 'topology' or orientation between the two membranes. I just find the addition of a second membrane very implausible, but it's not impossible to conceive. One would have not only to add the extra membrane but also to integrate its biogenesis with that of the plasma membrane, immediately and without harming the cell: how could this have occurred in a posibacterium with a thick cell, and what would be its selective advantage?

Orgel: The co-factors involved in some central biochemical pathways are different in the archaebacteria from elsewhere. How would that fit in with your model of a late divergence?

Wächtershäuser: Deazaflavin has been found in eubacteria as well.

Cavalier-Smith: One of the methanogenic coenzymes (coenzyme F) is also found in a eubacterium, *Streptomyces*, and in a posibacterium, which is interesting.

Orgel: So you think that both were present before the divergence?

Cavalier-Smith: Probably some evolved before the divergence, and some after. Differences don't necessarily mean independent origins: they very often arise through divergence. There are so many things in common in archaebacteria and eubacteria that we must suppose the common ancestor to have been a highly developed cell with many hundreds of different genes; you can't explain all these similarities in terms of separate origins from a precellular organism.

Orgel: But if it was something that was fairly fundamental to the biochemistry of all organisms, it would be surprising if you had to postulate a switch of co-enzyme. You could instead postulate that both were present before the separation.

Cavalier-Smith: One of the biggest switches is in the lipids of the bacterial membrane, from acyl esters to isoprenoid ethers. It is sufficiently drastic to justify treating archaebacteria and eubacteria as separate kingdoms, but it's not an impossible switch, if there was an intermediate with both types of lipids.

Wächtershäuser: There have been reports on sizeable amounts of fatty lipids being found in archaebacteria; and in eukaryotes (sponges) we find isoprenoid lipids.

Cavalier-Smith: With ester links?

Wächtershäuser: Yes. Of course, there may be archaebacteria living inside the sponges, but one also finds isoprenoid lipids in liver and brain cells; so it is quite common that eukaryotic cells contain isoprenoid acid esters.

Bu'Lock: Those esters have been known for many years. What we are concerned with in the archaebacteria are isoprenoid *ethers.*

Wächtershäuser: Yes, but you find ethers and esters in all kingdoms. You find a lot of ether lipids in eubacteria and in eukaryotes.

Cavalier-Smith: They are different; they are not isoprenoid ethers. Certainly some fungi have acyl ethers, but these are different.

Bu'Lock: The plasmalogens of mammals etc. are enol ethers and quite different from the true ether lipids that we are discussing. The interesting point is that only one section of the archaebacteria evolved the biterminal diether type of membrane-spanning lipid. It's interesting that certain true bacteria using ester lipids have also evolved biterminal diester lipids to fulfil a very similar task of membrane stabilization. This is an extremely good example of parallel evolutionary convergence in chemical terms. It fits with older ideas about convergence quite happily, because there's no way in which one type of lipid could be a descendant or ancestor of the other.

Orgel: What is your opinion about Norman Pace's claim (1991) that you can safely assume that the last common ancestor of all extant organisms lived at high temperatures?

Cavalier-Smith: His claim is based on the fact that in the archaebacterial part of the phylogenetic tree, both major branches ('crenarchaeotes' and 'euryarchaeotes') have thermophilic members; most of them in one branch (Crenarchaeota) are thermophilic, as are some of those in the Euryarchaeota. It is therefore probable that the common ancester of all archaebacteria was a thermophile. To that extent I agree with the Woese (1987) and Pace (1991) interpretation. But what I have always disputed (Cavalier-Smith 1991) is their assumption that the archaebacteria are in fact 'old'. Both the cytological and fossil evidence and the molecular gene duplication evidence suggest that they are more recent than eubacteria.

What then was the ancestral eubacterium like? Was it a thermophile? It could have been, but I think Woese's assumption that the thermophilic *Thermotoga* is near the base of the eubacterial tree is probably wrong. However, some of the green bacteria near the bottom of my tree are highly thermophilic, although most are mesophiles.

The tree I gave you is my present interpretation, but it may be incorrect. We need to find many more genes that underwent duplication within the eubacteria, and to get sequences for them from each of the different phyla, in order to work out the direction of evolution between them and therefore to locate the root of the tree with more confidence. Only then can we say what the latest common ancestor was really like.

Orgel: So you think the evidence that the common ancester was a thermophile is not conclusive.

Cavalier-Smith: That is my view. On general grounds of the stability of molecules, I would favour a non-thermophilic ancestor.

Baldwin: Do you have any thoughts on the possible origin of the murein sacculus?

Cavalier-Smith: I have suggested that the murein sacculus (cell wall) originated in the inside-out cell, before the origin of the first cell. One reason is that such an organism must have needed a mechanism of division early on, other than purely random fragmentation. Modern bacteria use murein as their division mechanism. It has D-amino acids, and two of them are among the easiest to form in Miller-type prebiotic mixes. At the bottom of my phylogenetic tree all the organisms contain murein. All those without murein seem to be derivative, so I think it was phylogenetically ancient.

Baldwin: There must have been some very ancient peptide-synthesizing mechanism preceding it?

Cavalier-Smith: I wouldn't suggest that murein originated before the evolution of translation and ribosomes; I would argue that a 'symbiosis' first occurred between a primitive RNA/protein world and a lipid world; and at that stage murein developed, to control the division of the lipid and to help attach the genes to it better.

References

Cavalier-Smith T 1987 The origin of eukaryote and archaebacterial cells. Ann NY Acad Sci 503:17–54

Cavalier-Smith T 1990 Microorganism megaevolution: integrating the living and fossil evidence. Rev Micropaleontol 33:145–154

Cavalier-Smith T 1991 The evolution of cells. In: Osawa S, Honjo T (eds) Evolution of life. Springer-Verlag, Tokyo, p 271–304

Cohen G, Shiffman D, Mevarech M, Aharonowitz Y 1990 Microbial isopenicillin N synthase genes: structure, function, diversity and evolution. Trends Biotechnol 8:105–111

Dawes IW 1981 Sporulation in evolution. In: Carlile MJ, Collins JF, Moseley BEB (eds) Molecular and cellular aspects of microbial evolution. Cambridge University Press, Cambridge, p 85–130

Kurr M, Huber R, König H et al 1991 *Methanopyrus kandleri*, gen. and sp. nov. represents a novel group of hyperthermophilic methanogens, growing at 110 °C. Arch Microbiol 156:239–247

Pace NR 1991 Origin of life—facing up to the physical setting. Cell 65:531–533

Singh PD, Young MG, Johnson JH, Cimarusti CM, Sykes RB 1984 Bacterial production of 7-formamidocephalosporins: isolation and structure determination. J Antibiot (Tokyo) 37:773–780

Woese CR 1987 Bacterial evolution. Microbiol Rev 51:221–271

Genes for polyketide secondary metabolic pathways in microorganisms and plants

David A. Hopwood and Chaitan Khosla

John Innes Institute, John Innes Centre, Norwich NR4 7UH, UK

Abstract. Recent advances in molecular genetics have led to the isolation, sequencing and functional analysis of genes encoding synthases that catalyse the formation of several classes of polyketides. The structures of the genes and their protein products differ strikingly in the various examples. For *Streptomyces* aromatic polyketides, exemplified by granaticin and tetracenomycin, the synthases correspond to Type II (bacterial and plant) fatty acid synthases in consisting of distinct proteins for such processes as condensation, acyl carrier function and ketoreduction. In contrast, for actinomycete macrolides such as erythromycin, similar catalytic functions are performed by a set of multifunctional proteins resembling Type I (animal) fatty acid synthases, but with every step in chain-building being catalysed by a different enzymic domain. *Penicillium patulum* has a simple Type I synthase for 6-methylsalicylic acid. For plant chalcones and stilbenes, a single small polypeptide acts as a condensing enzyme for carbon chain-building and may be unrelated to any of the other polyketide and fatty acid synthases. Thus, although these systems share a common general mechanism of chain assembly, they must differ in the ways that synthase 'programming' has evolved to determine chain length, choice of chain starter and extender units, and handling of successive keto groups during chain assembly, and so control the great diversity of possible chemical products.

1992 Secondary metabolites: their function and evolution. Wiley, Chichester (Ciba Foundation Symposium 171) p 88–112

Polyketides represent perhaps the largest family of secondary metabolites, especially abundant among actinomycete bacteria and eukaryotic fungi and higher plants, but also present in many other groups, including higher animals. They are interesting in this symposium because of recent evidence that synthetic enzymes for most polyketide *secondary* metabolites and those for fatty acids, an almost universal class of *primary* metabolites, have a common evolutionary origin. Fatty acids, as components of prokaryotic and eukaryotic cell membranes, are essential for the integrity of these organisms. Fatty acid biosynthesis probably evolved in a progenitor of modern prokaryotes and eukaryotes and was inherited, with largely invariant chemistry, by all its modern descendants. In contrast, polyketides are typical of specific stages in the life cycle of an organism, or of specific organs. Although a common underlying biochemistry is discernible, the genes that control it have diverged to code for enzymes that generate an almost infinite variety of chemical structures appropriate for particular ecological niches. Thus, many actinomycete polyketides are antibiotics that may help to defend the biomass accumulated by the producer during vegetative growth against loss to competing microorganisms during the reproductive phase; numerous fungal polyketides are toxins, such as patulin and the aflatoxins; mammalian examples include the prostaglandin and leukotriene hormones; and a large class of plant polyketides are flower pigments important for insect pollination.

After reviewing the chemistry that unites, and distinguishes, fatty acid and polyketide biosynthesis, we describe several alternative modes of organization of the DNA that codes for fatty acid and polyketide synthases, implying different strategies for the control of chemical structure by synthase 'programming', and speculate on the evolutionary relationships between them. Recent reviews, for further reading, include those by Hopwood & Sherman (1990), O'Hagan (1991) and Robinson (1991).

Fatty acid and polyketide biosynthesis

The enzyme system that catalyses the synthesis of fatty acids is termed the fatty acid synthase (FAS), and that involved in polyketide synthesis, the polyketide synthase (PKS). The essence of fatty acid and polyketide biosynthesis is an iterated *decarboxylative condensation* in which simple carboxylic acids are joined in head-to-tail fashion to produce a chain of 6–50 carbon atoms, each building unit contributing two carbons to the backbone of the chain. Figure 1(a) represents the simplest case, in which a malonyl 'extender unit' is condensed with an acetyl 'starter unit' to produce a four-carbon chain and CO_2. Both FASs and PKSs require that the acetyl and malonyl groups taking part in the condensation are not the free acids, but thioesters of them. The free acids are activated by attachment to the thiol of the 4′-phosphopantetheine moiety of CoA, which acts as a carrier. Two essential thiols are found within the FAS and PKS systems, one provided by a 4′-phosphopantetheine 'arm' bound to a serine hydroxyl of a synthase component distinct from the condensing enzyme, the acyl carrier protein (ACP), and the other by an active site cysteine residue on the condensing enzyme itself.

Because each condensation brings a keto group to the growing chain, the product would have keto groups on alternate carbon atoms if chain-building proceeded simply by a succession of condensations. This is precisely what happens at one end of the spectrum of polyketide synthesis—hence the name polyketide. However, at the other extreme, represented by the synthesis of

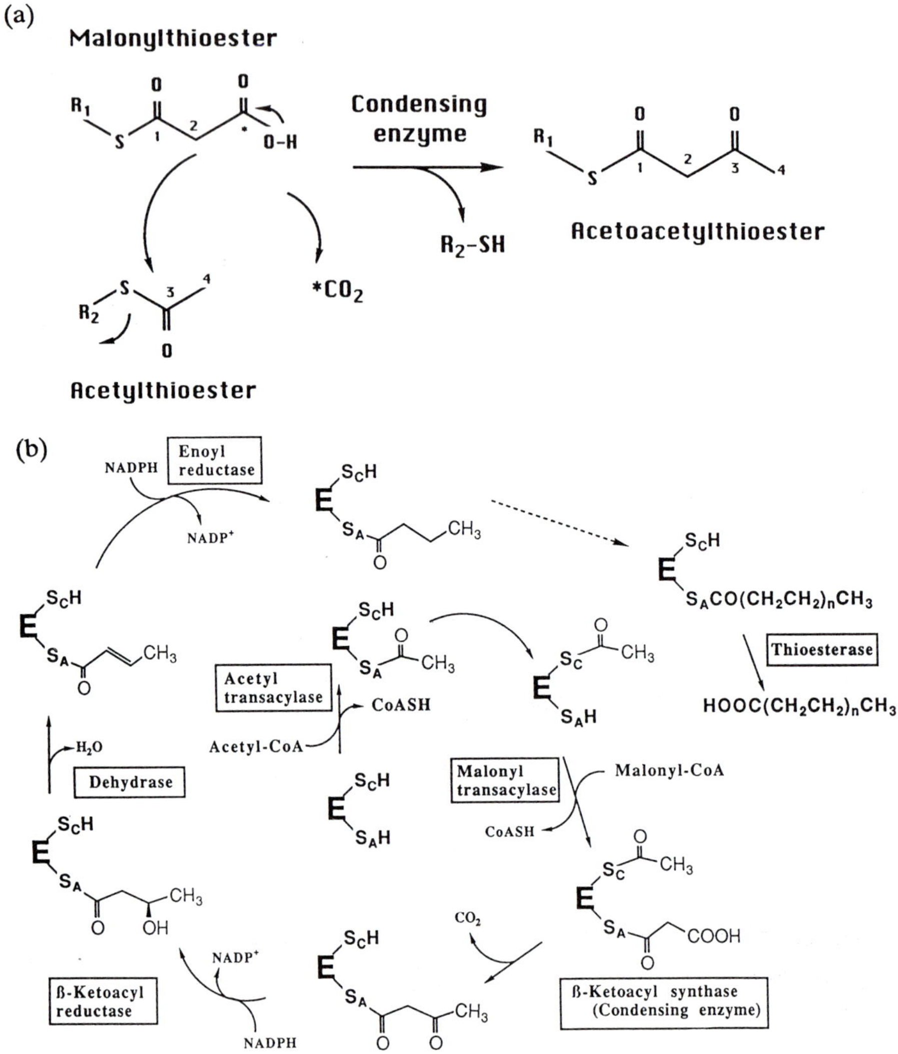

FIG. 1. Chemistry of fatty acid and polyketide biosynthesis. (a) The decarboxylative condensation reaction, illustrated by the first condensation in the synthesis of an unbranched fatty acid or polyketide. R_1 is typically the 4′-phosphopantetheine prosthetic group of the acyl carrier protein (ACP), and R_2 is the active site cysteine residue of the condensing enzyme. (b) The series of reactions in the synthesis of a typical unbranched, long-chain fatty acid. The starting point is in the centre of the cyclical part of the diagram, where acetyl-CoA enters. S_AH is the thiol of the ACP's 4′-phosphopantetheine prosthetic group, and S_CH is the thiol of the active site cysteine of the condensing enzyme. (Part (b) is reproduced from Robinson 1991, with minor modifications, by permission of the author and the Royal Society.)

saturated fatty acids like palmitic acid, all keto groups (except that of the terminal carboxyl) are removed during carbon chain assembly. This happens between successive condensations by a cycle of three reactions: reduction of keto to hydroxyl, elimination of the elements of water to produce a double bond in the carbon chain, and saturation of the double bond. Keto group removal thus requires three further enzymic capacities of the synthase: a ketoreductase, a dehydrase and an enoyl reductase activity. Furthermore, the complexity of the reduction–dehydration–reduction cycle may account for the requirement of an ACP to marshal the growing carbon chain around the active centres of the three enzymes.

Figure 1(b) shows the roles of the five components of a typical synthase named so far, and introduces the others—the acyltransferases (transacylases). For condensation to occur, an acetyltransferase typically moves the starter acetyl unit from the 4′-phosphopantetheine thiol of CoA to that of the ACP. The acetyl is then translocated, by an unknown mechanism, to the thiol of the condensing enzyme, leaving the ACP thiol free to receive a malonyl extender unit, transferred from malonyl-CoA by malonyltransferase. Decarboxylative condensation now occurs, to produce a four-carbon chain which is still attached to the ACP; the reduction–dehydration–reduction cycle operates, the growing chain returns to the thiol of the condensing enzyme, and a further extender unit is transacylated to the thiol of the ACP. After the correct number of cycles, the completed carbon chain leaves the FAS, either by transfer back to CoA or by hydrolysis to the free acid catalysed by a thioesterase. Thus, the FAS/PKS has up to eight constituent functions: condensing enzyme (β-ketoacyl-ACP synthase), β-ketoacyl-ACP reductase, β-hydroxyacyl-ACP dehydrase, enoyl-ACP reductase, ACP, and up to three acyltransferases (or two acyltransferases and a thioesterase). (We say ‘up to’ because sometimes the same transferase handles acetyl *and* malonyl groups, or malonyl groups *and* the mature carbon chain; and in the FAS from *Escherichia coli* there may be no role for acetyl transferase, because the special condensing enzyme III that carries out the first condensation in carbon chain building uses acetyl-CoA rather than acetyl-ACP as its substrate; Jackowski et al 1989.)

There are probably no fundamental differences between the biosynthetic capabilities of FASs and PKSs: together, their repertoires cover a spectrum. At one extreme, the FAS for a simple, saturated fatty acid such as palmitic acid builds a straight 16-carbon chain by condensing eight acetyl residues, reducing every keto group to alkyl functionality. Near the opposite extreme, the highly ‘programmed’ PKS for the *Streptomyces* ionophore monensin uses an acetyl starter unit, but the extender units vary and are chosen in a defined sequence (Fig. 2). They include not only acetate residues (A) from malonyl-CoA, but also propionate (P) from methylmalonyl-CoA (thus introducing a methyl side group onto the carbon chain) and butyrate (B) from ethylmalonyl-CoA, leaving an ethyl side group. We can write the sequence of origin of the

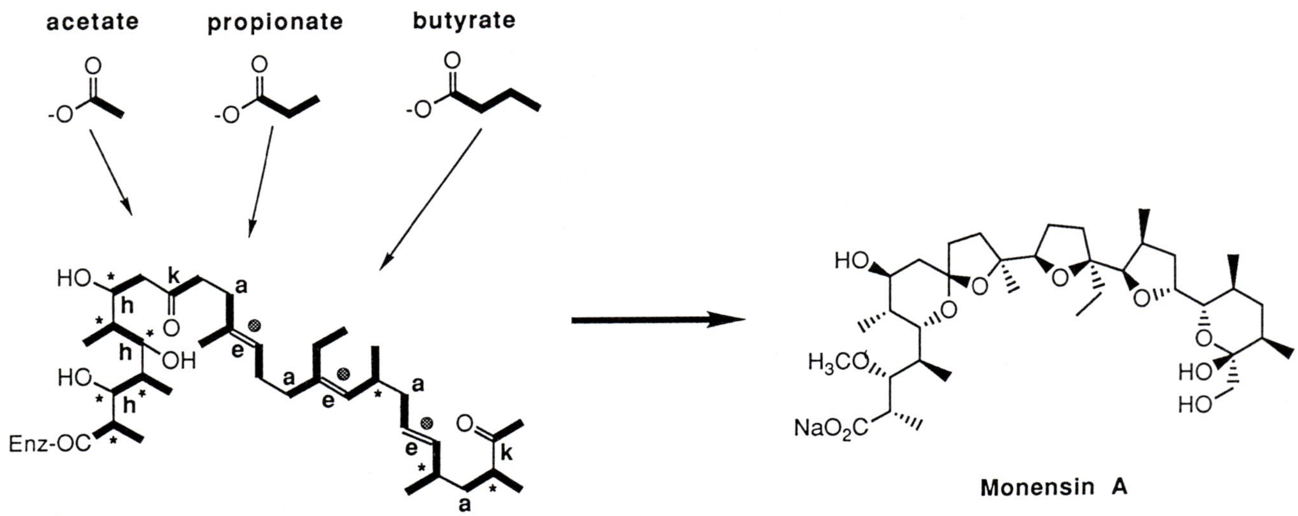

FIG. 2. Biosynthesis of monensin A, a 'complex' polyketide. Monensin A arises from a hypothetical polyketide precursor (*left*) derived by condensation of five acetyl and seven propionyl residues and one butyryl residue. k, h, e and a signify keto, hydroxy, enoyl or alkyl functionality at the indicated carbon atoms, established by presumed variations in the reductive cycle after each successive condensation of a new building unit onto the growing carbon chain. The asterisks indicate nine chiral carbon atoms with a specific *R* or *S* stereochemistry of the methyl or hydroxyl; the cross-hatched symbol indicates the three double bonds with *E* (*trans*) rather than *Z* (*cis*) configuration. (From Hopwood & Sherman 1990.)

pairs of chain carbons in shorthand as $A_1P_2P_3A_4P_5B_6A_7P_8A_9A_{10}P_{11}P_{12}P_{13}$. Moreover, many of the side chains occur on a chiral carbon, so the PKS must make the correct choice of *R* or *S* conformation. A further remarkable aspect of the programming of the monensin PKS is the correct handling of 12 keto groups: two remain, implying that there is abortion of the entire reduction–dehydration–reduction cycle after the corresponding condensation; three are reduced to hydroxyls, by curtailment of the cycle before the dehydration step; there are three double bonds, indicating ketoreduction and dehydration but no enoyl reduction, and these double bonds have specific stereochemistry—*E* rather than *Z*—which demands a correct choice by the PKS between these two theoretical possibilities; and four keto groups are modified by a complete cycle to alkyl functionality. Thus, the programme for the synthesis of monensin carries four kinds of instruction: one for chain length, one for the choice of the 13 starter and extender units, one for the handling of 12 keto groups in a specific sequence, and the fourth for the induction of stereochemistry at nine chiral carbons and three double bonds (Fig. 2). Or, expressed another way, assembly of the carbon chain requires the PKS to perform at least 47 catalysed steps, more than a quarter of them with a correct stereochemical course!

Not all FASs are programmed as simply as palmitate synthase, nor are the programmes of all PKSs so complex as that for monensin. For example, synthesis of branched fatty acids requires the FAS to use alternative starter units such as isobutyryl-CoA or 2-methylbutyryl-CoA to introduce methyl side chains at the 'starter' end of the molecules, or methylmalonyl extenders to afford methyl groups at other positions, as in the FAS of the uropygial gland of waterfowl. The unsaturated fatty acids of bacteria arise by abortion of enoyl reduction at a specific step in chain assembly to leave a double bond in the carbon chain (O'Hagan 1991). Conversely, the actinorhodin PKS (see below) assembles an unbranched 16-carbon chain from eight acetate residues, with only one keto group being modified by reduction–dehydration. However, for these aromatic and largely unreduced polyketides, also represented by the classical fungal example, 6-methylsalicylic acid, the synthase is apparently responsible for an additional process, the cyclization which probably accompanies release of the nascent carbon chain. Cyclization is doubtless essential to stabilize the highly reactive keto groups that were probably stabilized during chain-building by non-covalent bonding to the PKS itself.

Structures of FAS and PKS enzymes and organization of the genes that encode them

FAS genes

A fundamental advance in the understanding of fatty acid biochemistry came from the classification of FASs into Type I and Type II categories (reviewed by

McCarthy & Hardie 1984). Type I enzymes, typified by vertebrate FASs, consist of a single class of polypeptide carrying sites for all the catalytic activities and the attachment point of the ACP prosthetic group. The Type II FASs, found in certain bacteria and plants (*E. coli* being the best-studied example), consist instead of at least eight monofunctional polypeptides, each responsible for one of the reaction steps of the FAS, with a separate ACP. Fungal

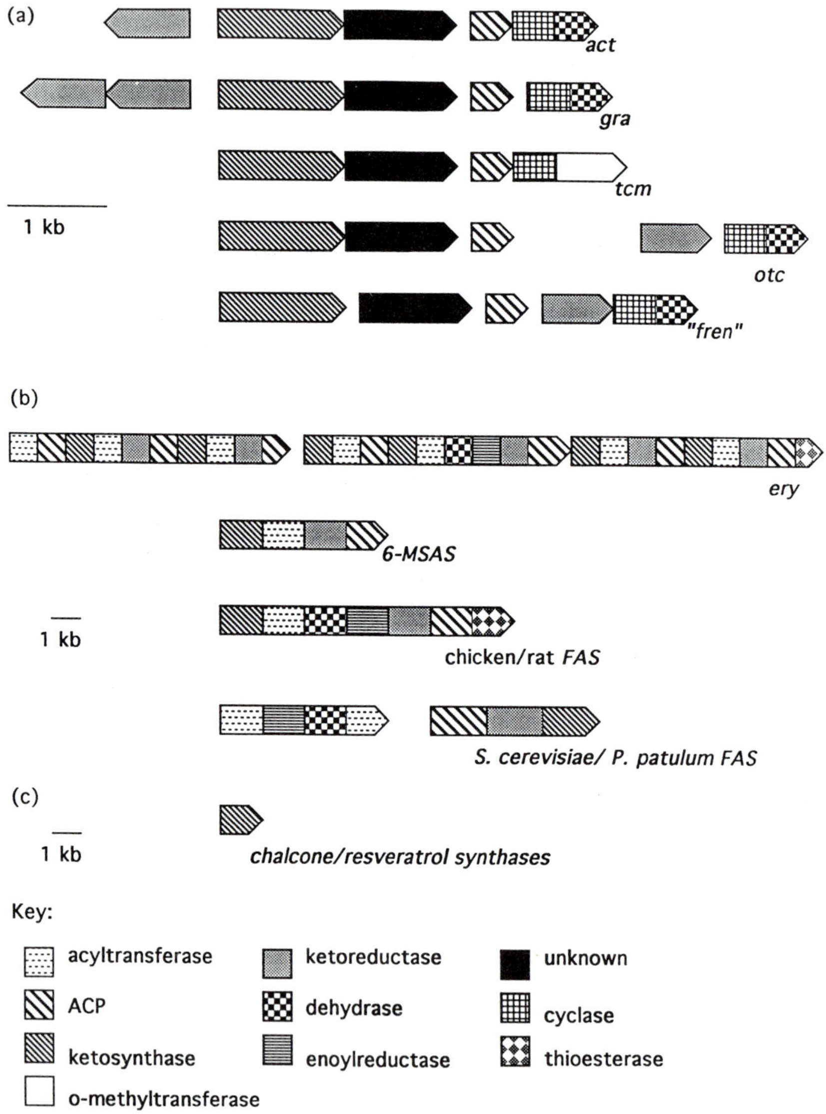

FASs, although classified as Type I, have the functionalities distributed between two large polypeptides, α and β (Fig. 3b). Interestingly, some bacteria have a Type I FAS—for example, that of *Brevibacterium ammoniogenes*, on which at least four active sites have so far been located (Morishima & Ikai 1987).

The genes for two vertebrate FASs (Amy et al 1989, Holzer et al 1989, Schweizer et al 1989, Yuan et al 1988) and the pairs of genes for two fungal FASs (Chirala et al 1987, Mohamed et al 1988, Schweizer et 1986, 1987, Wiesner et al 1988) have been sequenced and genetic and biochemical studies have generated maps of the active sites (Fig. 3b). For *E. coli* the picture is incomplete; a cluster carrying at least five genes, in the order *fabH* (for condensing enzyme III), *fabD* (malonyltransferase), *fabG* (ketoreductase), *acpP* (ACP) and *fabF* (condensing enzyme II), has been identified (J. E. Cronan Jr, personal communication), but several further genes remain to be defined.

PKS genes

Molecular genetics has greatly increased our understanding of the primary structures of representative PKSs from bacteria, fungi and higher plants. These

---

FIG. 3. Architecture of genes encoding fatty acid and polyketide synthases. (a) Type II PKSs. The linear arrangement of the three adjacent open reading frames (ORFs), including the ketosynthase and the ACP, is conserved for all Type II PKS gene clusters so far characterized. The product of the ORF of unknown function resembles the ketosynthases but lacks a putative active site thiol and may function with the ketosynthase protein in a heterodimer. Note that the *tcm* cluster has no ketoreductase or dehydrase functions but appears instead to have an O-methyltransferase function (Sherman et al 1991). The clusters are drawn approximately to scale except for the gap between the ACP and ketoreductase genes of the *otc* cluster, which is longer than shown. (b) Type I FASs and PKSs, arranged to show the conservation of the linear order of active sites between vertebrate Type I FASs and the PKSs. This arrangement is not conserved in the fungal FASs, which are encoded by two unlinked genes for the α and β subunits of the FAS. The domains within the Type I genes are arbitrarily shown of equal lengths, with no interdomain spaces. (c) Chalcone/resveratrol synthases, drawn to the same scale as the Type I synthases in (b). They each represent a single condensing enzyme that may be evolutionarily unrelated to other ketosynthases.
Source of information: *act* (actinorhodin synthase), M. A. Fernández-Moreno et al (in preparation); *gra* (granaticin synthase), Sherman et al (1989); *tcm* (tetracenomycin C synthase), Bibb et al (1989); *otc* (oxytetracycline synthase), D. H. Sherman & I. S. Hunter (personal communications); '*fren*' (presumptive frenolicin synthase), Maureen J. Bibb (personal communication); *ery* (erythromycin synthase), Cortes et al (1990), Donadio et al (1991); *6-MSAS* (6-methylsalicylic acid synthase), Beck et al (1990); chicken/rat *FAS*, Amy et al (1989), Holzer et al (1989), Schweizer et al (1989), Yuan et al (1988); *S. cerevisiae*/*P. patulum FAS*, Chirala et al (1987), Mohamed et al (1988), Schweizer et al (1986, 1987); *chalcone/resveratrol synthases*, Niesbach-Klösgen et al (1987), Schröder et al (1988).

advances are tantalizing, because so much more information will come from comparisons between the primary structures of a whole series of differently programmed PKSs, and from analysis of the three-dimensional architecture of the synthases, about which very little is known.

The biochemical demonstration (Dimroth et al 1976) that 6-methylsalicylic acid synthase (6-MSAS) of *Penicillium patulum* is a Type I protein has been confirmed by sequencing of the gene (Beck et al 1990). Moreover, the sequence of its active sites (Fig. 3b) resembles those of *vertebrate* FASs, rather than that of *P. patulum* itself.

For several aromatic *Streptomyces* polyketides, sequencing has instead revealed a Type II architecture, with a cluster of genes (Fig. 3a) coding for the components of each PKS, the condensing enzyme (which is perhaps also responsible for acyl transfer), ACP, ketoreductase and presumptive cyclase/dehydrase (Bibb et al 1989, Hopwood & Sherman 1990, Sherman et al 1989, 1991). Notable was the finding of genes for discrete ACPs, because of prior knowledge about the chalcone and resveratrol synthases of higher plants, which consist of two identical small ($\approx$40 kDa) polypeptide chains (Niesbach-Klösgen et al 1987, Schröder et al 1988) that catalyse three condensations of acetate residues from malonyl-CoA onto a coumaroyl-CoA starter, with no reduction or dehydration and in which no ACP domain is evident (and no separate ACP is needed). The building units for chain assembly are transferred from CoA directly to a cysteine thiol on the synthase, but this thiol is not part of a motif that resembles the active site of other FAS or PKS condensing enzymes (Lanz et al 1991).

In sharp contrast to the situation for the *Streptomyces* aromatic polyketides, the PKS for the macrolide antibiotic erythromycin from *Saccharopolyspora* (formerly *Streptomyces*) *erythraea* is a Type I enzyme (Donadio et al 1991, Cortes et al 1990, Bevitt et al 1992) with an amazingly complex primary structure. On each of three giant polypeptides are two 'modules' of information, each carrying sites for ketosynthase and acyltransferase activity, a (usually incomplete) set of sites for the reductive cycle of keto group modification, and an ACP domain. There would be a series of 28 sites (Fig. 3b), co-linear with their presumptive sequence of reactions with the growing carbon chain, and each used for one only of the reactions that convert seven propionate residues into the product of the PKS, 6-deoxyerythronolide B (Fig. 4).

The 'programming' of FAS and PKS systems

The central question in fatty acid and polyketide biosynthesis is: how does the structure of the FAS or PKS determine the programmed assembly and correct processing of the fatty acid or polyketide chain?

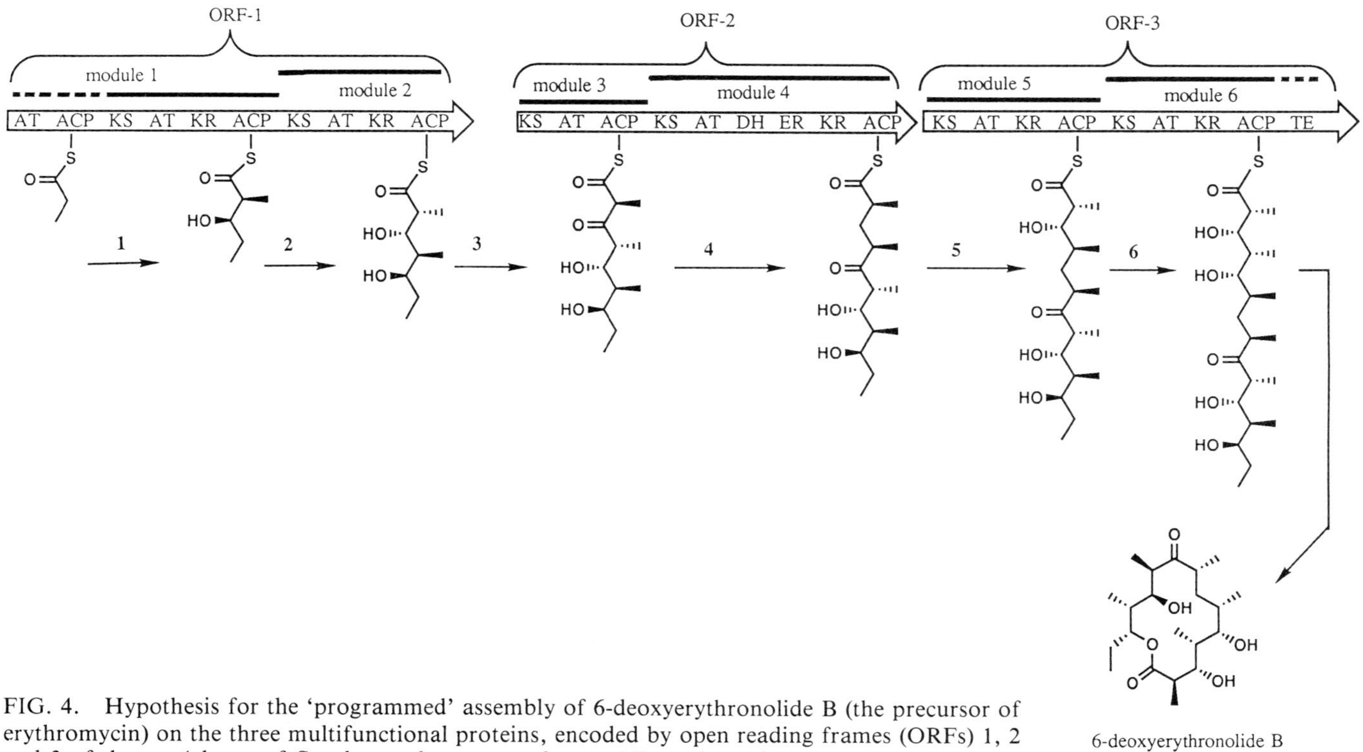

FIG. 4. Hypothesis for the ‘programmed’ assembly of 6-deoxyerythronolide B (the precursor of erythromycin) on the three multifunctional proteins, encoded by open reading frames (ORFs) 1, 2 and 3 of the *eryA* locus of *Saccharopolyspora erythraea*. AT, acyltransferase; ACP, acyl carrier protein; KS, β-keto-ACP synthase; KR, β-keto-ACP reductase; DH, dehydrase; ER, enoyl reductase; and TE, thioesterase (cyclase?). (The figure is based on the data of Cortes et al (1990) and Donadio et al (1991). Reproduced with permission from *The Polyketide Metabolites* by O’Hagan, published in 1991 by Ellis Horwood Ltd, with minor modifications.)

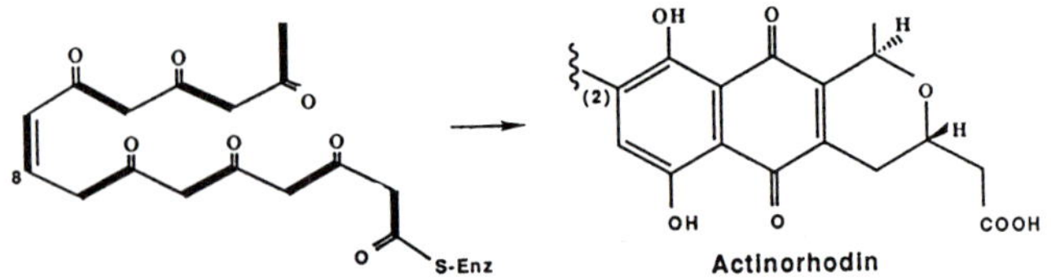

Actinorhodin

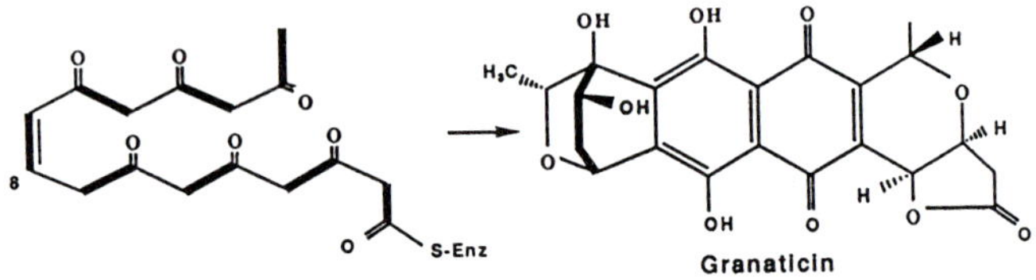

Granaticin

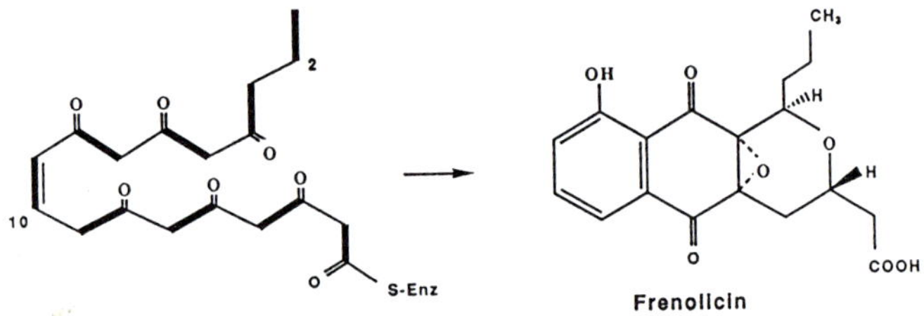

Frenolicin

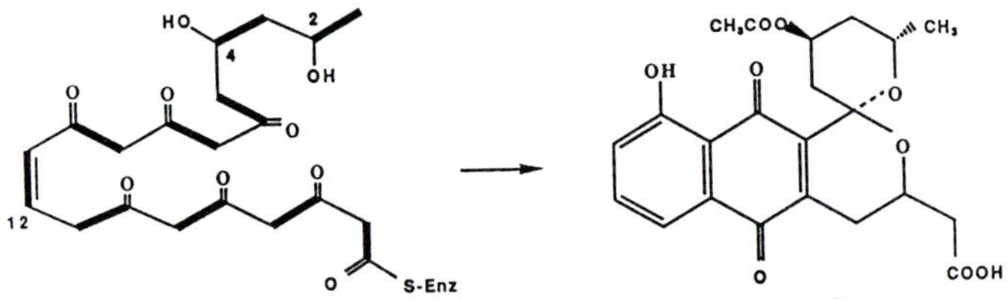

Griseusin B

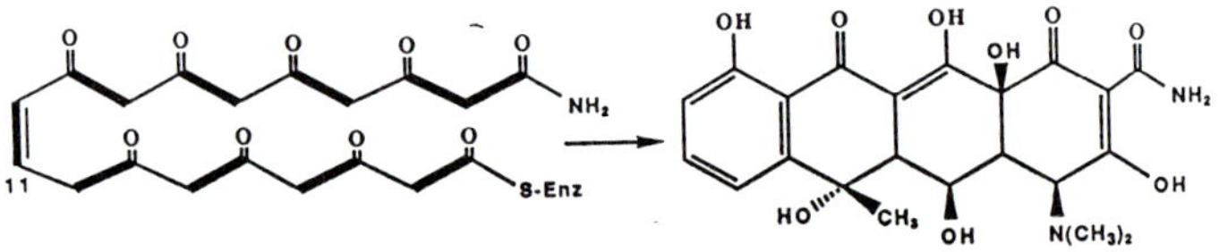

Oxytetracycline

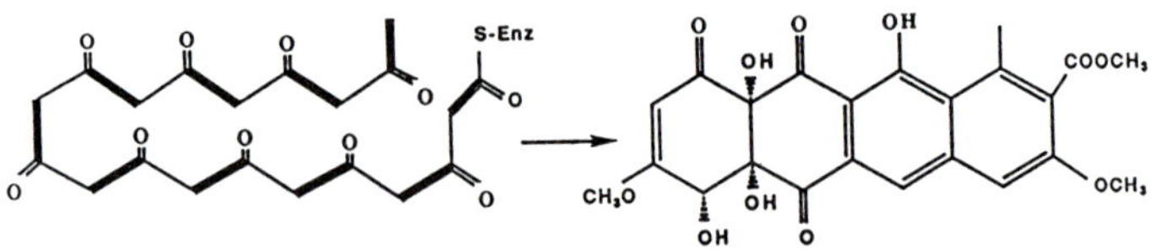

Tetracenomycin C

Type II systems

For the *E. coli* Type II FAS, little information exists about how the polypeptide subunits are associated: it might be only a loose association, with limited protein–protein interactions, a view supported by the ability of heterologous ACPs to replace the *E. coli* ACP in a functional complex (Simoni et al 1967), and by the isolation of individual condensing enzymes as homodimers (Garwin et al 1980). A three-dimensional structure derived for the *E. coli* FAS ACP indicates a strongly hydrophobic cleft as a possible site of fatty acid chain assembly (Kim & Prestegard 1990). This has led to speculation that the ACP conceivably controls chain length through contacts made between specific features of its hydrophobic surface and the hydrocarbon chain. Understanding of other aspects of programming—choice of alternative starter units for branched fatty acid synthesis (a characteristic of some FASs of Gram-positive bacteria), and control of the position of the double bond in unsaturated fatty acids—is obscure.

For the *Streptomyces* aromatic PKSs, we are addressing the programming question by constructing a series of hybrid PKS gene clusters with each gene in one cluster (for actinorhodin) replaced *in cis* by the corresponding gene from the others (Fig. 3a). This approach was initiated following successful complementation *in trans* of mutations in at least four of the five genes of the *act* (actinorhodin) PKS cluster by the corresponding *gra* (granaticin) genes (Sherman et al 1992). The set of parental compounds, and the presumptive

FIG. 5. 'Programming' of aromatic polyketide synthases. *Opposite*: Hypothetical polyketide precursors (*left*), synthesized by the products of the genes shown in Fig. 3a, of each of six aromatic antibiotics (*right*). The bold bonds on the left connect carbon atoms derived from individual monomeric building units. *Below*: Starter unit, extender units and keto group modifications for the biosynthesis of the set of hypothetical polyketide precursors. (For frenolicin and griseusin, the nature of the starter unit—and therefore some of the required keto group modifications—has not been verified chemically.)

Antibiotic	*Starter residue*	*Extender residues*	*Keto group modifications*
Actinorhodin monomer	Acetate	7 × acetate	Reduction/dehydration at C-8
Granaticin	Acetate	7 × acetate	Reduction/dehydration at C-8
Frenolicin	Acetate	8 × acetate	Reduction/dehydration/reduction at C-2; reduction/dehydration at C-10
Griseusin	Acetate	9 × acetate	Reduction at C-2 and C-4; reduction/dehydration at C-12
Oxytetracycline	Malonamide	8 × acetate	Reduction/dehydration at C-11
Tetracenomycin	Acetate	9 × acetate	None

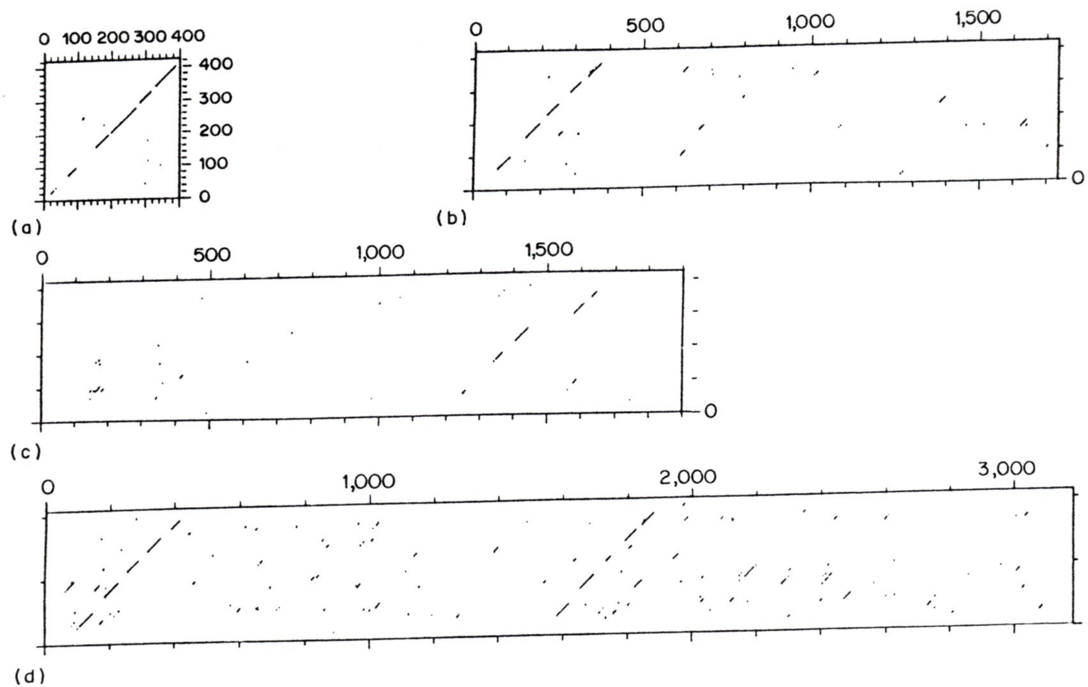

FIG. 6. Comparisons of the amino acid sequence of the actinorhodin PKS condensing enzyme (vertical axis) with those of (a) the *E. coli* fatty acid synthase condensing enzyme (FabB: Kauppinen et al 1988); (b) the *Penicillium patulum* 6-methylsalicylic acid synthase (Beck et al 1990); (c) the α subunit of the *Saccharomyces cerevisiae* fatty acid synthase (Schweizer et al 1987) and (d) the ORF-3 gene product of the *Saccharopolyspora erythraea* 6-deoxyerythronolide B synthase (Cortes et al 1990). These COMPARE/DOTPLOT analyses used a window of 40 residues and a stringency of 20. Interrupted lines with 45° slope indicate regions of similar amino acid sequence; note the nearly end-to-end alignment of the actinorhodin PKS sequence with the complete FabB protein, with one domain in each of the fungal PKS and FAS enzymes, and with two domains in the 6-deoxyerythronolide B ORF-3 (see Fig. 3b for the positions of these domains in each enzyme).

programming of each PKS, are shown in Fig. 5. Although all the programmes are relatively simple, there is variation in starter unit, number of extender units, and handling of various keto groups. Provided that isolable metabolites are produced by hosts carrying at least a reasonable proportion of the hybrid PKS clusters (preliminary results with ACP replacements are promising: Khosla et al 1992), their structures should implicate particular components of the synthases in various aspects of programming. Further refinements will involve construction of hybrid genes and targeted changes in specific amino acid residues.

We also hope to throw light on the three-dimensional architecture of these Type II PKS complexes. It is interesting that components of the tetracenomycin PKS appear to be membrane associated (Gramajo et al 1991). Perhaps significant cross-talk between them is facilitated by membrane anchorage, rather than through direct protein–protein interactions?

Type I systems

A model for the vertebrate Type I FAS (Witkowski et al 1991) has head-to-tail juxtaposition of the identical subunits to create two equivalent reaction centres in which the ACP and condensing enzyme thiols of *different* subunits would work together; however, control of chain length is not addressed. It has also been suggested that there might be interactions between monomers in the Type I fungal FAS, which has a heteromultimeric structure, $\alpha_6\beta_6$ (Schweizer et al 1989). Again, the programming of chain length is quite obscure.

Although the primary structure of the modular erythromycin PKS is, on the face of it, amazingly complex, understanding the programming of such synthases may not be so daunting as would at first appear. In a sense, the logic of the programming is apparent in the primary structure of the enzyme (assuming that the sequence of active sites really is the same as the order in which they are used; Fig. 4), in contrast to that of aromatic synthases, in which there is only one active site, used iteratively, for each class of biochemical reaction. The challenge is to understand how the three-dimensional structure of the PKS causes acyl chains to move, and be modified, at successive steps in their assembly (Bevitt et al 1992).

Evolution of fatty acid and polyketide synthases

Isotopic labelling studies in diverse systems have demonstrated the chemical equivalence of fatty acid and polyketide biosynthesis (O'Hagan 1991), in that both processes involve *decarboxylative* condensation of carboxylic acid thioesters. In virtually all cases a pantetheinylated ACP is involved and acyl transferases are required for transesterification of substrates and products between the synthase and CoA. It is these three features that set PKSs and FASs apart from other enzymes that catalyse condensation of carboxylic acid thioesters, such as acetoacetyl-CoA synthase (in polyhydroxybutyrate biosynthesis; Davis et al 1987, Anderson et al 1990) and 3-hydroxy-3-methylglutaryl (HMG)-CoA synthase (in steroid biosynthesis).

That most FAS and PKS genes must have had a common origin is shown by significant sequence identities between them, irrespective of their host or architecture. For the ketosynthases (Fig. 6) and ACPs, there is end-to-end similarity, as revealed by a diagonal alignment between the sequences of Type II proteins or extended domains of Type I enzymes. For ketoreductase, dehydrase, enoylreductase and acyltransferase sequences, the regions of identity/similarity appear to be more restricted, being found primarily around the putative active sites or cofactor binding sites. These are sometimes more similar to a non-FAS/PKS enzyme catalysing a similar reaction than to other FAS/PKS enzymes, raising the possibility that such components may have been recruited via transfer from another organism instead of co-evolving with the rest of the synthase. Perhaps the more conserved components have evolved to control the sequence of enzyme-catalysed reactions common to all pathways, and the less conserved components to control the selectivity of catalysis specific to particular pathways. Alternatively, the more conserved components might control both reaction sequence and selectivity, with the divergent components merely providing active sites for the additional functions. These speculations lead to several interesting questions which have important implications for designing recombinant organisms producing novel polyketides.

Phylogeny

What is the phylogenetic relationship between homologous FAS and PKS systems? Do PKSs such as plant chalcone and resveratrol synthases, which have few genetic and mechanistic similarities to other PKSs, belong to the same phylogenetic tree, or are they products of independent evolution? Are enzyme systems such as acetoacetyl-CoA synthase/reductase and HMG-CoA synthase/reductase related to FASs and PKSs? Do the archaebacteria, whose membranes contain isoprenoids (products of acetoacetyl-CoA synthase/reductase and HMG-CoA synthase/reductase) instead of fatty acids (De Rosa & Gambacorta 1988), also contain (cryptic) FAS/PKS genes, or are such genes completely absent from this group of organisms? Determination of further sequences for PKS and FAS genes from diverse systems, together with correlations between the structure and function of PKSs, will undoubtedly go far in addressing such questions.

Evolution of gene fusions

What were the driving forces behind the evolution of multifunctional PKS and FAS enzymes? Although identical arrangements of active sites can be observed in diverse multifunctional enzymes (such as vertebrate FAS and module 4, the only 'complete' module, of the erythromycin PKS), such arrangements are by no means universally conserved (for example, the *P. patulum* PKS is closer to the vertebrate FAS than the FAS of the fungus itself), suggesting the occurrence of several independent gene fusion events. In fungi and vertebrates perhaps gene fusion reflects the need for coordinate and/or equimolar synthesis, in the absence of operon organization, or for covalent (i.e. concentration-independent) association of various FAS/PKS components. In complex prokaryotic PKSs such as that for erythromycin, where corresponding components of different modules are extensively homologous, multifunctional enzymes may be necessary to prevent components from associating in incorrect combinations which would lead to unwanted side reactions. A potentially useful way of gaining insights into the rationale for gene fusions may be through genetic decoupling of domains in a multifunctional system such as the erythromycin PKS, and/or through genetic coupling of individual proteins in discrete systems such as aromatic *Streptomyces* PKSs.

Evolution of programming complexity

How did complex PKSs evolve from (presumably simple) primitive ones? The structural similarities between bacterial aromatic PKSs and Type II FASs on one hand, and fungal aromatic PKSs and vertebrate Type I FASs on the other, suggest that such PKS systems, which show only a modest increase in programme complexity over FASs, arose directly through mutations occurring in primitive FAS genes. Here, the challenge for Nature seems to have been how to have the same active sites acting repeatedly on the same or similar carboxylic acid monomers, with the occasional introduction of additional enzymic activities. The evolution of macrolide PKSs, however, presumably involved repeated gene duplication of a PKS module that resembled a vertebrate FAS. Perhaps this obviated the need for active site recycling, and at the same time opened up the possibility for an independent choice of monomers for each condensation reaction. The challenge here was probably to devise a way of channelling intermediates sequentially between divergent modules. Perhaps the head-to-tail spatial organization of active sites in vertebrate FASs offers a clue to how this problem was solved (Bevitt et al 1992). Yet again, genetic engineering could provide answers to such questions; functional expression of subsets of modules from the erythromycin PKS, and deliberate duplication of the actinorhodin PKS module, followed by 'matching' of the chain length control mechanism of one module with the substrate selectivity of the other, are possible approaches.

Acknowledgements

We thank Keith Chater and Peter Revill for useful comments on the manuscript. C.K. acknowledges support from the Ramsay Foundation. Work in the authors' laboratory was supported by grants-in-aid from the Agricultural and Food Research Council and the John Innes Foundation and by NIH grant GM39784.

References

Amy CM, Witkowski A, Naggert J, Williams B, Randhawa Z, Smith S 1989 Molecular cloning and sequencing of cDNAs encoding the entire rat fatty acid synthase. Proc Natl Acad Sci USA 86:3114–3118

Anderson VE, Bahnson BJ, Wlassics ID 1990 The reaction of acetyldithio-CoA, a readily enolized analog of acetyl-CoA with thiolase from *Zoogloea ramigera*. J Biol Chem 265:6255–6261

Beck J, Ripka S, Siegner A, Schiltz E,Schweizer E 1990 The multifunctional 6-methylsalicylic acid synthase gene of *Penicillium patulum*: its gene structure relative to that of other polyketide synthases. Eur J Biochem 192:487–498

Bevitt DJ, Cortes J, Haydock SF, Leadlay PF 1992 6-Deoxyerythronolide B synthase 2 from *Saccharopolyspora erythraea*: cloning of the structural gene, sequence analysis and inferred domain structure of the multifunctional enzyme. Eur J Biochem 204:39–49

Bibb MJ, Biró S, Motamedi H, Collins JF, Hutchinson CR 1989 Analysis of the nucleotide sequence of the *Streptomyces glaucescens tcmI* genes provides key information about the enzymology of polyketide antibiotic biosynthesis. EMBO (Eur Mol Biol Organ) J 8:2727–2736

Chirala SS, Kuziora MA, Spector DM, Wakil SJ 1987 Complementation of mutations and nucleotide sequence of FAS1 gene encoding β subunit of yeast fatty acid synthase. J Biol Chem 262:4231–4240

Cortes J, Haydock SF, Roberts GA, Bevitt DJ, Leadlay PF 1990 An unusually large multifunctional polypeptide in the erythromycin-producing polyketide synthase of *Saccharopolyspora erythraea*. Nature (Lond) 348:176–178

Davis JT, Moore RN, Imperiali B et al 1987 Biosynthetic thiolase from *Zoogloea ramigera*. I. Preliminary characterization and analysis of proton transfer reaction. J Biol Chem 262:82–89

De Rosa M, Gambacorta A 1988 The lipids of archaebacteria. Prog Lipid Res 27:153–175

Dimroth P, Ringelmann E, Lynen F 1976 6-Methylsalicylic acid synthetase from *Penicillium patulum*. Some catalytic properties of the enzyme and its relation to fatty acid synthetase. Eur J Biochem 68:591–596

Donadio S, Staver MJ, McAlpine JB, Swanson SJ, Katz L 1991 Modular organization of genes required for complex polyketide biosynthesis. Science (Wash DC) 252: 675–679

Garwin JL, Klages AL, Cronan JE 1980 Structural, enzymatic, and genetic studies of β-ketoacyl-acyl carrier protein synthases I and II of *Escherichia coli*. J Biol Chem 24:11949–11956

Gramajo HC, White J, Hutchinson R, Bibb MJ 1991 Overproduction and localization of components of the polyketide synthase of *Streptomyces glaucescens* involved in the production of the antibiotic tetracenomycin C. J Bacteriol 173: 6475–6483

Holzer KP, Liu W, Hammes GG 1989 Molecular cloning and sequencing of chicken liver fatty acid synthase cDNA. Proc Natl Acad Sci USA 86:4387–4391

Hopwood DA, Sherman DH 1990 Molecular genetics of polyketides and its comparison to fatty acid biosynthesis. Annu Rev Genet 24:37–66

Jackowski S, Murphy CM, Cronan JE, Rock CO 1989 Acetoacetyl-acyl carrier protein synthase: a target for the antibiotic thiolactomycin. J Biol Chem 264: 7624–7629

Kauppinen S, Siggaard-Andersen M, von Wettstein-Knowles P 1988 β-Ketoacyl-ACP synthase I of *Escherichia coli*: nucleotide sequence of the *fabB* gene and identification of the cerulenin binding residue. Carlsberg Res Commun 53:357–370

Khosla C, Ebert-Khosla S, Hopwood DA 1992 Targeted gene replacements in a *Streptomyces* polyketide synthase gene cluster: role for the acyl carrier protein. Mol Microbiol, in press

Kim Y, Prestegard JH 1990 Refinement of the NMR structures for acyl carrier protein with scalar coupling data. Proteins Struct Funct Genet 8:377–385

Knoche HW, Koths KE 1973 Characterization of a fatty acid synthetase from *Corynebacterium diphtheriae*. J Biol Chem 248:3517–3519

Lanz T, Tropf S, Marner F-J, Schröder J, Schröder G 1991 The role of cysteines in polyketide synthases. J Biol Chem 266:9971–9976

McCarthy AD, Hardie DG 1984 Fatty acid synthase—an example of protein evolution by gene fusion. Trends Biochem Sci 9:60–63

Mohamed AH, Chirala SS, Mody NH, Huang W-Y, Wakil SJ 1988 Primary structure of the multifunctional α subunit protein of yeast fatty acid synthase derived from *FAS2* gene sequence. J Biol Chem 263:12315–12325

Morishima N, Ikai A 1987 Active site organization of bacterial Type I fatty acid synthetase. J Biochem 102:1451–1457

Niesbach-Klösgen U, Barzen E, Bernhardt J et al 1987 Chalcone synthase genes in plants: a tool to study evolutionary relationships. Mol Evol 26:213–225

O'Hagan D (ed) 1991 The polyketide metabolites. Ellis Horwood, Chichester

Robinson JA 1991 Polyketide synthase complexes: their structure and function in antibiotic biosynthesis. Phil Trans R Soc Lond B Biol Sci 332:107–114

Schröder G, Brown JWS Schröder J 1988 Molecular analysis of resveratrol synthase cDNA genomic clones and relationship with chalcone synthase. Eur J Biochem 172:161–169

Schweizer E, Müller G, Roberts LM et al 1987 Genetic control of fatty acid synthetase biosynthesis and structure in lower fungi. Fat Sci Technol 89:570–577

Schweizer M, Roberts LM, Höltke H-J et al 1986 The pentafunctional FAS1 gene of yeast: its nucleotide sequence and order of catalytic domains. Mol & Gen Genet 203:479–486

Schweizer M, Takabayashi K, Laux T, Beck K-F, Schreglmann R 1989 Rat mammary gland fatty acid synthase: localization of the constituent domains and two functional polyadenylation/termination signals in the cDNA. Nucleic Acids Res 17:567–586

Sherman DH, Malpartida F, Bibb MJ, Kieser HM, Bibb MJ, Hopwood DA 1989 Structure and deduced function of the granaticin-producing polyketide synthase gene cluster of *Streptomyces violaceoruber* Tü22. EMBO (Eur Mol Biol Organ) J 8:2717–2725

Sherman DH, Bibb MJ, Simpson TJ et al 1991 Molecular genetic analysis reveals a putative bifunctional polyketide cyclase/dehydrase gene from *Streptomyces coelicolor* and *Streptomyces violaceoruber*, and a cyclase/*O*-methyltransferase from *Streptomyces glaucescens*. Tetrahedron 47:6029–6043

Sherman DH, Kim E-S, Bibb MJ, Hopwood DA 1992 Functional replacement of genes for individual polyketide synthase components in *Streptomyces coelicolor* A3(2) by heterologous genes from a different polyketide pathway. J Bacteriol, in press
Simoni RD, Criddle RS, Stumpf PK 1967 Fat metabolism in higher plants. XXXI. Purification and properties of plant and bacterial acyl carrier proteins. J Biol Chem 242:573–581
Wiesner P, Beck J, Beck K-F, Ripka S, Müller G, Lücke Schweizer E 1988 Isolation and sequence analysis of the fatty acid synthetase *FAS2* gene from *Penicillium patulum*. Eur J Biochem 177:69–79
Witkowski A, Rangan VS, Randhawa ZI, Amy CM, Smith S 1991 Structural organization of the multifunctional animal fatty-acid synthase. Eur J Biochem 198:571–579
Yuan Z, Liu W, Hammes GG 1988 Molecular cloning and sequencing of DNA complementary to chicken liver fatty acid synthase mRNA. Proc Natl Acad Sci USA 85:6328–6331

DISCUSSION

Leadlay: It is certainly striking that the 6-methylsalicylic acid synthase of *Penicillium patulum* has the same order of active sites, and about the same spacing, as you would expect to find in a vertebrate fatty acid synthase, and differs from its very own fatty acid synthase, which is a typical fungal type. It is also clear that each component of the erythromycin-producing polyketide synthase of *Saccharopolyspora erythraea* has the same order of active sites and the same general organization as a vertebrate fatty acid synthase, and not at all that of a typical bacterial fatty acid synthase. In the context of what we have already heard in the symposium, these similarities are extremely interesting.

One of the first things we have been concerned to do is to show that the three proposed genes for the erythromycin-producing polyketide synthase are really giving rise to protein. In fact, all three genes are expressed, and the resulting giant polypeptides have all been detected in *S. erythraea* and purified in roughly stoichiometric amounts (Caffrey et al 1992), so we needn't worry about pseudogenes and can start to think about how the growing polyketide chain gets from one end of the multienzyme complex to the other.

Cane: May I pick up on this theme of the ordering of the activities, particularly in module 4 and its similarity to the avian and mammalian fatty acid synthase gene. There is also a warning here: the naive conclusion, when we see that the linear sequence of the gene exactly mimics the proposed linear sequence of biosynthetic events, may be misleading, if one thinks about the way the avian protein is organized; the keto reductase and enoyl reductase at one end of one multifunctional polypeptide appear to be associated in a functional unit with the other end of the other polypeptide strand (Tsukamoto et al 1983). So it's not necessarily so that a given polyketide is processed all along the strand, in the fatty acid synthase. This leaves open the question of whether in the 6-deoxyerythronolide B synthase complex, there is more than one subunit and, if so, how the subunits are organized.

Hopwood: I agree that this is interesting. With the vertebrate fatty acid synthase, two reaction centres participate in this cross-talk that you are discussing, which is possible to envisage with a homodimer such as the vertebrate fatty acid synthase; but in the case of the three big gene products for erythromycin, it is quite a challenge to imagine how the proteins fit together in a three-dimensional way.

Cane: The other point to consider is the extent to which programming is implicit in the organization of the 6-deoxyerythronolide B synthase (*eryA*) gene, and what the consequences of that would be, and whether there is evidence to support it. One simple-minded interpretation of the molecular genetic data is that a starting unit is simply welded on the front ACP moiety of the synthase, and then what happens to it is inevitable, because any given ketosynthase is essentially stupid: that is, it is not particularly good at molecular recognition, but processes whatever is handed to it by the neighbouring protein. This would explain much about the structures of closely related macrolides and why you could have modular arrangements; it would also explain why there are families of structurally related polyethers, in which central portions of the synthetic chain assembly have been deleted and yet later condensations don't seem to be bothered by the change in substrate structure (Cane et al 1983). However, this view is at odds with the experimental facts, from studies carried out in a number of laboratories, including our own (Cane & Yang 1987, Cane & Ott 1988, Yue et al 1987).

We can feed precursors which are thioester analogues of chain-building intermediates; they are apparently recognized in phase. The biggest piece to have gone in now is a tetraketide (D. E. Cane & W.-T. Tan, unpublished work; Li et al 1992, Staunton & Sutkowski 1991a,b). This basically means that the synthase can look at a branched-chain polyhydroxylated thioester and say 'this is a tetraketide, therefore add such-and-such an acetate or propionate equivalent to it', rather than simply saying 'this is an acetate analogue with some strange piece stuck on at the beginning of the chain', thereby generating an analogue of the natural product. That must mean that recognition of structure and functionality is built into each of these modular catalytic units, and so all the microscopic reactions are in fact substrate specific; there must be a sorting-out of exactly what they have to do. It would be very nice to see what degree there is of fidelity—to what extent can a synthase from, say, module 3 or module 4 be fooled by variations in the structure of its natural substrate. It is impossible to do this with a whole-cell feeding experiment because the anomalous precursors would have to compete with formation of the natural product.

Hopwood: I am glad you mention those very nice experiments. They say that, if the system is really like a railway where you load things on at the beginning and they move to the end, they can also join at intermediate stations, and are recognized as such. As you say, a tetraketide will join at a 'tetraketide' station, and so on.

Bu'Lock: Mechanistically, it may be rather more sophisticated! One difficulty about polyketide assembly processes that people cannot shrug off is the way in which attachment to the ACP is always by the acyl unit that has been most recently attached. So, at some stage, when the next acyl unit goes in, the whole assembly has to come off the ACP; in other words, it is an *insertion* reaction. Such reactions involve continuing recognition of what the new acyl unit is being inserted into.

There is another point that would be accommodated by the idea of subunits in the more complex synthases. If you consider the structures of a large number of macrolides or polyene macrolides, they consist fundamentally of an annular structure, with organization into sectors which are apparently interchangeable. It seems that you can change one sector and the rest of the molecule stays the same, from one polymacrolide to another. This has always suggested to me that the process must be governed by some multiple template structure. What we have been told is that at least in some instances the process is determined by piling up different sequential, multifunctional proteins, which fits the idea very well.

Cane: These seem to be different from the polypeptide synthetases, where specificity derives from the amino acid-activating portions, and once the amino acid is loaded onto the synthetase the sequence is predetermined.

Bu'Lock: You can do that with polyketides to a limited extent, too. We made over 200 different avermectins this way (Dutton et al 1991).

Cane: But that involves only changes in the starter unit, and this appears to be very permissive, in that you can make many substitutions which are not tolerated in the middle of the polyketide chain. Comparing avermectins to milbemycins, there clearly has been an insertion or a deletion at the place where the C_{13} hydroxyl group is introduced.

There are a number of places where specificity may come in. It may be that the ketosynthases themselves are again stupid, but the acyltransferases, which load a growing chain onto the core cysteine of the ketosynthase, determine what gets transferred to the methylmalonyl or malonyl unit. There's an acyltransferase which will choose to put a methyl malonate or a malonate on a particular pantothenate, and another acyltransferase puts a particular partially grown chain on the core residue. Maybe, when those sites are loaded up properly, the condensing enzyme reaction then just takes place. Or it may be that the condensing enzyme accepts only certain substrates.

Bu'Lock: There is an example in the streptomycetes, where you can get competitive incorporation of alternative substrates in mid-chain. This was in

Vaněk's work on the biosynthesis of the various monensins (Pospíšil et al 1983).

Cane: One doesn't know whether that is a single enzyme system controlling the formation of all these products, where there is just substrate pressure; or whether in fact there are multiple synthases, each responsible for the formation of a characteristic monensin derivative.

Bu'Lock: Circumstantially, it looks like competitive binding of alternative substrates.

Cane: The lasalocids are like that too.

Hopwood: The point that David Cane and John Bu'Lock were making about the polyethers occurring in 'families', and about macrolides containing 'sectors', where you don't find all possible sequences of individual building units, is very interesting. It could be a reflection of the particular gene fusions that have happened, so that only certain possibilities have occurred, or it could reflect structural constraints. We can't tell which, yet.

Bu'Lock: Not until we have done some of the gene fusions.

Beppu: I think that differences in the composition of the synthetic enzyme complex may be important in determining the specificity of the condensation. I would like to know if there is any difference in the catalytic activity for the condensation, between these two different types of synthase?

Hopwood: We don't know. This is the point David Cane was making, when he was suggesting that the condensing enzyme might be 'stupid'; it will condense any thioesters that are correctly presented to it. The other alternative is the possibility that the ketosynthase may be 'clever' enough to choose the correct building units to assemble.

Cane: Someone has to isolate an active polyketide synthase.

Beppu: Could a 'stupid' enzyme with a lower activity have been the ancestral enzyme, in evolution?

Hopwood: You mean a more generalized activity that could then become specialized in different lines of evolution? I think that's quite possible.

Cavalier-Smith: For the fatty acid synthase, there can't be an overriding reason why in eukaryotes a fused version of the gene was favoured, because the higher plants have the dispersed gene, like the cyanobacteria, and it's likely that higher plants gave up the fused version that would have been in the host and replaced it by the dispersed, cyanobacterial version. If there was a strong reason for having the fused gene, that loss would have been selected against.

Hopwood: Doesn't that prove the rule, because in plants fatty acid synthesis takes place in the plastids, which are vestigial bacteria?

Cavalier-Smith: Yes, fatty acid synthase occurs in plastids, but the relevant genes are in the nucleus, so they would have had to evolve transit sequences for importing the fatty acid synthase proteins into the chloroplast. And it isn't necessarily true of all organisms with chloroplasts that fatty acid synthesis occurs in them. In the unicellular alga, *Euglena*, fatty acids are made in the cytoplasm. I think euglenoids have the same type of fatty acid synthase as animals and fungi?

Hopwood: I believe they have both, the animal type of fatty acid synthase in the cytoplasm and a bacterial form in the plastid (Hendren & Bloch 1980).

Cavalier-Smith: This supports the view that there is no overriding reason why fatty acid synthesis should be done one way rather than the other.

Davies: To what extent can these comparisons be carried over into the peptidyl synthetases?

Turner: There are parallels. I shall be referring to this in my talk (p 113).

Demain: I gather there is some relationship between the amino acid sequence of the ACV synthetase and the acyl carrier proteins of the fatty acid synthases?

Hopwood: The amino acid sequence similarity between the 4′-phosphopantetheine-binding regions of the ACV synthetases and the acyl carrier proteins of the fatty acid synthases was enough for the computer to accept them into the 'pile-up' analysis that we used to compare all of the sequences, suggesting an evolutionary relationship.

Demain: What about peptide synthetases? Would it accept the tyrocidin or gramicidin synthetases, and so on? The ACV synthetases of β-lactam-producing microorganisms are supposed to be similar to these.

Hopwood: Yes, the two peptide synthetases were also included in the analysis.

Cane: The ACV and peptide synthetases are similar in that they have similar pantothenyl domains; but that's the limit of their similarity.

Hunter: Professor Hopwood suggested that acyl carrier proteins are very 'forgiving', in that they can substitute functionally for each other in the experiments he described. But he managed to knock out the ACP for actinorhodin biosynthesis in his strain and antibiotic was no longer made; presumably there was still an intact copy of the ACP for fatty acid biosynthesis. It may not be working, but does that suggest that one reason these polyketide synthase genes are clustered is that they need to be expressed in space and time together? They may need to be assembled as they are translated.

Hopwood: That's a very interesting point. We haven't done the experiment yet, but when we get round to putting the fatty acid synthase ACP from *Streptomyces coelicolor* into the actinorhodin gene cluster of the same organism, it might very well function. As you say, it obviously doesn't function in the organism normally, or we would not have seen an actinorhodin-negative phenotype in our 'knockout' experiment. So it may well be a question of space–time organization.

Hunter: So are all your constructions made by insertion *in cis* in the cluster?

Hopwood: In the ACP replacements that I described, yes (Khosla et al 1992).

Chater: Could you say anything about that question by asking which fragments of the *act* cluster, when carried on a plasmid, are able to complement defined *act* mutations? In that sense, can you reshuffle pathways when you don't have the genes *in cis*?

Hopwood: Mutations in four of the actinorhodin polyketide synthase genes (for the ketosynthase, ketoreductase, and cyclase/dehydrase) have been successfully complemented by the corresponding ones in the granaticin cluster, *in trans* (Sherman et al 1992). So they don't have to be *in cis* in the cluster in order to function. However, in this experiment, presumably they are being expressed in the cells in which actinorhodin is normally produced, whereas at that time those cells may conceivably not be making lipids any longer. This may be why the cell's own fatty acid synthase genes cannot complement the mutations in the actinorhodin genes.

Cane: When complementation results in the formation of the blue pigment, the presumption is that it's actinorhodin, but when it results in the formation of 'a pigment', you need to be sure that the pigment produced is either an actinorhodin precursor, or a shunt metabolite.

Hopwood: Absolutely! We may be sitting on a goldmine here. We started this work to try to get different compounds: if we put the frenolicin gene in, we might get two extra carbons in the chain, or with the griseusin gene we might get four. We haven't been able to investigate the chemistry yet to know what the pigments produced in the complementation experiments are, but I hope that some of them are indeed different from actinorhodin and so will give us clues as to which components of the polyketide synthase are responsible for carbon chain length and the other variables that distinguish the different compounds.

References

Caffrey P, Bevitt DJ, Staunton J, Leadlay PF 1992 Identification of DEBS1, DEBS2 and DEBS3 of the multienzyme polypeptides of the erythromycin-producing polyketide synthase from *Saccharopolyspora erythraea*. FEBS (Fed Eur Biochem Soc) Lett 304:225–228

Cane DE, Ott WR 1988 Macrolide biosynthesis. 5. Intact incorporation of a chain elongation intermediate into nargenicin. J Am Chem Soc 110:4840–4841

Cane DE, Yang C 1987 Macrolide biosynthesis. 4. Intact incorporation of a chain elongation intermediate into erythromycin. J Am Chem Soc 109:1255–1257

Cane DE, Celmer WD, Westley JW 1983 A unified stereochemical model of polyether structure and biogenesis. J Am Chem Soc 105:3594–3600

Dutton CJ, Gibson SP, Goodie AC et al 1991 Novel avermectins produced by mutational biosynthesis. J Antibiot (Tokyo) 44:357–365

Hendren RW, Bloch K 1980 Fatty acid synthetases from *Euglena gracilis*. J Biol Chem 255:1504–1508

Khosla C, Ebert-Khosla S, Hopwood DA 1992 Targeted gene replacements in a *Streptomyces* polyketide synthase gene cluster: role for the acyl carrier protein. Mol Microbiol, in press

Li Z, Martin M, Vederas JC 1992 Biosynthetic incorporation of labeled tetraketide intermediates into dehydrocurvularin, a phytotoxin from *Alternaria cinerariae*, with assistance of β-oxidation inhibitors. J Am Chem Soc 114:1531–1533

Pospísil S, Sedmera P, Hávranek V, Krumphanzl V, Vanék Z 1983 Biosynthesis of monensins A and B. J Antibiot (Tokyo) 36:617–619

Sherman DH, Kim E-S, Bibb MJ, Hopwood DA 1992 Functional replacement of genes for individual polyketide synthase components in *Streptomyces coelicolor* A3(2) by heterologous genes from a different polyketide pathway. J Bacteriol, in press

Staunton J, Sutkowski AC 1991a Biosynthesis of aspyrone, a metabolite of *Aspergillus melleus*: advanced precursor studies to identify the product of the polyketide synthase. J Chem Soc Chem Commun, p 1108–1110

Staunton J, Sutkowski AC 1991b The polyketide synthase (PKS) of aspyrone biosynthesis: evidence for the enzyme bound intermediates from incorporation studies with *N*-acetylcysteamine thioesters in intact cells of *Aspergillus melleus*. J Chem Soc Chem Commun, p 1110–1112

Tsukamoto Y, Wong H, Mattick JS, Wakil SJ 1983 The architecture of the animal fatty acid synthetase complex. IV. Mapping of active centers and model for the mechanism of action. J Biol Chem 258:15312–15322

Yue S, Duncan JS, Yamamoto Y, Hutchinson CR 1987 Macrolide biosynthesis. Tylactone formation involves the processive addition of three carbon units. J Am Chem Soc 109: 1253–1255

Genes for the biosynthesis of β-lactam compounds in microorganisms

Geoffrey Turner

Department of Molecular Biology and Biotechnology and Krebs Institute for Biomolecular Research, University of Sheffield, Sheffield S10 2UH, UK

Abstract. Rapid progress has recently been made in the characterization of genes and gene clusters involved in the biosynthesis of β-lactam antibiotics such as penicillins, cephalosporins and cephamycins. The biosynthetic pathways are found in a wide range of microorganisms, including fungi, actinomycetes and Gram-negative bacteria. Comparisons of gene sequences (particularly the genes encoding isopenicillin N synthetase) and gene organization in these different microorganisms have led to proposals about the evolution of this group of pathways, and how they might have been transferred from prokaryotes to eukaryotes. The isolation and characterization of the genes encoding ACV (tripeptide) synthetase, the first step in the β-lactam biosynthetic pathway, have revealed the presence of three partly repeated domains, most likely responsible for the recognition, adenylation and activation of the three amino acid precursors of the penams and cephems. This has confirmed their classification as peptide synthetases, distantly related to enzymes responsible for the synthesis of peptide antibiotics in *Bacillus brevis* and other bacteria and fungi.

1992 Secondary metabolites: their function and evolution. Wiley, Chichester (Ciba Foundation Symposium 171) p 113–128

β-Lactam compounds are made by a wide variety of microorganisms, including filamentous fungi, actinomycetes, and Gram-negative bacteria (Elander 1983). While structures vary, the common feature is the presence of the β-lactam ring, which may be alone, or fused to a second ring structure. The best-understood biosynthetic pathways are those for the penicillins and cephalosporins (Fig. 1), where the β-lactam ring is fused to either a thiazolidine (penicillins) or a dihydrothiazine ring (cephalosporins). Both of these structures contain sulphur in the second ring (derived from cysteine), but β-lactam compounds are also known where this is replaced by carbon (carbapenems, e.g. thienamycin) or oxygen (oxapenems, e.g. clavulanic acid). Sometimes, the ring even exists on its own, as is the case for monocyclic β-lactams (e.g. nocardicin and sulfazecin).

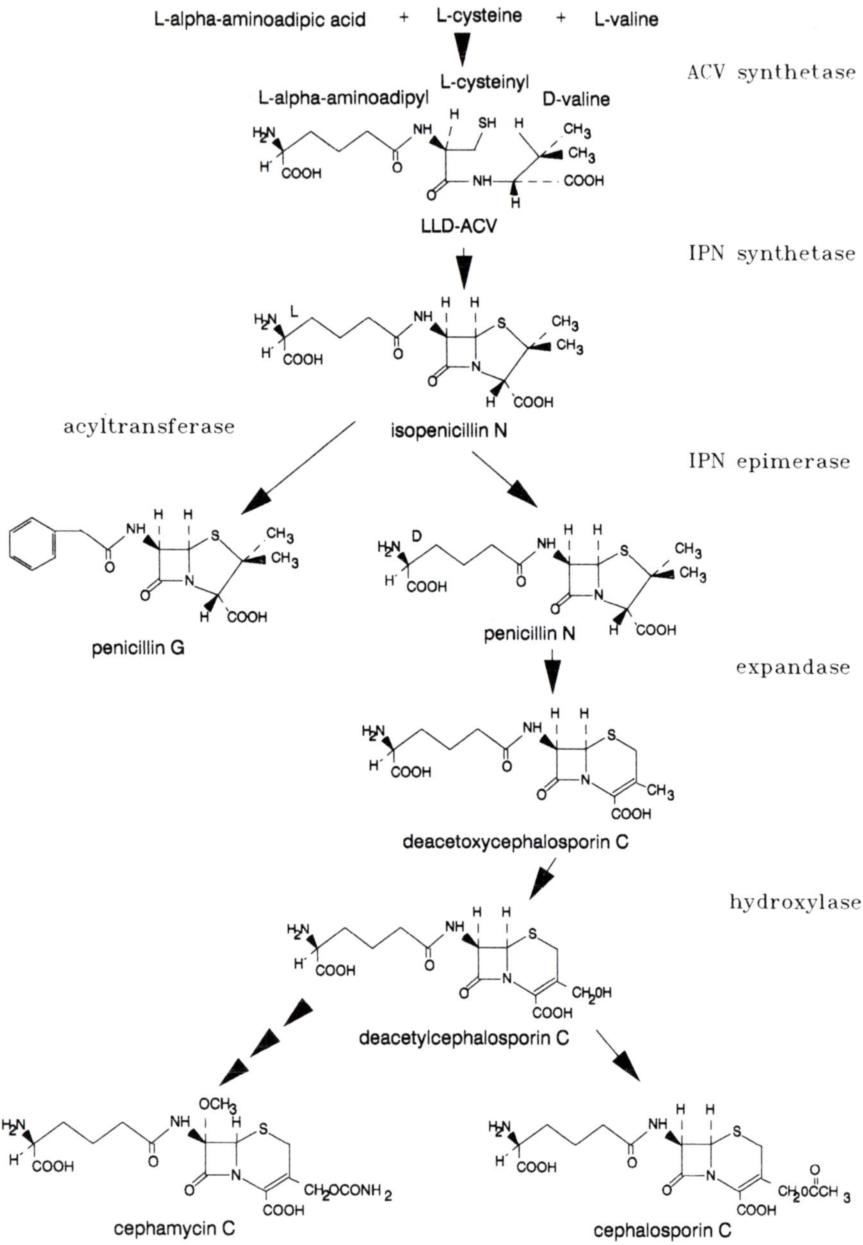

FIG. 1. Biosynthesis of penams and cephems in microorganisms. No organism possesses all the pathways shown. Enzymes for which genes have been isolated are shown.

Since the biosynthetic routes for carbapenems, oxapenems and monobactams appear to be distinct from the penam/cephem pathways (e.g. glycerol and ornithine are precursors of the oxapenem clavulanic acid), and only partly characterized at present (Baldwin et al 1991a), this discussion will be confined to the pathways shown in Fig. 1. The penam/cephem pathway has been extensively reviewed recently (Queener 1990, Martin et al 1991).

The aim of this paper is to summarize the recent findings from a number of laboratories, particularly with respect to the structure and organization of the genes encoding the steps in the biosynthetic pathways, and how these have provided new ideas about the function and origin of the enzymes involved.

The biosynthetic pathway

Ever since penicillin and cephalosporin became medically important there has been an interest in understanding the chemistry and biochemistry of their synthesis. The biosynthesis of the penams (penicillins) and cephems (cephalosporins, cephamycins) is shown in Fig. 1, where it can be seen that these pathways share the first two steps, leading to isopenicillin N.

The primary metabolites which are the starting point for the penams/cephems are the three amino acids L-α-aminoadipic acid, L-cysteine and L-valine. The first step is the formation of the tripeptide δ-(L-α-aminoadipyl)-L-cysteinyl-D-valine (ACV), catalysed by the single large enzyme, ACV synthetase (ACVS).

The tripeptide is cyclized by isopenicillin N synthetase (IPNS), also called ACV cyclase or cyclase, in an oxidative reaction in which four hydrogen atoms are removed from ACV, and one molecule of oxygen is consumed (Baldwin & Abraham 1988). The fate of the isopenicillin N now depends on the organism. Only the fungi seem to possess the enzyme acyltransferase (acyl-CoA:6-aminopenicillanic acid acyltransferase, ACT), which removes the aminoadipyl side chain. Depending on the compounds available to the organism, this can be replaced by an acyl derivative such as phenyl acetyl (penicillin G; Fig. 1) or phenoxyacetyl. 6-Aminopenicillanic acid (6-APA), lacking a side chain, is also found in fermentation medium (Batchelor et al 1959).

Some fungi, and many bacterial species, further modify the penam ring by the sequential action of expandase and hydroxylase activities to form deacetylcephalosporin C. This is converted into either cephalosporins or cephamycins, depending on the organism.

Biosynthetic genes

A fuller understanding of these pathways has been greatly helped by the isolation of the cognate genes. The first success was as recent as 1985, when the gene encoding IPNS of *Acremonium chrysogenum* (formerly *Cephalosporium acremonium*) was isolated (Samson et al 1985). Two factors which have greatly

TABLE 1 β-Lactam-producing microorganisms

		Bacteria	
β-Lactam	*Fungi*	*Actinomycetes*	*Eubacteria*
Penam	*Aspergillus nidulans* *Penicillium chrysogenum* *Epidermophyton* *Trichophyton* *Polypaecilum* *Malbranchea* *Microsporum*		
Cephem	*Acremonium chrysogenum* *Paecilomyces* *Scopulariopsis* *Diheterospora* *Spiroidium*	*Streptomyces* *Nocardia*	*Flavobacterium* *Lysobacter* *Xanthomonas*
Clavam		*Streptomyces*	
Carbapenem		*Streptomyces*	*Erwinia* *Serratia*
Monocyclic β-lactams		*Nocardia*	*Pseudomonas* *Chromobacterium* *Acetobacter* *Gluconobacter* *Agrobacterium*

Adapted from Elander 1983, Cohen et al 1990.

aided the isolation of genes from a wide variety of organisms (Table 1) are the clustering of genes, in both bacteria and fungi, and the surprisingly high degree of sequence identity between organisms which are phylogenetically unrelated. The significance of this will be discussed later.

Genes common to penam and cephem pathways

ACV synthetase gene(s). The ACV synthetase, catalysing the first step, is by far the most complex enzyme of the pathway. Whereas earlier studies detected the formation of the dipeptide L-α-aminoadipyl-L-cysteine, suggesting a two-step process, possibly requiring two enzymes, more recent data suggested that direct synthesis of a tripeptide was the most likely scenario (Banko et al 1987). The idea that two genes were required to encode two enzymes led to their naming as *pcbA* and *pcbB* (Ingolia & Queener 1989). Unfortunately, this premature naming of the gene(s) before characterization has persisted as *pcbAB*, though it has also been called *acvA* to avoid this problem (Smith et al 1990a, MacCabe et al 1991). *pcbAB* suggests the presence of two separable functions that can be identified by intragenic complementation. Such complementation might be possible, but has yet to be demonstrated.

Another idea, that the ACV synthetase was a peptide synthetase of the same class as the gramicidin synthetases (GS1 and GS2) of *Bacillus brevis*, was directly testable on purification of the ACVS from *Aspergillus nidulans* (Van Liempt et al 1989). Although the enzyme is very unstable, it was possible to demonstrate synthesis of ACV by a single polypeptide, ATP/pyrophosphate exchange, and covalent binding of L-[^{14}C]valine to the protein. This strong evidence was confirmed by sequencing the isolated gene.

One method of isolating the ACVS gene was by using DNA fragments from a 7-formamidocephalosporin-producing *Flavobacterium* sp. as probes against *Penicillium chrysogenum* DNA (Smith et al 1990b). A cosmid insert of *Flavobacterium* DNA, known to carry the biosynthetic gene cluster, was the source of this DNA. The only steps common to the β-lactam biosynthetic pathways of the bacterium and the fungus are the ACVS and the IPNS steps (Fig. 1). After elimination of the IPNS cross-hybridization signal, the remaining probes identified three adjacent regions in the *P. chrysogenum* DNA. Sequence analysis of these regions showed them to be three partly repeated sequences, encoding domains A, B and C (Fig. 2), contained within a large open reading frame of 11.3 kb, potentially encoding a single polypeptide of 424 kDa. Each domain consists of about 570 amino acids.

These domains have close identity with similar domains in the *Bacillus brevis* peptide synthetase genes, *tycA*, *grs1*, *grs2*. One of the most conserved regions of the domains also has similarity to certain other enzymes which carry out ATP/PP_i exchange reactions, including firefly luciferase and coumarate CoA ligase. Taken together, these results strongly suggest that each domain activates and binds a specific amino acid. An amino acid would be activated by adenylation, and bound to the enzyme as a thioester. Successive transfer of amino acids, probably via a pantetheine cofactor, would lead to the formation of peptide bonds between the attached amino acids. Recent data have suggested a stoichiometry of one pantetheine per enzyme molecule (Baldwin et al 1991b), though sequence analysis has identified three putative pantetheine binding sites (Gutierrez et al 1991). Finally, the tripeptide must be released from the protein, presumably by a thioesterase; a putative thioesterase domain, showing some similarity with those of animal fatty acid synthetases, can be identified in the C-terminal end of the protein.

ACVS genes from a number of β-lactam producers have now been sequenced, and all have a similar structure to the *P. chrysogenum* gene (Coque et al 1991, MacCabe et al 1991).

IPN synthetase gene. Since the isolation of the *Acremonium chrysogenum* IPNS gene (Samson et al 1985), the high degree of sequence conservation has permitted the isolation of IPNS genes from some of the organisms shown in Table 1, representing a good cross-section of species. Since these genes are relatively small, and were isolated before the ACVS gene (*acvA*), a large amount of sequence information has now been accumulated and compared. These data have been used to formulate ideas about the evolution of the β-lactam biosynthetic pathways, and are discussed below.

Unlike the ACV synthetase, the IPNS protein is not related to any other known class of enzymes. Cyclases are also involved in the biosynthesis of other β-lactams such as clavulanic acid (Baldwin et al 1991a), and the isolation and analysis of the genes encoding them should throw more light on the evolution of these other pathways.

The availability of the *A. chrysogenum* gene has also permitted the production of large amounts of pure enzyme, using *Escherichia coli* expression systems. This has facilitated detailed studies on the chemistry of the reaction (Baldwin & Abraham 1988).

A gene specific to hydrophobic penicillin producers

The gene encoding the acyltransferase is found in hydrophobic penicillin producers (e.g. penicillin G) such as *P. chrysogenum* and *A. nidulans* (i.e. some filamentous fungi), but is absent from cephem producers, and has not yet been detected in prokaryotes, suggesting that it is a later addition to the repertoire. In both *P. chrysogenum* and *A. nidulans* the genes are interrupted by three introns in equivalent positions (Barredo et al 1989, Montenegro et al 1990).

The enzyme can use IPN or 6-APA as a substrate for the substitution or addition of different side chains. While the gene encodes a 40 kDa polypeptide, post-translational processing gives rise to two non-identical peptides of 11.5 and 28.5 kDa. The active enzyme appears to require the association of both subunits (Whiteman et al 1990).

Small amounts of 6-APA accumulate in the culture medium of penicillin producers as a 'natural penicillin', but acyltransferase normally adds a side chain which depends on the availability of suitable precursors in the medium, such as phenyl acetic acid or phenoxyacetic acid. These are presented to the enzyme as acyl-CoA derivatives. How the CoA is added to these compounds is not known, and no specific CoA ligases have yet been characterized in the fungi.

The three genes so far described are clustered in *P. chrysogenum* and *A. nidulans* (Fig. 2), and seem to be sufficient for the synthesis of penicillin from amino acid precursors. This has been demonstrated by laboratory transfer (by transformation) of the *P. chrysogenum* gene cluster to *Aspergillus niger* and *Neurospora crassa*. These fungi do not naturally make β-lactam antibiotics, nor do they contain any detectable related genes, but make penicillin after introduction of the cluster (Smith et al 1990c).

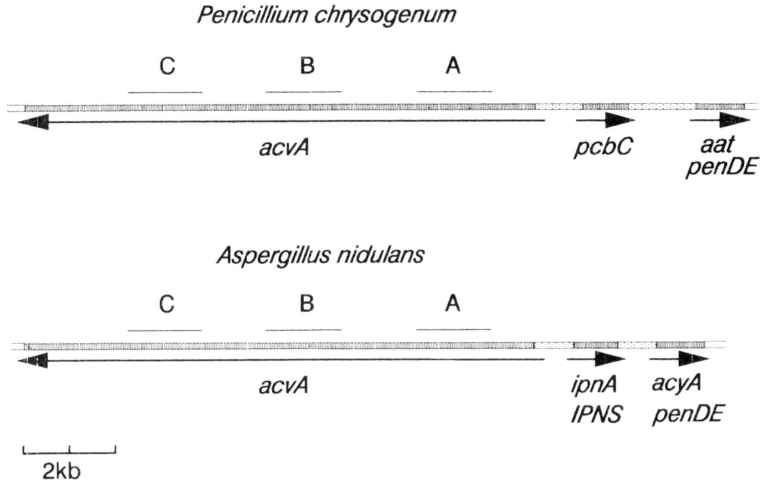

FIG. 2. Penicillin biosynthetic gene clusters in *Aspergillus nidulans* and *Penicillium chrysogenum*. Different gene designations have been used in the literature. Domains A, B and C of the genes encoding δ-(L-α-aminoadipyl)-L-cysteinyl-D-valine synthetase (ACVS) are shown.

Genes specific to the cephem pathway

Epimerase. Isopenicillin N is converted to penicillin N, the substrate for ring expansion, by epimerase, an enzyme not found in organisms that produce hydrophobic penicillins. The epimerase gene of *Streptomyces clavuligerus* is upstream of the expandase, and these two genes are expressed as a single transcript (Kovacevic et al 1990).

Expandase/hydroxylase. Study of the expandase and hydroxylase genes has revealed more interesting differences between a prokaryote and a eukaryote. These reactions are catalysed by a single protein in the fungus *A. chrysogenum*, but *Streptomyces clavuligerus* possess two separate genes. Sequence comparison of the *A. chrysogenum* expandase/hydroxylase and *S. clavuligerus* expandase genes revealed a high degree of amino acid sequency identity (57%) between them, the fungal gene having a C-terminal extension of 20 amino acids (Queener 1990). Both expandase and hydroxylase are α-ketoglutarate-linked dioxygenases. Cloning and sequencing of the hydroxylase gene of *S. clavuligerus* have now shown that the hydroxylase and expandase genes of this bacterium are also related, with 59% amino acid sequence identity (Kovacevic & Miller 1991). It has been suggested that an ancestral dual function expandase/hydroxylase, inherited by the fungus, was later duplicated in some bacteria, one copy taking on the expandase role, and one the hydroxylase role, though each enzyme retains some of the other activity.

Horizontal gene transfer and evolution

Taxonomic criteria and 5S rRNA sequences can be used to draw a phylogenetic tree depicting the approximate times of separation of β-lactam-producing organisms (Cohen et al 1990, Landan et al 1990). However, the sequence divergences observed for IPNS of these same species indicate a much closer relatedness. It can be suggested that either functional constraints on the IPNS protein have resulted in a slower rate of evolution, or that there has been horizontal, interspecies transfer of the IPNS gene after the separation of prokaryotes and eukaryotes (two billion years ago) and close to the divergence of the Gram-positive and Gram-negative bacteria (1.0–1.5 billion years). The main argument against structural constraints is that within the filamentous fungal and streptomycete branches of the tree, the divergence of these genes is similar to that seen for the 5S rRNA genes. It still remains possible that the rate of change of the gene sequences has been more rapid since the divergence of bacteria and fungi.

The horizontal transfer proposed for the IPNS gene is generally extrapolated to suggest that a complete antibiotic cluster may have transferred from a prokaryote to a eukaryote. Although data on the ACVS are just beginning to accumulate, full analysis of this information might lead to a revision of the hypothesis, at least with respect to the dating of the transfer. The degree of identity of the amino acids encoded by the ACVS genes of *P. chrysogenum* and *Nocardia lactamdurans* is only 43% overall (Coque et al 1991), compared to 57% for the IPNS genes of the same species. If individual domains are compared, then the identity is higher, sometimes approaching the level of IPNS identity.

The gene organization in bacteria and fungi is also interesting to compare. The β-lactam genes of *Streptomyces* and *Flavobacterium* are organized in clusters, where at least the ACVS, IPNS, epimerase, expandase and hydroxylase are located. Depending on the species, other steps are required for the final product, and it remains to be seen whether these will be part of the same cluster (Fig. 3).

Interestingly, the *S. clavuligerus* (Madduri et al 1991) and *N. lactamdurans* (Coque et al 1991) clusters also contains a gene encoding lysine 6-aminotransferase, which converts lysine to aminoadipic acid. It has long been known that while filamentous fungi can synthesize lysine via α-aminoadipic acid, bacteria use the diaminopimelic acid pathway which does not involve aminoadipate (Martin & Aharonowitz 1983). An equivalent lysine 6-aminotransferase gene has not yet been detected in fungi.

The only remnants in filamentous fungi of a proposed cluster transfer are the first two genes of the pathway, encoding ACVS and IPNS, which lack introns. The gene for acyltransferase (*act*) is also part of the fungal penicillin cluster (Fig. 3), but contains introns, and has not yet been found in a prokaryote.

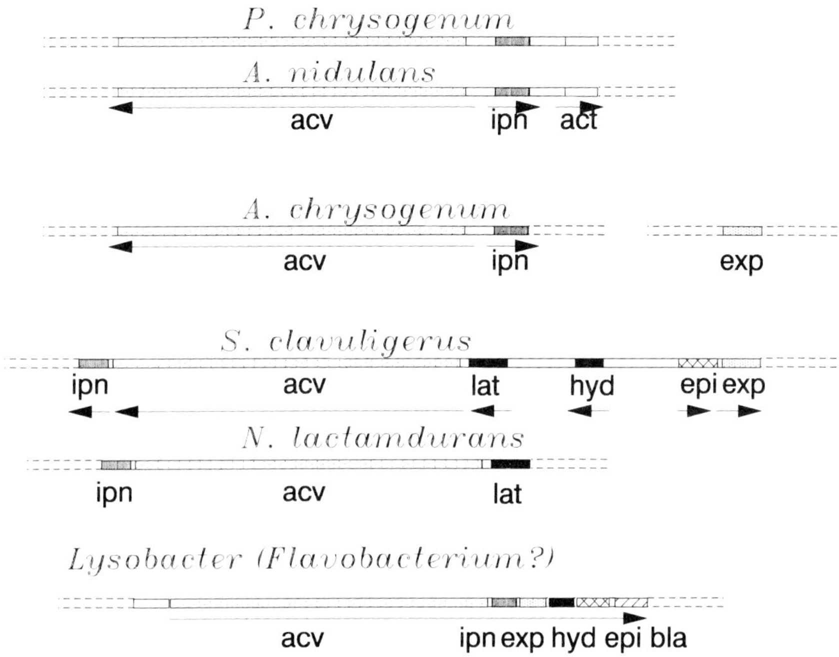

FIG. 3. Clustering of biosynthetic genes in fungi and bacteria. Genes encoding: acv, ACVS; ipn, IPNS; act, acyltransferase; exp, expandase; hyd, hydroxylase; epi, epimerase; lat, lysine aminotransferase; bla, β-lactamase. References for the data are in the text, except for *Lysobacter* (Kimura et al 1990) (which is very closely related to *Flavobacterium*; D. Caswell, personal communication). Arrows show the orientation of the coding regions. The expandase gene of *A. chrysogenum* is on a different chromosome from the ACVS and IPNS genes.

Conversely, the expandase/hydroxylase of *A. chrysogenum* is located on a different linkage group from the ACV and IPNS gene 'cluster'.

In each of the three fungal species analysed so far, the gene encoding ACVS is in the opposite orientation from that encoding IPNS, and the coding regions are separated by a short region (approximately 1 kb) which presumably contains the promoters for divergent transcription. The prokaryotic genes, on the other hand, lie in the same orientation. The reason for the clustering of pathway genes in fungi is not at all clear. In prokaryotes, clustered genes are often transcribed in a single unit to give a polycistronic (multiple gene) mRNA, which also provides an explanation for a common orientation of the clustered genes. This ensures that all the enzymes for a particular pathway are produced simultaneously, under a common control. Clustering of pathway genes is quite common in fungi (e.g. Hawkins et al 1988), but each gene has its own promoter, so orientation is not important.

Concluding remarks

Similarities between the IPNS genes of streptomycetes and fungi led to an original proposal that horizontal transfer occurred some 370 million years ago (Weigel et al 1988). Consideration of the additional data from *Flavobacterium*, a Gram-negative bacterium, pushed this date back by a billion years (Cohen et al 1990). Recent data suggest that the ACVS genes might not be so closely related, implying a different rate of sequence substitution from IPNS, particularly in certain regions of the large ACVS genes. The horizontal transfer hypothesis needs to take into account these new data.

Whatever the mechanism of pathway transfer, the widespread occurrence of functional pathways implies that they confer some selective advantage on strains carrying them. In this context, it is interesting to note that a wild-type, non-penicillin-producing strain of *Aspergillus nidulans* isolated from the soil lacks the entire gene cluster (MacCabe et al 1990). This strain is able to cross with laboratory strains of *A. nidulans*, showing it to be the same species. It appears that loss of penicillin production in a given species is not greatly disadvantageous.

Further back in time from the evolution of diversity in the β-lactam biosynthetic pathways, the evolution of the peptide synthetases is now a fascinating topic. It has long been known that biologically active, small peptides containing unusual amino acids are produced by a wide variety of microbes, and most of these are probably made by peptide synthetases which belong to the same ancient family. It is over 20 years since Lipmann (1971) discussed the evolution of peptide synthetases, but only recently have sequence data begun to provide new insights into this field. The availability of the genes should make a significant impact over the next few years on both the enzymology and the evolution of these complex proteins.

Acknowledgements

Work in the author's laboratory is supported by grants from SmithKline Beecham Pharmaceuticals and the Science and Engineering Research Council.

References

Baldwin JE, Abraham EP 1988 The biosynthesis of penicillins and cephalosporins. Nat Prod Rep 5:129–145

Baldwin JE, Adlington RM, Bryans JS et al 1991a Isolation of dihydroclavaminic acid, an intermediate in the biosynthesis of clavulanic acid. Tetrahedron 47:4089–4100

Baldwin JE, Bird JW, Field RA, O'Callaghan NM, Schofield CJ, Willis AC 1991b Isolation and partial characterization of ACV synthetase from *Cephalosporium acremonium* and *Streptomyces clavuligerus*—evidence for the presence of phosphopantothenate in ACV synthetase. J Antibiot (Tokyo) 44:241–248

Banko G, Demain AL, Wolfe S 1987 δ-(-L-α-Aminoadipyl)-L-cysteinyl-D-valine synthetase (ACV synthetase): a multifunctional enzyme with broad substrate specificity for the synthesis of penicillin and cephalosporin precursors. J Am Chem Soc 109:2858–2860

Barredo JL, Van Solingen P, Diez B et al 1989 Cloning and characterization of the acyl-coenzyme A:6-aminopenicillanic acid-acyltransferase gene of *Penicillium chrysogenum*. Gene 83:291–300

Batchelor FR, Doyle FP, Nayler JHC, Rolinson GN 1959 Synthesis of penicillin-6-aminopenicillanic acid in penicillin fermentations. Nature (Lond) 183:257–258

Cohen G, Shiffman D, Mevarech M, Aharonowitz Y 1990 Microbial isopenicillin N synthase genes: structure, function, diversity and evolution. Trends Biotechnol 8:105–111

Coque JJR, Martin JF, Calzada JG, Liras P 1991 The cephamycin biosynthetic genes pcbAB, encoding a large multidomain peptide synthetase, and pcbC of *Nocardia lactamdurans* are clustered together in an organization different from the same genes in *Acremonium chrysogenum* and *Penicillium chrysogenum*. Mol Microbiol 5:1125–1133

Coque JR, Liras P, Laiz L, Martin JF 1991 A gene encoding lysine 6-aminotransferase which forms the β-lactam precursor α-aminoadipic acid, is located in the cluster of cephamycin biosynthetic genes in *Nocardia lactamdurans*. J Bacteriol 173: 6258–6264

Elander RP 1983 Strain improvement and preservation of β-lactam-producing microorganisms. In: Demain AL, Solomon NA (eds) Antibiotics containing the β-lactam structure. Springer-Verlag, Berlin, vol 1:97–146

Gutierrez S, Diez B, Montenegro E, Martin J-F 1991 Characterization of the *Cephalosporium acremonium pcbAB* gene encoding α-aminoadipyl-cysteinyl-valine synthetase, a large multidomain peptide synthetase: linkage to the *pcbC* gene as a cluster of early cephalosporin biosynthetic genes and evidence of multiple functional domains. J Bacteriol 173:2354–2365

Hawkins AR, Lamb HK, Smith M, Keyte JW, Roberts CF 1988 Molecular organization of the quinic acid utilisation gene cluster in *Aspergillus nidulans*. Mol & Gen Genet 214:224–231

Ingolia TD, Queener SW 1989 Beta-lactam biosynthetic genes. Med Res Rev 9:245–264

Kimura H, Miyanoshita H, Sumino Y 1990 DNA and applications thereof. Japan Patent Application No. 2–3762

Kovacevic S, Miller JR 1991 Cloning and sequencing of the β-lactam hydroxylase gene (*cefF*) from *Streptomyces clavuligerus*: gene duplication may have led to separate hydroxylase and expandase activities in the actinomycetes. J Bacteriol 173:398–400

Kovacevic S, Tobin MB, Miller JR 1990 The β-lactam biosynthesis genes for isopenicillin N epimerase and deacetoxycephalosporin C synthetase are expressed from a single transcript in *Streptomyces clavuligerus*. J Bacteriol 172:3952–3958

Landan G, Cohen G, Aharonowitz Y, Shuali Y, Graur D, Shiffman D 1990 Evolution of isopenicillin N synthase genes may have involved horizontal gene transfer. Mol Biol Evol 7:399–406

Lipmann F 1971 Attempts to map a process evolution of peptide biosynthesis. Science (Wash DC) 173:875–884

MacCabe AP, Riach MBR, Unkles SE, Kinghorn JR 1990 The *Aspergillus nidulans npeA* locus consists of three contiguous genes required for penicillin biosynthesis. EMBO (Eur Mol Biol Organ) J 9:279–287

MacCabe AP, van Liempt H, Palissa H et al 1991 δ-(L-α-Aminoadipyl)-L-cysteinyl-D-valine synthetase from *Aspergillus nidulans*. Molecular characterization of the acvA gene encoding the first enzyme of the penicillin biosynthetic pathway. J Biol Chem 266:12646–12654

Madduri K, Stuttard C, Vining LC 1991 Cloning and location of a gene governing lysine ϵ-aminotransferase, an enzyme initiating β-lactam biosynthesis in *Streptomyces* spp. J Bacteriol 173:985–988

Martin JF, Aharonowitz Y 1983 Regulation of biosynthesis of β-lactam antibiotics. In: Demain AL, Solomon NA (eds) Antibiotics containing the β-lactam structure. Springer-Verlag, Berlin, vol 1:229–254

Martin JF, Ingolia TD, Queener SW 1991 Molecular genetics of penicillin and cephalosporin antibiotic biosynthesis. In: Leong SA, Berka RM (eds) Systems and applications for filamentous fungi. Marcel Dekker, New York, Basel & Hong Kong, p 149–196

Montenegro E, Barredo JL, Gutierrez S, Diez B, Alvarez E, Martin JF 1990 Cloning, characterization of the acyl-CoA:6-amino penicillanic acid acyltransferase gene of *Aspergillus nidulans* and linkage to the isopenicillin N synthase gene. Mol & Gen Genet 221:322–330

Queener SW 1990 Molecular biology of penicillin and cephalosporin biosynthesis. Antimicrob Agents Chemother 34:943–948

Samson SM, Belagaje R, Blankenship DT et al 1985 Isolation, sequence determination and expression in *Escherichia coli* of the isopenicillin N synthetase gene from *Cephalosporium acremonium*. Nature (Lond) 318:191–194

Smith DJ, Earl AJ, Turner G 1990a The multifunctional peptide synthetase performing the first step of penicillin biosynthesis in *Penicillium chrysogenum* is a 421,073 dalton protein homologous to *Bacillus brevis* peptide synthetases. EMBO (Eur Mol Biol Organ) J 9:2743–2750

Smith DJ, Burnham MKR, Bull JH et al 1990b β-Lactam antibiotic biosynthetic genes have been conserved in clusters in prokaryotes and eukaryotes. EMBO (Eur Mol Biol Organ) J 9:741–747

Smith DJ, Burnham MKR, Edwards J, Earl AJ, Turner G 1990c Cloning and heterologous expression of the penicillin biosynthetic gene cluster from *Penicillium chrysogenum*. Bio/Technology 8:39–41

Van Liempt H, Von Döhren H, Kleinkauf H 1989 δ-(L-α-Aminoadipyl)-L-cysteinyl-D-valine synthetase from *Aspergillus nidulans*. The first enzyme in penicillin biosynthesis is a multifunctional peptide synthetase. J Biol Chem 264:3680–3684

Weigel BJ, Burgett SG, Chen VJ et al 1988 Cloning and expression in *Escherichia coli* of isopenicillin N synthetase genes from *Streptomyces lipmanii* and *Aspergillus nidulans*. J Bacteriol 170:3817–3826

Whiteman PA, Abraham EP, Baldwin JE et al 1990 Acylcoenzyme A:G-aminopenicillanic acid acyltransferase from *Penicillium chrysogenum* and *Aspergillus nidulans*. FEBS (Fed Eur Biochem Soc) Lett 262:342–344

DISCUSSION

Demain: You didn't mention the work of Peñalva, who postulates two different horizontal transfers of the β-lactam synthetic genes (Peñalva et al 1990). Why was that?

Turner: The reason is really that one wants to look for the simplest hypothesis to explain the close similarity of the genes.

Cane: In relation to the similarity between the ACV and gramicidin synthetase genes, there are a couple of provocative aspects. One is that when the ACV synthetase sequence was first determined, it was noted that the supposedly conserved cysteines that had been observed in the tyrocidin and gramicidin synthetases were absent from the ACV sequence, and so the notion that there was a thioester synthesis template was questioned (MacCabe et al 1991). Recently, Vater's group determined the amino acid to which a valine is apparently attached in gramicidin S synthetase, as well as the surrounding sequence, and discovered that it is attached to a serine rather than a cysteine, at least if one assumes that the protein has been directly aminoacylated (Schlumbohm et al 1991). The surrounding sequence is found not only in the tyrocidin and gramicidin synthetases but occurs three times in the ACV synthetase. The problem is that the amino acid now appears to be attached as an oxoester, not a thioester. It's not clear what each amino acid is really attached to, and there's no proof that they are attached directly to the serine.

Secondly, in relation to the question of doing block replacements, a recent patent application from the Berlin group refers to the replacement of the aminoadipyl domain of the ACV synthetase with the D-Phe domain of gramicidin S1 (Veenstra et al 1991). It wasn't clear whether this was a proposal that they could do this, or whether it had actually already been done.

Turner: It was a proposal only. There was no experimental evidence, although it may be possible.

Baldwin: You referred to the clavulanic acid pathway enzymes. The enzyme that has been sequenced is clavaminic acid synthetase, but the persons who have done that will not release the sequence for commercial reasons.

You noticed an apparently slower rate of change of IPNS than of ACVS. The IPNS reaction is of course a totally different chemical process from the condensation of three amino acids. It involves almost certainly an extraordinarily reactive intermediate; we postulated an iron(IV) oxo species. My view of the catalysis of IPNS is that the protein component of that catalyst is not acting in the sense of a Pauling enzyme, which lowers activation energy by stabilizing transition states; it actually is avoiding other reactions, because the intermediate has such a high chemical potential that there is no need for any transition state lowering to permit the reaction to go. What is needed is to stop other reactions going on, and, in order to do that, to make the penicillin structure, there may be a very strong limitation on what you can do with the substrate; whereas a condensation of three amino acids to make two peptide bonds is, I suspect intuitively, a much more easily achieved process which does not require this stringency of the active site to stop other reactions.

I have thought a lot about how IPNS could have evolved. What you have discussed about the evolution of this pathway is concerned with how the pathway may have got from one organism to another; but the real question is where it came from in the first case. In my view, IPNS takes a molecule of dioxygen, an atom of ferrous iron, and a thiol-containing substrate, and takes out four hydrogens and makes penicillin. I have always held that all such reactions of secondary metabolism biosynthesis can be achieved in the absence of enzymes, and I think there's enough evidence to show that it is so. They have intrinsically the desire to go along the pathway.

We have made a lot of experiments to see if it is possible to carry out the reaction of IPNS in the total absence of this enzyme, and to make penicillins *in vitro*. I have to say that none of these experiments has been successful, although using biological techniques we can detect about 10 molecules of penicillin if they pass by. I think the reason is that our methods have been wrong! Probably, from what we understand of the enzyme mechanism, to do it in the absence of IPNS we need to work at very low oxygen tension. There are various technical and chemical reasons why this could be the answer. If it turns out to be true and by going to low oxygen tension (as low as 0.1 torr), we can make penicillin from ACV, with iron and dioxygen, it would seem that the making of penicillin may be a very ancient reaction that might have evolved at a time when the atmospheric composition of the earth was changing from anoxic to oxygenic. Unfortunately, I can't yet report the success of the experiment!

Cavalier-Smith: For IPN synthetase, in the *fungi*, you have to postulate *higher* rates of evolution on any hypothesis, because fungi are relatively recently evolved compared with the divergence time between *Flavobacterium* and *Streptomyces*. I would suspect at least a seven-fold difference between the divergence times of the different fungi from each other and of *Flavobacterium* from *Streptomyces*. With such unequal rates of evolution in different branches of the phylogenetic tree, it is going to be difficult to use gene sequence trees to decide between the two hypotheses. Unequal rates can give highly misleading results. So it's a very open question.

Beppu: Professor Turner, is the exon–intron structure found in the fungal gene?

Turner: Yes; the acyltransferase genes of *Penicillium* and *Aspergillus*, the only ones that have been sequenced and published, have three introns in the acyltransferase in similar positions. The ACV has none.

Beppu: If you assume horizontal gene transfer from prokaryotes to eukaryotes, I don't understand why the exon–intron structure was introduced.

Turner: This is getting into the question of where introns come from, of course.

Demain: But no one is suggesting that the acyltransferase gene was transferred laterally. It is apparently a *Penicillium* gene that was recruited, and happens to be next to the others. However, the other two genes (encoding ACVS and IPNS) do not have introns and are candidates for lateral transfer.

Turner: I wouldn't be totally committed to that, because the question of how introns arise is a very interesting one. I am thinking of transposable elements in higher plants, such as the Ds/Ac elements in maize, which can hop into a gene, and sometimes do not prevent the gene expressing; they apparently generate splice sites and can be spliced out of the primary transcript while sitting in the gene (Wessler 1989). So the presence of introns does not rule out a prokaryotic origin.

Davies: With respect to the production of β-lactams by different organisms, there was a report that *E. coli* makes a β-lactam. Has this been confirmed?

Demain: This has been published (Allison 1990). The intracellular contents of many different organisms were examined using affinity columns. It was found that compounds like β-lactams are made by *E. coli*, *Enterobacter cloacae* and many other bacteria. I haven't seen a confirmation of this report.

Davies: Has anyone probed *E. coli* or *Salmonella* with an antibody to the β-lactam ring?

Baldwin: No.

Beppu: We have not done this. I have a question: if you isolate a mutant of a β-lactam-producing streptomycete, deficient for antibiotic synthesis, does it show any pleiotropic changes in its phenotype?

Turner: There is an interesting example in a eukaryote, *Aspergillus nidulans*, which normally makes penicillin. Certain soil isolates of *A. nidulans* made no penicillin and became known as natural non-penicillin producers (Cole et al 1976). Recently, Jim Kinghorn's group showed that the whole penicillin gene cluster was missing from one of these isolates (MacCabe et al 1990). This strain was isolated from soil a number of times, so it certainly continues to survive in Nature quite happily, without the antibiotic pathway.

Vining: Professor Turner, I think you raised a 'chicken-and-egg' type of question when you suggested that the *lat* gene had been recruited or had 'hopped' into the actinomycete gene cluster; without *lat*, α-aminoadipic acid cannot be made in prokaryotes, and, if there is no aminoadipic acid, then aminoadipylcysteinyl-D-valine, the key intermediate in β-lactam biosynthesis, also cannot be made. This argues for *lat* having been involved early in the evolution of β-lactam biosynthesis, rather than recruited later. Another interesting point is that *lat* is not present in most streptomycetes; you find it only in the β-lactam-producing species.

Turner: What is the sequence relationship between the *lat* genes? This enzyme is not confined to β-lactam-producing organisms, is it?

Vining: It is present in other bacteria, such as *Pseudomonas* and *Flavobacterium* species. I don't know the degree of nucleotide sequence similarity, but the streptomycete and flavobacterium enzymes have quite different protein structures.

Demain: It is interesting that the IPN synthetase has no sequence relationship with any other gene that has been looked at. What is the situation with the expansion enzyme? This catalyses a unique reaction, but is quite similar in its activity to other α-ketoglutarate-linked dioxygenases. Has anybody looked at the sequence to see if it resembles any of the genes encoding those enzymes?

Turner: I imagine that this has been looked at, but I don't know. It is only recently that the gene for the expandase was isolated.

Baldwin: I don't think any similarities have been recognized.

Demain: But has anyone made the statement that the expandase is a unique protein, in terms of sequence?

Baldwin: There are similarities with IPNS that people have construed as homologies.

Demain: I am thinking about enzymes that are *not* in the β-lactam pathway.

Turner: A similar question arises with the *Penicillium* acyltransferase. Barredo et al (1989) suggested that it may be related to the penicillin acylases of bacteria. These bacterial enzymes remove the side chain, which effectively is what the acyltransferase does before putting on another side chain. Functionally, you can see some potential similarity, but there is little discernible amino acid sequence similarity between penicillin acylases and acyltransferase.

Demain: The penicillin acylases have some relationship to the new cephalosporin acylase in their sequence, as I understand it.

References

Allison DG 1990 Purification of an intracellular bacterial β-lactam by affinity chromatography. J Microbiol Methods 12:83–89

Barredo JL, Van Solingen P, Diez B et al 1989 Cloning and characterization of the acyl-coenzyme A:6-aminopenicillanic acid-acyltransferase gene of *Penicillium chrysogenum*. Gene (Amst) 83:291–300

Cole DS, Holt G, Macdonald KD 1976 Relationship of the genetic determination of impaired penicillin production in naturally occurring strains to that in induced mutants of *Aspergillus nidulans*. J Gen Microbiol 96:423–426

MacCabe AP, Riach MBR, Unkles SE, Kinghorn JR 1990 The *Aspergillus nidulans npeA* locus consists of three contiguous genes required for pencillin biosynthesis. EMBO (Eur Mol Biol Organ) J 9:279–287

MacCabe AP, van Liempt H, Palissa H et al 1991 δ-(L-α-Aminoadipyl)-L-cysteinyl-D-valine synthetase from *Aspergillus nidulans*. Molecular characterization of the acvA gene encoding the first enzyme of the penicillin biosynthetic pathway. J Biol Chem 266:12646–12654

Peñalva MA, Moya A, Dopazo J, Ramon D 1990 Sequences of isopenicillin N synthase genes suggest horizontal gene transfer from prokaryotes to eukaryotes. Proc R Soc London B Biol Sci 241:164–169

Schlumbohm W, Stein T, Ullrich C et al 1991 An active serine is involved in covalent substrate amino acid binding at each reaction center of gramicidin-S synthetase. J Biol Chem 266:23135–23141

Veenstra AE, Martin JR, Garcia BD et al 1991 Cloning of aminoadipylcysteinylvaline synthetase gene of *Penicillium chrysogenum*, its mutation, and use in preparing new β-lactam antibiotics. Eur Pat Appl 91-200411 27 Feb 1991 (Chem Abstr 115(25):272696j)

Wessler SR 1989 The splicing of maize transposable elements from pre-mRNA—a minireview. Gene (Amst) 82:127–133

Regulation of gibberellin formation by the fungus *Gibberella fujikuroi*

B Brückner

Friedrich Schiller University Jena, Department of General Microbiology, Neugasse 24, 0-6900 Jena, Germany

Abstract. Gibberellins are a classic example of the production of plant growth regulators by microorganisms. They are important biotechnological products and are increasingly used in agriculture and horticulture. The economic importance of these plant hormones has led to an extensive study of the regulation of gibberellin biosynthesis. There have been reports of light, growth rate, inoculum size and carbon and ammonium sources acting as regulators of gibberellic acid biosynthesis. Besides light stimulation, nitrogen repression is a well-known regulatory principle of secondary metabolite formation. In *Gibberella fujikuroi* ammonium interferes with the production of gibberellic acid whereas phosphate does not influence the biosynthesis. It was found that the negative effect of ammonium ions is due to both the inhibition of activity and the repression of *de novo* synthesis of specific gibberellin-producing enzymes. Besides nitrogen control, the biosynthesis of gibberellins is suppressed by glucose. This glucose effect can be overcome by the addition of mevalonic acid. Therefore, the key enzyme of the isoprenoid pathway, the HMG-CoA reductase, seems to be the target of C-catabolite repression. A detailed knowledge of the regulation of gibberellin biosynthesis is important for fermentation processes. The biological function of gibberellin formation for the producing fungus is discussed.

1992 Secondary metabolites: their function and evolution. Wiley, Chichester (Ciba Foundation Symposium 171) p 129–143

The gibberellins (GA) are a group of plant growth hormones that were first isolated from the fungus *Gibberella fujikuroi* (imperfect stage, *Fusarium moniliforme*), a widespread phytopathogenic fungus often found on corn, rice, barley and other crops in different geographical regions throughout the world. Besides gibberellins, *G. fujikuroi* produces a number of interesting biologically active secondary metabolites, including pigments, phytotoxins and mycotoxins (Brückner et al 1989).

Mevalonic acid is the structural unit of many secondary metabolites, including the gibberellins, the carotenoids and the trichothecene toxins. Mevalonic acid is first converted via the isopentenyl, geranyl and farnesyl pyrophosphates and then further metabolized to the different secondary metabolites. Gibberellins have many functions in plants. They stimulate stem elongation, induce flowering and overcome seed dormancy. Because of the economic importance of the gibberellins as plant hormones, industrial researchers have established the conditions in which gibberellins can be produced and have looked for *G. fujikuroi* strains that synthesize increased amounts of these hormones.

Knowledge of the regulation of gibberellin biosynthesis is very important for the optimization of culture conditions. Up to now, little has been known about the regulation of gibberellins. There have been reports that light, ammonium ions, carbon source and growth rate all act as regulators of gibberellin biosynthesis in *G. fujikuroi*.

Zweig & De Vay (1959) first reported that 'cultures which were kept in darkness usually produced less gibberellins than cultures which were kept in the light'. Mertz & Henson (1967) and Mertz (1970) found a light-stimulated increase in acetate incorporation which could be prevented by AMO-1618, an inhibitor of GA biosynthesis which acts at the step of *ent*-kaurene synthetase. Johnson & Coolbaugh (1990) showed that the difference between levels of GA production in light- and dark-grown daughter cultures was greater if the two-day-old stock cultures had been maintained in darkness. Light-grown and older inocula (5–6 days) appeared to lose their capacity for light stimulation of GA synthesis. Furthermore, these authors found that the kaurenoic acid oxidation appears to be light stimulated. Rates of oxidation to products having GA-like properties were increased by up to 60% in enzyme preparations from cultures grown in light.

Interestingly, blue light has been found to stimulate isoprenoid biosynthesis in several of the Ascomycete fungi. Carotenogenesis in *G. fujikuroi, Blakeslea trispora* and *Neurospora crassa* is markedly increased, more than 300-fold, by blue light irradiation. Johnson & Coolbaugh (1990) suggested that light-stimulated GA biosynthesis also involves blue light reception.

Avalos and co-workers (1988) studied the biosynthesis of carotenoids and gibberellins in wild and mutant strains of *Gibberella fujikuroi*. They found that wild-type carotenoid formation is stimulated 2.9-fold, whereas kaurene synthesis is increased 438-fold. In contrast, enzymic activity of the regulatory mutant is even higher in dark-grown mycelia. Therefore, light affects the total amount of terpenoids formed by *G. fujikuroi*.

The negative control of gibberellin A_3 (GA_3; gibberellic acid) biosynthesis by rapidly utilized carbon sources was well established early on by Borrow et al (1961, 1964). An initial concentration of glucose in the culture of more than 3% results in a low yield of GA_3, and at this glucose level there is a lag phase that is otherwise absent (Vass & Jefferys 1979). These effects were overcome by feeding glucose in such a way that the concentration at any time was less than 4%. Mevalonic acid seems to be an initiator in the production of GA_3. Feeding of the prime precursor under conditions of catabolic repression overcomes the lag period. GA_3 production starts a short time after the addition of mevalonate. Therefore, the key enzyme in the isoprenoid pathway, 3-hydroxy-3-methylglutaryl-coenzyme A reductase (HMG-CoA reductase), is likely to be one, or the only, target of catabolic regulation.

The inhibitory effect of ammonium ions on gibberellin biosynthesis by *G. fujikuroi* has been known for a considerable time and was first reported by Borrow et al (1964).

We have studied the nitrogen regulation of GA synthesis in some detail (Brückner & Blechschmidt 1991). Figure 1 shows the influence of $(NH_4)_2SO_4$ concentration on the yield of GA_3 in defined and complex media. In defined medium, optimal production of GA_3 occurred at 22.5 mM $(NH_4)_2SO_4$. In a complex medium containing corn steep liquor, the conditions for *G. fujikuroi*

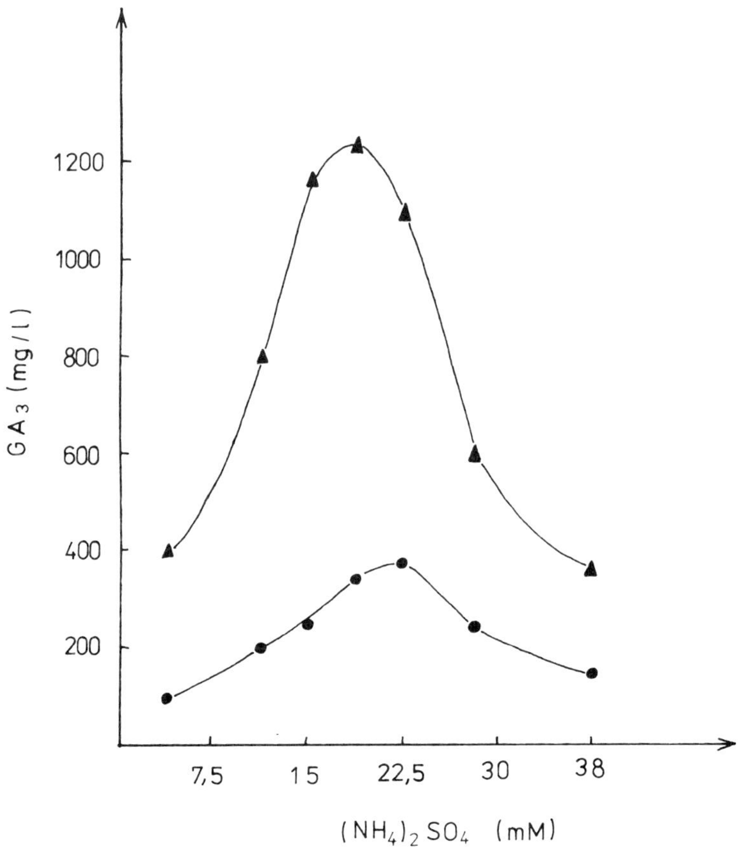

FIG. 1. The effect of $(NH_4)_2SO_4$ concentration on the maximum gibberellic acid (GA_3) titre in defined (filled circles) and complex (filled triangles) media.

growth and for the production of GA_3 are much better. The maximum GA_3 titre was measured in medium containing 19 mM $(NH_4)_2SO_4$. With a further increase in nitrogen concentration, the formation of GA_3 was decreased drastically. Figure 2 shows the effect of ammonium excess (38 mM $(NH_4)_2SO_4$) on fungal growth and gibberellin formation in comparison to optimal (19 mM $(NH_4)_2SO_4$) production conditions. The synthesis of GA_3 was reduced three-fold in the medium at high ammonium concentration, whereas growth was not greatly affected. Under these conditions, a significant amount of ammonium still remained in the culture fluid after 168 hours. Nevertheless, the formation of GA_3 began at the same time as it did in the medium containing the lower concentration of $(NH_4)_2SO_4$. These results are in contrast to the findings of other authors (Borrow et al 1961, 1964, Bu'Lock et al 1974, Vass & Jefferys 1979), who reported that GA_3 production began only when, or soon after, the nitrogen source in the medium had been exhausted.

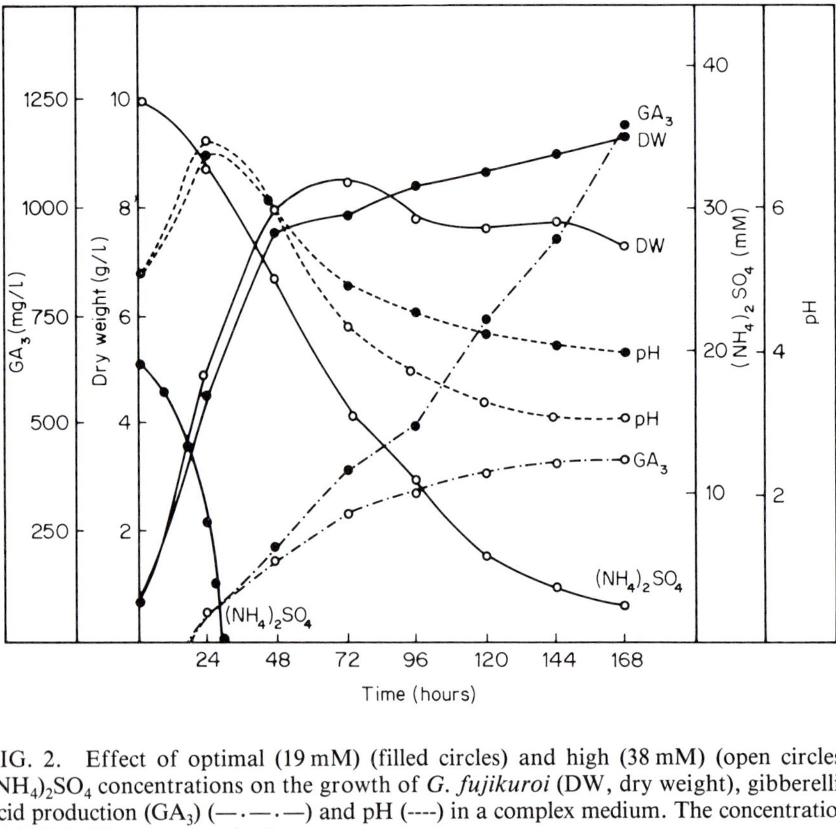

FIG. 2. Effect of optimal (19 mM) (filled circles) and high (38 mM) (open circles) $(NH_4)_2SO_4$ concentrations on the growth of *G. fujikuroi* (DW, dry weight), gibberellic acid production (GA_3) (—·—·—) and pH (----) in a complex medium. The concentration of $(NH_4)_2SO_4$ over time is shown.

To investigate further the effect of ammonium ions on biosynthetic enzyme activity, we added a high level of $(NH_4)_2SO_4$ (38 mM) to complex medium cultures at different times after GA_3 production had started. The medium already contained 19 mM $(NH_4)_2SO_4$. Figure 3 shows that the biosynthesis of gibberellic acid (GA_3) was nearly completely inhibited a few hours after the addition of ammonium sulphate. In another experiment, the onset of gibberellin biosynthesis in complex medium containing a high or low concentration of $(NH_4)_2SO_4$ was studied by adding cycloheximide (100 μg/ml) to batch cultures

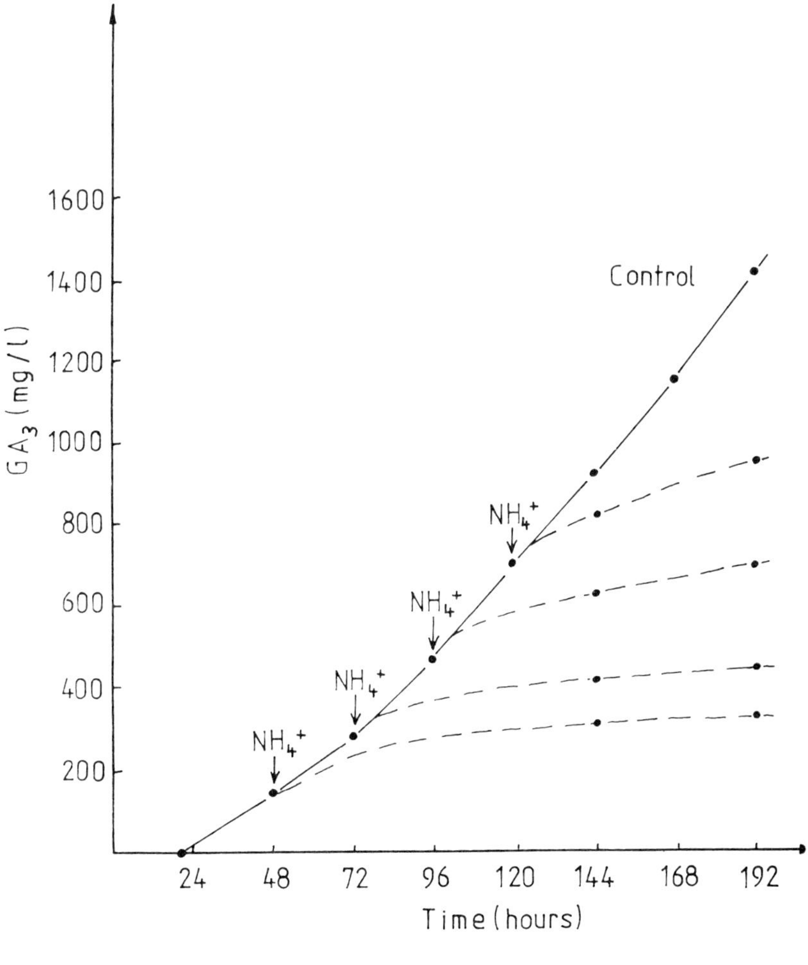

FIG. 3. Influence of the addition of 38 mM ammonium sulphate (arrows) on the production of gibberellic acid in a complex medium containing 19 mM $(NH_2)_2SO_4$ as the major nitrogen source.

of various ages. Addition of this protein synthesis inhibitor at 17 hours or 20 hours prevented a further increase in mycelial weight, and no gibberellins were formed. Addition of cycloheximide at 24, 30 or 41 hours allowed a short phase of gibberellin biosynthesis. Therefore, the enzymes of gibberellin pathway are not constitutive but were produced in the first 20 hours of cultivation.

To determine the specific gibberellin productivity during fermentation under different conditions, we prepared resting cell suspensions with washed mycelia previously grown on one of two concentrations (19 mM or 38 mM) of $(NH_4)_2SO_4$. The resting cell activity (mg GA_3 per mg mycelium per hour) of mycelia which had been grown at both concentrations of $(NH_4)_2SO_4$ increased during growth, reached a maximum after 72 hours, and fell during the stationary phase (Fig. 4). The activity of the gibberellin biosynthetic enzyme complex of cells previously grown in 38 mM $(NH_4)_2SO_4$ was lower than that in cells from an ammonium-limited medium (19 mM $(NH_4)_2SO_4$). This indicated that one or more enzymes in the gibberellin biosynthetic pathway were repressed—that is, their own production was reduced in the presence of ammonium ion. Moreover, the addition of $(NH_4)_2SO_4$ to the resting cell system led to a strong decrease in resting cell GA synthetic activity. This indicates that the negative effect of NH_4^+ on gibberellin synthesis is due to repression *and* to the direct inhibition of one or more enzymes in the gibberellin biosynthetic pathway.

The target of ammonium regulation is not clear. It was suggested by Bearder (1983) that the dehydrogenation of GA_4 to GA_7 is reduced by nitrogen excess, which inhibits the GA_4 1,2-dehydrogenase preferentially. To determine the influence of a high ammonium concentration on the activity of this enzyme, we used a mutant blocked at the last step in GA_3 biosynthesis, which therefore produced mostly GA_7. If the 1,2-GA_4 dehydrogenase is the major site of NH_4^+ inhibition, the GA_4:GA_7 ratio should increase sharply because of the accumulation of GA_4. Although the GA_4:GA_7 ratio rose slightly, the reduction of GA_7 formation was not associated with an increase in GA_4, the precursor of GA_7. Therefore, the 1,2-GA_4 dehydrogenase is probably not the major NH_4^+-sensitive enzyme of gibberellin synthesis in *G. fujikuroi*. The target of ammonium inhibition must lie earlier in the biosynthetic pathway, because of the decrease in all the analysed gibberellins by $(NH_4)_2SO_4$.

Bu'Lock et al (1974) studied the regulation of the biosynthesis of two different secondary metabolites—GA_3 and the polyketide bikaverin (a red pigment)—in relation to the nitrogen concentration. In batch culture with progressively more acute nitrogen limitation, bikaverin synthesis is initiated first, and then GA_3 synthesis follows. It is interesting that this series of events can be reproduced in chemostat cultures growing at different rates. These authors (Bu'Lock et al 1974) concluded that the gibberellin pathway is under the same type of overall regulation by growth rate as bikaverin synthesis,

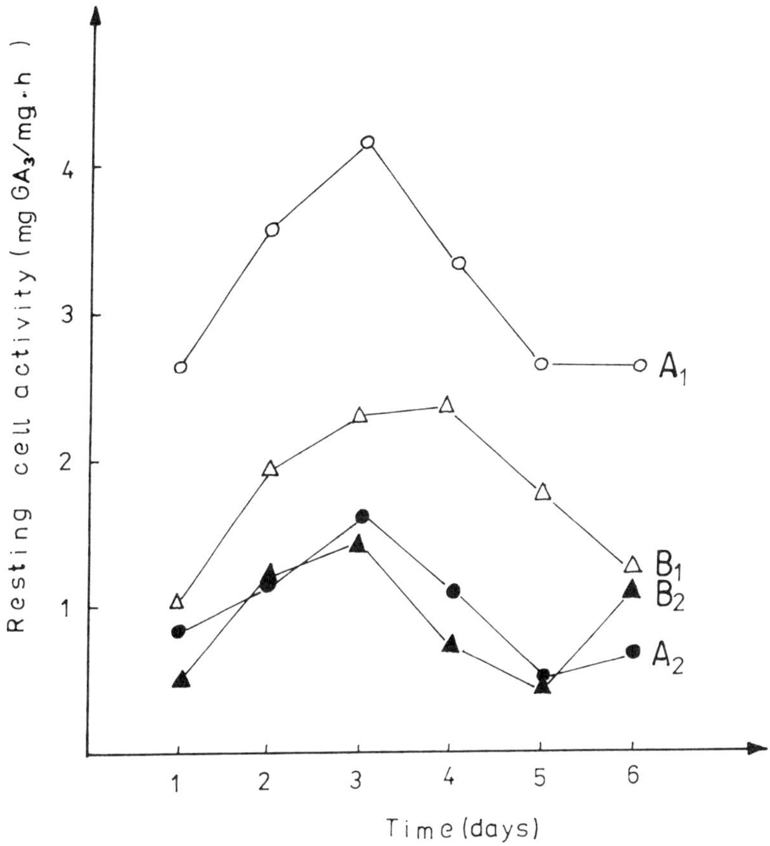

FIG. 4. Resting cell activity of cultures previously grown on (A) 19 mM or (B) 38 mM $(NH_4)_2SO_4$ in a resting cell system with 38 mM $(NH_4)_2SO_4$ or without $(NH_4)_2SO_4$. ○, △, resting cell system without $(NH_4)_2SO_4$; ●, ▲, resting cell system with $(NH_4)_2SO_4$.

but that GA synthesis begins at a lower level of limited nitrogen and a corresponding lower growth rate. In their opinion, a drastic decrease in growth rate caused by phosphate or other limitations should give the same onset of GA biosynthesis as nitrogen limitation. To test this suggestion, we examined the influence of phosphate limitation on the formation of gibberellin under conditions of nitrogen excess. Large amounts of ammonium prevent marked GA formation, although the growth rate is reduced drastically by phosphate limitation. Therefore, nitrogen controls gibberellin biosynthesis specifically. On the other hand, there are some new results on regulatory genes in *Fusarium* strains which control the biosynthesis of secondary metabolites produced on different pathways. Bu'Lock & De Gomez (1990) have described a pleiotropic regulatory mutation in *Fusarium graminearum* which prevents the expression of the zearalenone pathway and also prevents the expression of the isoprenoid pathway leading to trichothecene mycotoxins. Up to now, we have not been able to find a similar regulatory mutant which has at the same time lost the ability to produce bikaverins and gibberellins.

Finally, it should be pointed out that the distinction between trophophase (the period of growth and primary metabolism) and idiophase (the phase of secondary metabolism) is true for gibberellin production only under laboratory conditions. Zak (1976) suggested that gibberellins may play a role in the initial infection of the host plant. In contrast to submerged cultivation, where the main production of gibberellins begins only after fungal growth has ceased, gibberellin production in infected plants occurs during the exponential phase of fungal growth. Gibberellin-produing strains of *F. moniliforme* produced significantly more biomass than non-producing strains. Therefore, gibberellin formation seems to be the phytopathogenic principle of this fungus.

Interestingly, the main steps in the gibberellin biosynthetic pathway are identical or quite similar in plants and in the fungus. Chapman & Ragan (1980) have suggested that the occurrence of this pathway in *Gibberella* can be considered as an example of gene transfer from plants to the fungus. The close association of *G. fujikuroi* with host plants may have provided an environment for natural genetic engineering.

An answer to this interesting question can be found only by cloning the specific genes for gibberellin biosynthesis from the fungus and comparing them with the analogous genes of the host plant.

Acknowledgements

I thank B. Groß and S. Nickel for kind technical assistance.

References

Avalos J, Mackenzie A, Nelki DS, Bramley PM 1988 Terpenoid biosynthesis in cell extracts of wild-type and mutant strains of *Gibberella fujikuroi*. Biochim Biophys Acta 966:257–265

Bearder JR 1983 In vivo diterpenoid biosynthesis in Gibberella fujikuroi: the pathway after ent-kaurene. In: Crozier A (ed) The biochemistry and physiology of gibberellins, vol 1, chapter 5. Praeger, New York

Borrow A, Jefferys EG, Kessell RHJ, Lloyd EC et al 1961 The metabolism of *Gibberella fujikuroi* in stirred culture. Can J Microbiol 7:227–276

Borrow A, Brown S, Jefferys EG et al 1964 Metabolism of *Gibberella fujikuroi* in stirred culture. Can J Microbiol 10:407–444

Brückner B, Blechschmidt D 1991 Nitrogren regulation of gibberellin biosynthesis in *Gibberella fujikuroi*. Appl Microbiol Biotechnol 35:646–650

Brückner N, Blechschmidt D, Schubert B 1989 *Fusarium moniliforme* SHELD—a fungus producing a broad spectrum of bioactive metabolites. Zentralbl Bakteriol 144:3–12

Bu'Lock JD, De Gomez SC 1990 Effect of a regulatory mutation on trichothecene production by *Fusarium graminearum* strain NRRL 3198. Mycol Res 94:851–864

Bu'Lock JD, Detroy RW, Hoštálek Z, Munim-al-Shakarchi A 1974 Regulation of secondary biosynthesis in *Gibberella fujikuroi*. Trans Br Mycol Soc 62:377–389

Chapman DJ, Ragan MA 1980 Evolution of biochemical pathways: evidence from comparative biochemistry. Annu Rev Plant Physiol 31:639–678

Johnson SW, Coolbaugh RC 1990 Light-stimulated gibberellin biosynthesis in *Gibberella fujikuroi*. Plant Physiol 94:1696–1701

Mertz D 1970 Light-stimulated incorporation of L-leucine into the gibberellin of *Gibberella fujikuroi*. Plant Cell Physiol 11:273–279

Mertz D, Henson W 1967 The effect of the plant growth retardants Amo 1618 and CCC on gibberellin production in *Fusarium moniliforme*. Light stimulated biosynthesis of gibberellins. Physiol Plant 20:187–199

Vass RC, Jefferys EG 1979 Gibberellic acid. In: Rose AH (ed) Economic microbiology: secondary products of metabolism. Academic Press, New York & London, vol 3:421–435

Zak JC 1976 Pathogenicity of a gibberellin-producing and a nonproducing strain of *Fusarium moniliforme* in oats as determined by a colorimetric assay for *N*-acetyl glucosamine. Mycologia 68:151–155

Zweig G, De Vay JE 1959 On the biosynthesis of gibberellins from carbon-14-substrates by *Fusarium moniliforme*. Mycologia 51:877–886

DISCUSSION

Demain: Dr Brückner, you told us that a large number of secondary metabolites are made by *Gibberella fujikuroi*. Does that mean that there are many different strains of this organism, each of which make some of these metabolites, or that any particular strain makes many gibberellins, *and* makes all the other secondary compounds that are found?

Brückner: There are many strains of the species *G. fujikuroi*, and each of the strains will make some of these different compounds, but probably not all. We use one strain which was isolated from rice. It makes gibberellin, and it also makes carotenoids and bikaverin.

Davies: Is the formation of these other compounds also regulated by light?

Brückner: Light regulates especially the isoprenone compounds, which include the carotenoids, as well as the gibberellins.

Baldwin: You mentioned that glucose inhibits the terpene pathway through HMG-CoA reductase. Is that actually proved, at the enzyme level?

Brückner: No; this is only a suggestion, that when mevalonic acid overcomes the inhibitory effect of glucose, the inhibition must be the result of the lowering of mevalonic acid production with glucose, and so only when we give mevalonic acid can we overcome the glucose effect. This means that this key enzyme must be affected by glucose. But we don't know if it is an inhibition of the activity of this enzyme, or, for instance, an influence on the *de novo* synthesis of HMG-CoA reductase.

Demain: Of course, this suggestion would only be valid if, when you add mevalonic acid to the oil fermentation, you do *not* see a stimulatory effect on gibberellin production.

Brückner: There is really no effect on gibberellin yield; but the productivity is very high and it's not possible to find a further increase with mevalonic acid.

Cane: You want to be cautious here, because the main pathway for mevalonic acid metabolism will be for the production of ergosterol in this organism, at least before the resting stage. Unlike cholesterol production, which is closely linked to HMG-CoA reductase, in studies on the formation of metabolites such as sesquiterpene phytoalexins in tobacco plant tissue cultures, there is no simple correlation between HMG-CoA reductase activity and terpenoid phytoalexin production (Vögeli & Chappell 1988, Chappell et al 1991). It could be that the branch point is later on—for example, in the induction of the geranylgeranyl diphosphate synthase, or the kaurene cyclase. This has to be determined experimentally.

I am interested in the stimulation of metabolite production by light, because there are no light-harvesting organelles in the fungus, and only a very small number of enzymic activities are known to be driven by light. I am curious about the mechanism by which an organism like this would respond to light.

Turner: There are one or two examples now in fungi of light regulation. A gene for carotenoid biosynthesis in *Neurospora crassa* is regulated by blue light (Nelson et al 1989), and in *Aspergillus nidulans* light is required for conidiation (Mooney & Yager 1990). In *Aspergillus*, although the light receptor hasn't been identified, conidiation requires red light and is inhibited by far red light. It is interesting that a mutation was put into *Aspergillus nidulans* by geneticists so that it wouldn't need light for sporulation, the velvet mutation. Exactly how the velvet gene is involved in this light response is not known.

Chater: There has been molecular analysis of light induction of carotenoid synthesis in the myxobacteria, by a group in northern Spain (Martínez-Laborda et al 1990) and by Dave Hodgson at Warwick University. This may be a good model for looking at light effects.

Bu'Lock: There is a very marked division of opinion from laboratory to laboratory as to whether we should investigate light effects! The experimental set-ups required to demonstrate and define light effects are exceedingly tedious; they are also very fruitful, within certain limitations. Very often, people from the more botanical sciences are the ones prepared to do these things. Others of us decide just to leave the light on and to refrain from investigating light effects!

The same tends to happen in relation to trace metals. You can decide to follow Weinberg's approach and investigate trace metal effects in every system you look at, which is a minefield and can give you quite extraordinary effects. My view is that we are not going to investigate trace metal effects; use the tap! Providing you don't move from one part of the country to another, you are all right. But in fact, occasionally you cannot escape profound trace metal effects. One such is an effect on bikaverin synthesis, which, in unpublished work from our laboratory, was shown to be highly zinc determined. But once you enter into the world of trace metal effects, you have to you start a whole lot of experiments that are very difficult to finish. So, most of the time, we tend not to investigate them.

Dr Brückner, is there any evidence for a phosphate limitation of gibberellin production?

Brückner: No, we have no evidence. We compared two conditions, excess of phosphate and limited phosphate. Under conditions of excess, we see normal growth; with a low phosphate concentration (0.1 g/l KH_2PO_4) we saw a reduced growth rate, but nevertheless the rate of production of gibberellin was very low in both cases; this was because of the excess of ammonium ion in this study. Therefore, gibberellin formation is under the specific control of nitrogen but not of phosphate or of growth rate.

Chater: Dr Brückner, is there any information about whether the plant in which *G. fujikuroi* grows contributes any inducers of gibberellin synthesis?

Brückner: I know of only one publication on that. Dr Zak studied the growth of non-producing strains and a gibberellin-producing strain on rice plants. He found that non-producing strains have a lower growth rate and that the gibberellin-producing strain was at an advantage and grew faster and produced GA_3 from the beginning of exponential growth (Zak 1976). This is in contrast to laboratory conditions, where GA formation starts only after growth has been finished.

Chater: Could this be because the plant provides a non-specific inducer?

Brückner: It is possible. On the other hand, I think that the nutrient compounds are in limited concentrations and no inhibitory effects can occur.

Chater: Or there might also be a *specific* response of the plant to the fungus. Dr Brückner, are the experiments you are describing with a strain that has already been developed as a high producing strain, or a wild-type strain?

Brückner: These experiments are done with a wild-type strain.

Piepersberg: I am wondering whether the mutants blocked in gibberellic acid biosynthesis would infect the rice plant, and whether you have any idea of what gibberellin does in the infection process?

Brückner: I think the blocked mutant should grow, but at a lower rate. A specific case is the so-called 'bakanae' disease of rice. Only the gibberellic acid-positive organisms produce the disease. The blocked mutants do not cause the super-elongation of rice seedlings. The pathogenic process of this disease involves the production of gibberellic acid.

Piepersberg: But the mutants can nevertheless still infect the natural host?

Brückner: Yes; we know a lot of diseases of *Fusarium* strains on plants, not only because of gibberellin production, but also simply because of fungal growth on the plants, such as in tomatoes and other crops. But I cannot say whether gibberellin plays a role in these other diseases, in plants such as lupins, tomatoes or carnations. I don't think so. In general, *Fusarium* strains proved to be aggressive phytopathogens for different reasons—for example, because of the formation of phytotoxins like fusaric acid.

Demain: Has anyone taken a fungal mutant that cannot make gibberellic acid and tried to infect a plant with it?

Brückner: Yes, and in both cases, mutant and wild-type, the organism can grow in the plant, but at a different rate. Gibberellic acid producers grow faster, which suggests that GA helps the infection of plants; because of the GA_3-induced elongation of the cells, maybe the fungus can more easily invade the plant tissues.

Chater: A possible experiment would be to feed a plant with gibberellic acid, after infecting it with a non-producing mutant; then you might get the full growth rate.

Davies: Does gibberellic acid enter the plant, through some lesion on the plant surface?

Brückner: Yes. In the plant, the gibberellins as natural plant hormones are in low concentrations, but when you give GA_3 from outside, at higher concentrations, this causes the hormonal balance to be destroyed.

Davies: In relation to that point, you suggested that GA_4 and GA_7 intermediates are the more agriculturally interesting. Why is that?

Brückner: There are several new publications on the functions of GA_4 and GA_7 and there may be some other effects apart from those of gibberellic acid—for instance, in the production of seeds by conifers, or in the treatment with GA_4 and GA_7 to reduce russet on 'golden delicious' apples. Another important field of application of GA_4 and GA_7 is the induction of male flower formation in monoecious and gynoecious cucumbers.

Demain: There are different agricultural effects of the different gibberellins. I have always thought of gibberellins as the equivalent of steroids in animals, but the problem has been that except for GA_3 (gibberellic acid), the gibberellins have not been available for large-scale testing in agriculture. As I recall from my industrial experience, the mixture of GA_4 and GA_7 has been shown to have specific effects on plants, not shown by gibberellic acid. But then, as you say, there are very many different gibberellins, perhaps as many as 79 known ones, and they are not available for individual testing.

Turner: The repression of production rates by glucose is reminiscent of penicillin production in *Penicillium* or *Aspergillus*. Nobody has really resolved this glucose repression effect. In *Penicillium* and *Aspergillus*, as in your organism, it's not an absolute repression. Glucose slows down the production, but will not stop it. I wonder to what extent one can separate a glucose effect from the simultaneous effect of altered growth rates, so it's actually that glucose is *speeding up* the growth, and the faster growth leads to the repression of production?

Bu'Lock: That is the usual impression. The key observation is that with *Penicillium chrysogenum* you can get round the inhibitory glucose effect by substituting a rate-determining slow feed with glucose for a big batch with a non-glucose source supplied from the beginning. To go beyond that and ask what is the actual mechanism of that glucose effect is more difficult, and there has been comparatively little work. People tend to be empirical and to get round it by using soluble starch or something like that as their carbon source.

Turner: We have been able to show that this 50% or so inhibition of penicillin production by glucose in *Aspergillus nidulans* does act at the level of the genes (Brakhage et al 1992). You can make reporter genes (by fusing the promoter of the ACV synthase and the IPNS, to appropriate reporter genes). With glucose, a pattern of a slow increase in production of the reporter gene product occurs, about 50% down on the growth on lactose. So there seems to be an effect from glucose acting at the gene level, again not an absolute effect.

In *A. nidulans* there are mutations called *cre*, which are catabolite-repression mutations. Where glucose acts on the other genes in a clearer way, such as on the gene for isocitrate lyase, when it will repress enzyme formation, or alcohol dehydrogenase, if you introduce a *creA* mutation, there is no repression with glucose. But one of these *cre* mutations that we tried doesn't affect the glucose repression (or whatever this repression is) in the *Aspergillus* antibiotic regulation mechanism; it doesn't seem to be mediated by the normal catabolite-repression system.

Demain: You mention two enzymes (IPNS and ACVS) that you made fusions with. Did they both behave in the same way?

Turner: We found that the IPNS is affected by glucose, whereas the ACVS is not.

Demain: In cephalosporin production we see an inhibitory effect of glucose but there is no glucose effect on the formation of ACVS, yet a strong one on IPNS formation. I don't want to extrapolate to *Penicillium* (because there is some indirect evidence obtained by others [Revilla et al 1986] of a glucose effect on *Penicillium* ACVS formation), but there is absolutely no repressive effect on ACVS synthesis in *Cephalosporium acremonium* (*Acremonium chrysogenum*) or *Streptomyces clavuligerus*. We do find a curious type of inhibition by glucose of the *activity* of ACVS, which may or may not have physiological importance (Zhang & Demain 1991). In cell-free extracts, it can be explained by a competition for ATP between the ACVS and the early steps of glycolysis; this inhibition can be reversed by increasing the ATP concentration. With regard to repression by glucose, it's mainly acting on later enzymes in the β-lactam pathway, especially the expandase enzyme. This enzyme is usually the most sensitive in terms of glucose repression, in both *Cephalosporium* and *Streptomyces*. The IPNS is also susceptible. It would be very interesting to know what the mechanisms are, but there has been almost no work on the basic mechanism of these glucose effects. In other organisms—for example, in the studies of George Jones (1985) on phenoxazinone synthase of actinomycin synthesis—there is an effect of glucose on transcription. With the β-lactams, there have been no molecular studies, until this work that Professor Turner mentions.

Haslam: Could I ask a rather more general question? One of the themes that is apparent in this area of the biosynthesis of secondary metabolites is that acetic acid is a prime building block; it's the major building block for many sorts of compounds. To what extent can one ascribe the effects of various additives in changing gibberellin production, which is a very specific example, to changes in the energy state of the cell that is producing these compounds? Acetyl-CoA is a key point in the generation of energy in the cell; to what extent are all these effects simply reflecting the partitioning of acetyl-CoA in different directions?

Bu'Lock: We certainly have some ideas on this. We have examined a polyketide-type product and an isoprenoid-type product in several different systems. The system in which we have done most work is the gibberellin-producing *Gibberella fujikuroi* (Bu'Lock et al 1974). If you use a nitrogen-limited medium with a fairly high K_S value (we used glycine as nitrogen source, which has a much higher K_S than ammonium), you can run a continuous culture which doesn't make either compound; you can also run such a culture at a lower growth rate which makes, predominantly, and within analytical limits, the bikaverin polyketide; and you need to reduce the growth rate even lower to get full expression of gibberellin synthesis. Gibberellin and bikaverin are using the same acetyl-CoA pool; but which of the two you get seems to depend, not on the magnitude of that pool, but on the competition between enzymes doing different things to it. That is also my impression from less recent definitive work with the two biosynthetic routes from acetyl-CoA in *Fusarium graminearum*, to the polyketide zearalenone and the isoprenoid trichothecenes.

Brückner: Yes, and we have studied the competition for mevalonic acid. When we make mutants which cannot produce carotenoids, they sometimes produce a higher yield of gibberellins, because the entire mevalonate pool goes into the gibberellin pathway. The effect of acetate is difficult to study because of pH effects and transport problems. It's not possible to get good results, but mevalonic acid acts as an effective precursor and increases the gibberellin yield.

References

Brakhage AA, Browne P, Turner G 1992 Regulation of *Aspergillus nidulans* penicillin biosynthesis and penicillin biosynthesis genes *acvA* and *ipnA* by glucose. J Bacteriol 174, in press

Bu'Lock JD, Detroy RW, Hoštálek Z, Munim-al-Shakarchi A 1974 Regulation of secondary biosynthesis in *Gibberella fujikuroi*. Trans Br Mycol Soc 62:377–389

Chappell J, Vonlanken C, Vögeli U 1991 Elicitor-inducible 3-hydroxy-3-methylglutaryl coenzyme-A reductase activity is required for sesquiterpene accumulation in tobacco cell suspension cultures. Plant Physiol 97:693–698

Jones GH 1985 Regulation of phenoxazinone synthase expression in *Streptomyces antibioticus*. J Bacteriol 163:1215–1221

Martínez-Laborda A, Balsalobre JM, Fontes M, Murillo FJ 1990 Accumulation of carotenoids in structural and regulatory mutants of the bacterium *Myxococcus xanthus*. Mol & Gen Genet 223:205–210

Mooney JL, Yager LN 1990 Light is required for conidiation in *Aspergillus nidulans*. Genes & Dev 4:1473–1482

Nelson MA, Morelli G, Carattoli A, Romano N, Macino G 1989 Molecular cloning of a *Neurospora crassa* carotenoid biosynthetic gene (albino-3) regulated by blue light and the products of the white collar genes. Mol Cell Biol 9:1271–1276

Revilla G, Ramos FR, Lopez-Nieto MJ, Alvarez E, Martin JF 1986 Glucose represses formation of δ-(L-α-aminoadipyl)-L-cysteinyl-D-valine and isopenicillin N synthase but not penicillin acyltransferase in *Penicillium chrysogenum*. J Bacteriol 168:947–952

Vögeli U, Chappell J 1988 Induction of sesquiterpene cyclase and suppression of squalene synthetase activities in plant cell cultures treated with fungal elicitor. Plant Physiol 88:1291–1296

Zak JC 1976 Pathogenicity of a gibberellin-producing and a nonproducing strain of *Fusarium moniliforme* in oats as determined by a colorimetric assay for *N*-acetyl glucosamine. Mycologia 68:151–155

Zhang J, Demain AL 1991 Regulation of ACV synthetase in penicillin- and cephalosporin-producing microorganisms. Biotechnol Adv 9:623–641

Genetic regulation of secondary metabolic pathways in *Streptomyces*

Keith F. Chater

John Innes Institute, John Innes Centre, Norwich NR4 7UH, UK

Abstract. *Streptomyces* species are (along with the fungi) the best-known antibiotic-producing organisms. Often, they make several different antibiotics. The biosynthesis of each antibiotic is encoded by a complex gene cluster that usually also contains regulatory and resistance genes. Typically, there may be more than one such pathway-specific regulatory gene per cluster. Both activator and repressor genes are known. Some of the regulatory genes for different pathways are related. In *S. coelicolor*, expression of several such biosynthetic gene clusters also depends on at least 11 globally acting genes, at least one of which is involved in the translation of a rare codon (UUA). A protein phosphorylation cascade also seems to be involved. Gene clusters closely similar to those for the biosynthesis of aromatic polyketide antibiotics determine spore pigment in some species. These genes show different regulation from antibiotic production genes. The evolution of gene clusters for polyketide antibiotics, and the possible adaptive benefits of secondary metabolism, are discussed.

1992 Secondary metabolites: their function and evolution. Wiley, Chichester (Ciba Foundation Symposium 171) p 144–162

A universally acceptable definition of secondary metabolites is elusive. Few, however, would argue against the inclusion of antibiotics produced by stepwise synthesis from the products of primary metabolism. Although many organisms produce antibiotics, one bacterial genus—*Streptomyces*—is numerically pre-eminent. The morphology of *Streptomyces* colonies growing on surfaces is unusually complex among bacteria, involving growth of branching hyphae to form a mycelium from which aerial branches grow and produce chains of spores. The start of aerial growth usually coincides with the onset of antibiotic synthesis, suggesting that there is some benefit in coordinating the two processes. This article reviews recent knowledge of how *Streptomyces* antibiotic biosynthesis is controlled at the genetic level, and explores its relevance to the evolution and role of secondary metabolites.

Genetic regulation of antibiotic production in *Streptomyces coelicolor* A3(2)

Identification and organization of genes influencing antibiotic production

The study of antibiotic biosynthesis by physiological and chemical techniques has recently been complemented by genetic approaches. *S. coelicolor* A3(2), which is particularly well characterized genetically, is suited to such investigations because it produces four distinct antibiotics, two of them coloured (actinorhodin, blue; undecylprodigiosin, red) and therefore easily studied. The other two are methylenomycin and 'calcium-dependent antibiotic' (Chater & Hopwood 1989). The biosynthetic enzymes of each pathway are specified by a gene cluster. Mutations in these genes affect production only of the cognate antibiotic. Mutations in at least 11 other genes impair production of several—often all—of these antibiotics. Many of these mutations have been mapped (Fig. 1). The designation of the widely acting (global) mutations reflects the phenotype sought during isolation of the mutants: *afs* mutants did not make the hormone-like A-factor, *abs* mutants were defective only in antibiotic production, and *bld* (bald) mutants did not make aerial mycelium. Such mutant phenotypes suggest that antibiotic production is subject to global, sometimes morphologically coupled, regulation.

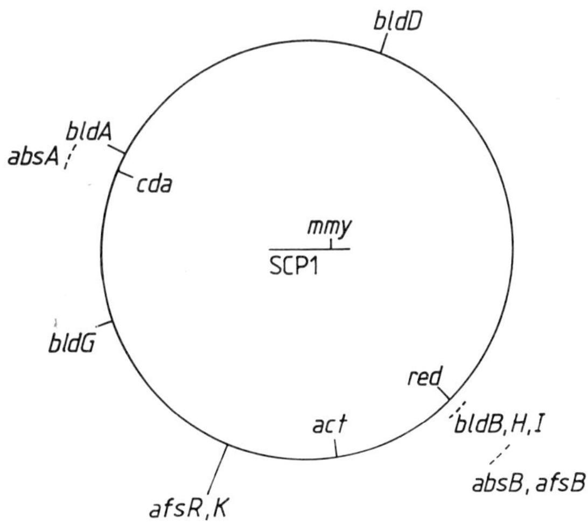

FIG. 1. Locations of genes affecting antibiotic production on the linkage map of *Streptomyces coelicolor* A3(2). Inside the circle, clusters of genes for antibiotic biosynthesis. Outside the circle, genes affecting the production of several antibiotics. The map is based on that of Hopwood (1988) with some modifications and the addition of *absA* (Adamidis et al 1990), *absB* and *afsR* (Champness 1991), *afsK* (Horinouchi & Beppu 1992) and *bldI* (Harasym et al 1990).

The *act* (actinorhodin), *red* (undecylprodigiosin) and *mmy* (methylenomycin) gene clusters have been cloned (see Chater 1990 for references). Each contains many individual genes, arranged into several transcription units (Fig. 2). Within each cluster at least some of the transcription units are controlled by the products of pathway-specific regulatory genes, some of which activate, whereas other repress, transcription. Thus, *act*II-ORF4 (ORF, open reading frame) and *redD* encode activators, and *act*II-ORF1 and *mmyR* encode repressors.

Global regulation and the act *genes*

Studies of how the global 'regulatory' *bld, afs* and *abs* genes may work have focused mainly on their effects on actinorhodin synthesis. These studies are

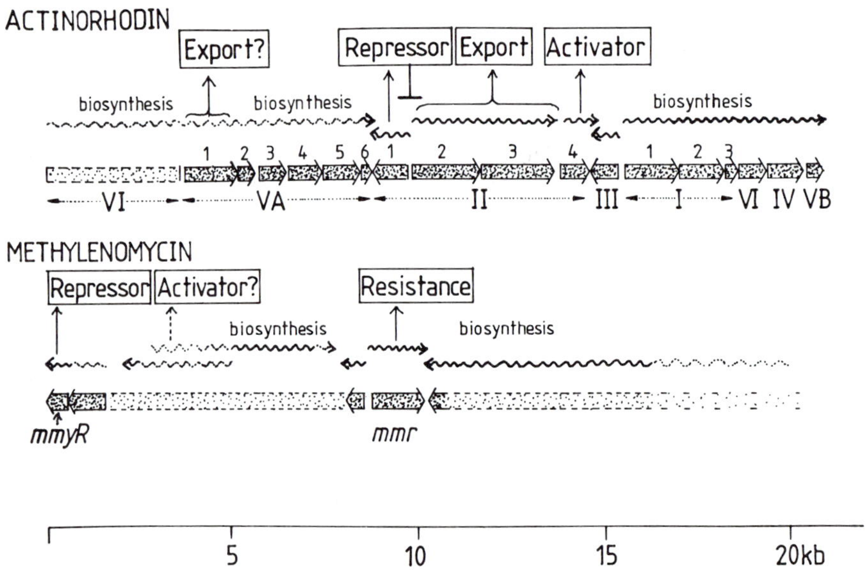

FIG. 2. The organization of two gene clusters for antibiotic production in *Streptomyces coelicolor* A3(2). The diagrams represent currently available information on the *act* (actinorhodin) cluster (Hallam et al 1988, Fernández-Moreno et al 1991, Caballero et al 1991; F. Malpartida, personal communication) and the *mmy* (methylenomycin) cluster (Chater & Bruton 1985, Neal & Chater 1987, 1991; A. Wietzorrek, C. J. Bruton & K. F. Chater, unpublished results). Dashed lines: regions for which detailed information is not available. Stippled arrows: the position and extent of open reading frames. Wavy arrows: the position and extent of transcripts (dotted where published information is not available). Functionally related regions of the *act* cluster are indicated by roman numerals, and the numbering of open reading frames within these regions is shown by small numbers above the genes.

tending to implicate a pivotal role for the pathway-specific regulatory gene *act*II-ORF4. This is particularly clear for *bldA*, which has been cloned and shown to specify the transfer RNA that translates the rare codon UUA as a leucine residue (Lawlor et al 1987, Leskiw et al 1991a). Circumstantial evidence strongly suggests that such codons (TTA in the DNA) are absent from all genes essential for vegetative growth in streptomycetes, and none have been reported in any structural genes encoding enzymes of antibiotic synthesis sequenced so far from this genus. Instead, the few known TTA codon-containing genes are nearly all resistance or regulatory genes from clusters of antibiotic biosynthetic genes (Leskiw et al 1991b), including *act*II-ORF4 (Fernández-Moreno et al 1991). When the single TTA codon of *act*II-ORF4 was changed to an alternative leucine codon and the altered gene introduced into a *bldA* mutant, actinorhodin was produced (Fernández-Moreno et al 1991). Thus, the UUA codon in *act*II-ORF4 mRNA is the major means by which *bldA* influences actinorhodin production. This codon is probably translated at a low efficiency even in a *bldA* mutant: when *act*II-ORF4 is present in such a mutant at 50–100 copies on a high copy number plasmid, presumptively causing a proportionate increase in the mRNA, enough activator is made to permit actinorhodin synthesis (Passantino et al 1991). Similar effects of extra copies of *act*II-ORF4 on actinorhodin production in *bldD, G* and *H* mutants may indicate that these genes are involved, along with *bldA*, in producing a functional $tRNA^{Leu}_{TTA}$ (Passantino et al 1991). The mode of action of the other seven known pleiotropic genes on actinorhodin synthesis is unknown, though effects on the transcription or translation of *act*II-ORF4 or on the stability or post-translational modification of its product are possible. Alternatively, some of the gene products might directly participate in initiating transcription of *act* structural genes, for example by encoding an RNA polymerase σ factor specific for antibiotic production genes (though the promoters of those *act* genes analysed are not similar to each other; Hallam et al 1988, Parro et al 1991).

TTA codons and regulatory genes associated with the production of other antibiotics

Although transcriptional defects for the *red* (Guthrie & Chater 1990) and *mmy* genes (A. Wietzorrek & K. F. Chater, unpublished results) have been demonstrated in a *bldA* mutant, no TTA codons have yet been found in the *red* or *mmy* gene clusters: the *redD* (activator) gene is TTA-free (Narva & Feitelson 1990) and no activator gene has been identified for the *mmy* cluster. Nevertheless, circumstantial evidence (see later) suggests that both clusters contain other, as yet unidentified, regulatory genes. Positively acting regulatory genes have also been identified in clusters of genes for antibiotic biosynthesis in other *Streptomyces* species. Those for streptomycin in *S. griseus* and the phylogenetically somewhat distant *S. glaucescens* (Distler et al 1987;

J. Distler, personal communication), and for bialaphos in *S. hygroscopicus* (Raibaud et al 1991) also contain TTA codons; however, the daunorubicin regulatory genes *dnrI* and *dnrJ* do not (Stutzman-Engwall et al 1992). There are no reports on whether mutations in *bldA* affect production of the relevant antibiotics in these strains.

It is attractive to postulate that UUA codons can be translated efficiently only in certain conditions (i.e. those suitable for antibiotic production), and that this is a major controlling mechanism (Leskiw et al 1991a,b). However, the evidence on this is equivocal. It may be that these codons have little regulatory significance, and that pressure for their loss during the evolution of *Streptomyces* DNA to its characteristic high content of G + C was less severe in poorly expressed regulatory genes. If so, streptomycetes might exist in which *bldA* was required for vegetative growth; but *bldA* mutants of the taxonomically distinct species *S. griseus* do exist, and resemble *bldA* mutants of *S. coelicolor* (J. Kwak & K. Kendrick, personal communication). Probably, therefore, the observed distribution of TTA codons is ancient, and likely to be significant.

A phosphorylated regulatory protein

Like *act*II-ORF4, the *afsR* gene of *S. coelicolor*, at high copy number, stimulates transcription of *act* genes (Horinouchi et al 1989, Stein & Cohen 1989). It has recently been shown that purified AfsR (the protein encoded by the *afsR* gene) becomes radiolabelled when incubated with [γ-^{32}P]ATP and cell-free extracts of *S. coelicolor* or *S. lividans*, through the action of membrane-bound AfsR-kinase activity (AfsK; Hong et al 1991, Horinouchi & Beppu 1992). This suggests that an uncharacterized extracellular signal is perceived by AfsK, stimulating it to phosphorylate, and thereby activate, AfsR. AfsK is specified by a gene (*afsK*) located immediately downstream of *afsR* (Horinouchi & Beppu 1992). Could other global regulatory genes also be involved in a protein phosphorylation cascade? This notion has been encouraged by the sequencing of a newly discovered DNA fragment that provokes antibiotic over-production when present at high copy number in *S. coelicolor*. This fragment contains a gene, *afsQ* that encodes a product homologous to the regulator components of protein kinase–response regulator pairs from other bacteria (Horinouchi et al 1991, Albright et al 1989). It is not clear why, as with *afsR*, extra copies of *afsQ* stimulate antibiotic production (a property also shown by another newly described gene, *abaA*, from *S. coelicolor*: Fernández-Moreno et al 1992).

Some of the proteins specified by antibiotic regulatory genes fall into families

Out of eleven known sequences of defined or presumptive regulatory genes associated with antibiotic biosynthetic gene clusters in various *Streptomyces* spp. and in *Saccharopolyspora erythraea*, a close relative of streptomycetes (Table 1),

TABLE 1 Pathway-specific regulatory genes for antibiotic biosynthesis in actinomycetes

Species	*Antibiotic*	*Regulatory gene*	*Reference*	*Action*	*TTA codons?*	*Related genes*
S. coelicolor	Actinorhodin	*act*II-ORF4	1	Positive	1	*redD*, *dnrI*, *afsR*
		*act*II-ORF1	1, 2	Negative	0	*tetR*, *mmyR*
	Undecylprodigiosin	*redD*	1, 3	Positive	0	*act*II-ORF4, *dnrI*, *afsR*
	Methylenomycin	*mmyR*	4	Negative	0	*tetR*, *act*II-ORF1
S. peucetius	Daunomycin	$dnrR_2$	5	Positive	Not sequenced	
		dnrI	5	Positive	0	*redD*, *act*II-ORF4, *afsR*
		dnrJ	5	Positive	0	*strS*, *eryC1*, *degT*, *prg*1
S. griseus	Streptomycin	*strR*	6	Positive	1	Lambda Q?
		strS (?)	5, 7	?	?	*dnrJ*, *eryC1*, *degT*, *prg*1
S. glaucescens	Streptomycin	*strR*	8	Positive	1	Lambda Q?
S. hygroscopicus	Bialaphos	*brpA*	9	Positive	1	Response regulator genes
Saccharopolyspora erythraea	Erythromycin	*eryC1* (?)	10	Positive (?)	0	*dnrJ*, *degT*, *strS*, *prg*1
S. alboniger	Puromycin	*prg*1	11	Positive (?)	DNA sequence not published	*dnrJ*, *degT*, *strS*

References: 1. Fernández-Moreno et al 1991; 2. Caballero et al 1991; 3. Narva & Feitelson 1990; 4. C.J. Bruton & K.F. Chater, unpublished results; 5. Stutzman-Engwall et al 1992; 6. Distler et al 1987; 7. K. Mansouri & W. Piepersberg, unpublished results; 8. J. Distler, personal communication; 9. Raibaud et al 1991; 10. Dhillon et al 1989; 11. Lacalle et al 1992.

three encode proteins with similarities to the N-terminal region of AfsR. These are the *act*II-ORF4 and *redD* proteins (Fernández-Moreno et al 1991) and the *dnrI* gene product, excess of which stimulates *S. peucetius* to overproduce ϵ-rhodomycinone, an intermediate in the biosynthesis of the aromatic polyketide daunorubicin (Stutzman-Engwall et al 1992). The *dnrI* gene can substitute for *act*II-ORF4 in stimulating actinorhodin synthesis in *S. coelicolor*, a remarkable observation in view of the considerable divergence of these genes (their products show about 39% identity in the region of extensive overlap). The regions conserved among all four proteins do not include the postulated DNA-binding or ATP-binding motifs tentatively identified in RedD or AfsR (Narva & Feitelson 1990, Horinouchi et al 1990), and show no resemblance to other known proteins. Thus, these proteins may have evolved from a common ancestor that was already devoted to the control of something like secondary metabolism, rather than evolving independently from a family of proteins having diverse physiological roles. *act*II-ORF4 and *dnrI* are associated with genes for polyketide synthesis, and undecylprodigiosin, synthesis of which is controlled by *redD*, could also originate in part from a polyketide intermediate (giving rise to the side chain), which suggests that there may have been an ancient association of regulatory genes of this type with genes for polyketide synthesis. Some of the other positive regulatory proteins for which sequences are available do show resemblances to regulatory proteins from other organisms. The *dnrJ* gene, which is apparently needed for the production of daunorubicin in *S. peucetius*, and is located immediately downstream of *dnrI*, specifies a protein that resembles the products of the functionally little-characterized *prg*1, *strS* and *eryC1* genes from the gene clusters for, respectively, puromycin (Lacalle et al 1992), streptomycin (in *S. griseus*; K. Mansouri & W. Piepersberg, unpublished results quoted by Stutzman-Engwall et al 1992) and erythromycin (in *Saccharopolyspora erythraea*; Dhillon et al 1989), and that of the regulatory gene *degT* from *Bacillus stearothermophilus* (Takagi et al 1990) (mode of action unknown). Stutzman-Engwall et al (1992) speculated that the *dnrI–dnrJ* gene pair may provide another example of a two-component kinase-regulator system involved in regulating secondary metabolism.

The negatively acting *mmyR* and *act*II-ORF1 gene products associated with the *mmy* and *act* clusters of *S. coelicolor* are both related to repressors of tetracycline resistance genes in plasmids and transposons of Gram-negative bacteria (C. J. Bruton & K. F. Chater, unpublished results; Fernández-Moreno et al 1991). There may have been long conservation of linkage of such repressors with genes for antibiotic export, because the methylenomycin-resistance protein (Neal & Chater 1987) and one of the products of the transcription unit regulated by *act*II-ORF1 (Caballero et al 1991, Fernández-Moreno et al 1991) are homologous to tetracycline resistance proteins.

Regulation of streptomycin biosynthesis in *Streptomyces griseus* by a low molecular mass compound

Many actinomycetes produce extracellular hormone-like butyrolactone compounds (reviewed by Horinouchi & Beppu 1990). The function of some of these is obscure, and their production may be lost without detectable phenotypic effects (this is true for *S. coelicolor*). However, in certain species, notably *S. griseus* and *S. virginiae*, these compounds are required for antibiotic production. Thus, *S. griseus* mutants unable to make A-factor (2-*S*-isocapryloyl-3-*R*-hydroxymethyl-γ-butyrolactone) fail to make—or be resistant to—streptomycin (or to make a yellow pigment), unless supplied exogenously with as little as 10^{-9} M A-factor. The lipophilic nature of A-factor should allow it to diffuse freely across membranes, so the intracellular and extracellular concentrations are likely to be the same. An intracellular A-factor-binding protein with a K_d of about 10^{-9} M is present at a few tens of molecules per genome, and it is postulated that in the absence of bound factor this binding protein represses the expression of a gene encoding another regulatory protein: a candidate protein has been discovered which binds near to, and may activate, the promoter of *strR*, a pathway-specific activator gene for streptomycin biosynthesis and resistance (Vujaklija et al 1991). (In *S. virginiae*, the gene encoding virginiae butyrolactone-binding protein has been sequenced, and its product shows homology to part of an *Escherichia coli* transcriptional anti-terminator, NusG; Okamoto et al 1992.)

The functional significance of A-factor-like compounds in these cascades is presumably to allow a bacterial population to respond synchronously either to population density, or to some physiological signal—perhaps connected with nutritional limitation—that causes increased butyrolactone production. Different recognition specificities between the factors and their binding proteins may aid different adaptation and speciation. In some strains the compounds could even serve to disorganize the behaviour of competing strains, not necessarily only in respect of secondary metabolism: in *S. griseus* the formation of the aerial mycelium also depends on A-factor (Horinouchi & Beppu 1990).

Multiple pathway-specific regulatory genes

Pathway-specific regulation of antibiotic biosynthesis is beginning to seem more complex than was recognized even in recent reviews. The positive regulatory gene *act*II-ORF4 probably controls most or all of the *act* biosynthetic genes, but not the genes for actinorhodin export; instead, these are regulated by the *act*II-ORF1 product, a repressor (Fernández-Moreno et al 1991, Caballero et al 1991). Methylenomycin is over-produced either when the repressor gene is disrupted or when an extra copy of a region near the repressor gene is added

(Chater & Bruton 1985). The latter region may perhaps contain an activator gene (a potential target for *bldA* action, which would explain why the *mmy* genes are not transcribed in a *bldA* mutant; see above). Undecylprodigiosin synthesis is positively regulated by the *redD* gene (Narva & Feitelson 1990), but mutations activating the transcription of *red* genes in a *bldA* mutant map elsewhere in the *red* gene cluster (E. P. Guthrie & K. F. Chater, unpublished results). In addition to the well-known *strR* gene (Distler et al 1987), circumstantial evidence (see above) suggests that *strS* also regulates streptomycin biosynthesis. The daunorubicin gene cluster includes two separate regulatory loci, $dnrR_1$ and $dnrR_2$, each capable at multiple copy number of inducing over-production of daunorubicin or its precursor ϵ-rhodomycinone; moreover, $dnrR_1$ itself consists of two genes, *dnrI* and *dnrJ* (Stutzman-Engwall et al 1992).

Why this complexity? Several (non-exclusive) answers are possible. (1) The clusters are assembled from older gene cluster subsets containing their own regulatory elements (which would perhaps also account for the multiplicity of transcription units that form each biosynthetic cluster). (2) Converging biosynthetic pathways, using different primary metabolites as starting materials for each branch, may require complex regulation. (3) Regulation by multiple global and pathway-specific elements may be necessary to allow integration of the various developmental and environmental signals that influence secondary metabolism.

Spore pigments: antibiotic-like secondary metabolites subject to distinctive regulation

Recently, it has become clear that some (perhaps many) *Streptomyces* spore pigments are determined by sets of genes closely similar to the *act* genes and to other genes for the synthesis of aromatic polyketides (Davis & Chater 1990, Blanco et al 1992; P. Brian, personal communication, T. McNeil, personal communication). First results suggest that these gene sets may be less complex than, for example, the *act* cluster, though this will not be very easy to prove until the pigments have been chemically defined. Visually, it seems that in colonies growing on agar, spore pigment is present only in sporulating hyphae, and actinorhodin only in the ageing substrate mycelium. P. Brian & G. Blanco (personal communication) recently showed that several promoters from the spore pigment clusters of *S. coelicolor* and *S. halstedii* have different transcriptional control from that of *act* genes: they are expressed independently of *bldA*, depending instead on certain sporulation-specific genes (*whiA, whiB* and *whiH*) that are not needed for expression of *act* genes.

The function of spore pigments is unknown, but possibilities include the regulation of sporulation, contributing to the surface properties (hydrophobicity for example) of spores, or shutting down activity of the chromosome at the end of sporulation. The widespread occurrence of quite similar gene clusters (mostly of unknown function, but many probably specifying spore pigments), suggests that such clusters may have occurred very early in the evolution of streptomycetes, and that the more well-known and diverse determinants of polyketide antibiotic synthesis might even have evolved subsequently from them. (Alternatively, selection pressure for divergence may have been greater for genes for antibiotic production.) The apparent comparative simplicity of the spore pigment clusters allows for the possibility that the insertion of a suitably regulated promoter near one end of the cluster could cause them to be regulated in an entirely different manner. Indeed, this situation has been artifically produced: *in vitro* insertion of a randomly cloned promoter upstream of what subsequently turned out to be a large part of the *S. coelicolor* spore pigment cluster resulted in the production of a brown polyketide pigment by the substrate mycelium (Horinouchi & Beppu 1985). Over evolutionary time, new adaptive benefits, such as those associated with antibiotic activity (see below), would be selected. In future, this hypothesis may be testable in part by phylogenetic analysis of the DNA sequences of genes for polyketide antibiotic and pigment synthesis. Of course, this model does not provide any insight into the evolution of the many antibiotics that are not synthesized by the polyketide route.

Concluding remarks

This article has focused on only a few sets of genes. The synthesis of many other antibiotics also coincides with the onset of aerial mycelium formation, and global mutants lacking both aerial mycelium and antibiotic production are found in many different species. One general potential adaptive benefit of this interlinking of morphological and physiological differentiation could be the protection of the colony against invasion and overgrowth by other bacteria at the stage in colony development when aerial hyphae are growing parasitically on the lysing substrate mycelium (Chater & Merrick 1979). Many antibiotics and other secondary metabolites may fulfil different, perhaps more specific roles, for example in developmental regulation. Consideration of the factors regulating the production of these compounds, whether in *Streptomyces* species or in other organisms, may help us to understand their biological roles.

Acknowledgement

I am grateful to David Hopwood for helpful comments on the manuscript.

References

Adamidis T, Riggle P, Champness W 1990 Mutations in a new *Streptomyces coelicolor* locus which globally block antibiotic synthesis but not sporulation. J Bacteriol 172:2962–2969

Albright LM, Huala E, Ausubel F 1989 Prokaryotic signal transduction mediated by sensor and regulator protein pairs. Annu Rev Genet 23:311–336

Blanco G, Pereda A, Méndez C, Salas JA 1992 Cloning and disruption of a fragment of *Streptomyces halstedii* DNA involved in the biosynthesis of a spore pigment. Gene (Amst) 112:59–66

Caballero JL, Malpartida F, Hopwood DA 1991 Transcriptional organisation and regulation of an antibiotic export complex in the producing *Streptomyces* culture. Mol & Gen Genet 228:372–380

Champness WC 1988 New loci required for *Streptomyces coelicolor* morphological and physiological differentiation. J Bacteriol 170:1168–1174

Champness W 1991 Genetic analysis of regulation of *Streptomyces coelicolor* antibiotic synthesis. Proceedings of Seoul Conference on Actinomycetes, Seoul National University, Seoul, Korea, 1991, p 45–52

Chater KF 1990 The improving prospects for yield increase by genetic engineering in antibiotic producing streptomyces. Bio/Technology 8:115–121

Chater KF, Bruton CJ 1985 Resistance, regulatory and production genes for the antibiotic methylenomycin are clustered. EMBO (Eur Mol Biol Organ) J 4: 1893–1897

Chater KF, Hopwood DA 1989 Antibiotic biosynthesis in *Streptomyces*. In: Hopwood DA, Chater KF (eds) Genetics of bacterial diversity. Academic Press, London, p 129–150

Chater KF, Merrick MJ 1979 *Streptomyces*. In: Parish JH (ed) Developmental biology of prokaryotes. Blackwell Scientific Publications, Oxford p 93–114

Davis NK, Chater KF 1990 Spore colour in *Streptomyces coelicolor* A3(2) involves the developmentally regulated synthesis of a compound biosynthetically related to polyketide antibiotics. Mol Microbiol 4:1679–1692

Dhillon N, Hale RS, Cortes J, Leadlay PF 1989 Molecular characterization of a gene from *Saccharopolyspora erythraea* (*Streptomyces erythraeus*) which is involved in erythromycin biosynthesis. Mol Microbiol 3:1405–1414

Distler J, Ebert A, Mansouri K, Pissowotzki K, Stockmann M, Piepersberg W 1987 Gene cluster for streptomycin biosynthesis in *Streptomyces griseus*: nucleotide sequence of three genes and analysis of transcriptional activity. Nucleic Acids Res 15:8041–8056

Fernández-Moreno MA, Caballero JL, Hopwood DA, Malpartida F 1991 The *act* cluster contains regulatory and antibiotic export genes, direct targets for translational control by the *bldA* tRNA gene of Streptomyces. Cell 66:769–780

Fernández-Moreno MA, Martín-Triana AJ, Martinez E et al 1992 *abaA*, a new locus for antibiotic production in *Streptomyces coelicolor*. J Bacteriol 174: 2958–2967

Guthrie EP, Chater KF 1990 The level of a transcript required for production of a *Streptomyces coelicolor* antibiotic is conditionally dependent on a tRNA gene. J Bacteriol 172:6189–6193

Hallam SE, Malpartida F, Hopwood DA 1988 Nucleotide sequence, transcription and deduced function of a gene involved in polyketide antibiotic synthesis in *Streptomyces coelicolor*. Gene (Amst) 74:305–320

Harasym M, Zhang L-H, Chater K, Piret J 1990 The *Streptomyces coelicolor* A3(2) *bldB* region contains at least two genes involved in morphological development. J Gen Microbiol 136:1543–1550

Hong S-K, Kito M, Beppu T, Horinouchi S 1991 Phosphorylation of the AfsR product, a global regulatory protein for secondary metabolite formation in *Streptomyces coelicolor* A3(2). J Bacteriol 173:2311–2318

Hopwood DA 1988 Towards an understanding of gene switching in *Streptomyces*, the basis of sporulation and antibiotic production. Proc R Soc Lond B Biol Sci 235:121–138

Horinouchi S, Beppu T 1985 Construction and application of a promoter-probe plasmid that allows chromogenic identification in *Streptomyces lividans*. J Bacteriol 162:406–412

Horinouchi S, Beppu T 1990 Autoregulatory factors of secondary metabolism and morphogenesis in actinomycetes. Crit Rev Biotechnol 10:191–204

Horinouchi S, Beppu T 1992 Regulation of secondary metabolism and cell differentiation in *Streptomyces*: A-factor as a microbial hormone and the AfsR protein as a component of a two-component regulatory system. Gene (Amst) 115:167–172

Horinouchi S, Malpartida F, Hopwood DA, Beppu T 1989 *afsB* stimulates transcription of the actinorhodin biosynthetic pathway in *Streptomyces coelicolor* A3(2) and *Streptomyces lividans*. Mol & Gen Genet 215:355–357

Horinouchi S, Kito M, Nishiyama M et al 1990 Primary structure of *afsR*, a global regulatory gene for secondary metabolite formation in *Streptomyces coelicolor* A3(2). Gene (Amst) 95:49–56

Horinouchi S, Miyake K, Hong S-K, Vujaklija D, Ueda K, Beppu T 1991 Regulation by A-factor and *afsR* of secondary metabolism and morphogenesis in *Streptomyces*. Actinomycetologica 5:119–125

Lacalle RA, Tercero JA, Jiménez A 1992 Cloning of the complete biosynthetic gene cluster for an aminonucleoside antibiotic, puromycin, and its regulated expression in heterologous hosts. EMBO (Eur Mol Biol Organ) J 11:785–792

Lawlor EJ, Baylis HA, Chater KF 1987 Pleiotropic morphological and antibiotic deficiencies result from mutations in a gene encoding a tRNA-like product in *Streptomyces coelicolor* A3(2). Genes & Dev 1:1305–1310

Leskiw BK, Lawlor EJ, Fernandez-Abalos JM, Chater KF 1991a TTA codons in some genes prevent their expression in a class of developmental, antibiotic-negative *Streptomyces* mutants. Proc Natl Acad Sci USA 88:2461–2465

Leskiw BK, Bibb MJ, Chater KF 1991b The use of a rare codon specifically during development? Mol Microbiol 5:2861–2867

Narva KE, Feitelson JS 1990 Nucleotide sequence and transcriptional analysis of the *redD* locus of *Streptomyces coelicolor* A3(2). J Bacteriol 172:326–333

Neal RJ, Chater KF 1987 Nucleotide sequence analysis reveals similarities between proteins determining methylenomycin A resistance in *Streptomyces* and tetracycline resistance in eubacteria. Gene (Amst) 58:229–241

Neal RJ, Chater KF 1991 Bidirectional promoter and terminator regions bracket *mmr*, a resistance gene embedded in the *Streptomyces coelicolor* A3(2) gene cluster encoding methylenomycin production. Gene (Amst) 100:75–83

Okamoto S, Nihira T, Kataoka H, Suzuki A, Yamada Y 1992 Purification and molecular cloning of a butyrolactone autoregulator receptor from *Streptomyces virginiae*. J Biol Chem 267:1093–1098

Parro V, Hopwood DA, Malpartida F, Mellado RP 1991 Divergent transcription of genes involved in polyketide antibiotic synthesis in *Streptomyces coelicolor*. Nucleic Acids Res 19:2623–2627

Passantino R, Puglia AM, Chater KF 1991 Additional copies of the *act*II regulatory gene induce actinorhodin production in pleiotropic *bld* mutants of *Streptomyces coelicolor* A3(2). J Gen Microbiol 137:2059–2064

Raibaud A, Zalacain M, Holt TG, Tizard R, Thompson CJ 1991 Nucleotide sequence analysis reveals linked N-acetyl hydrolase, thioesterase, transport, and regulatory genes encoded by the bialaphos biosynthetic gene cluster of *Streptomyces hygroscopicus*. J Bacteriol 173:4454–4463

Stein D, Cohen SN 1989 A cloned regulatory gene of *Streptomyces lividans* can suppress the pigment deficiency phenotype of different developmental mutants. J Bacteriol 171:2258–2261

Stutzman-Engwall KJ, Otten S, Hutchinson CR 1992 Regulation of secondary metabolism in *Streptomyces* spp and overproduction of daunorubicin in *Streptomyces peucetius*. J Bacteriol 174:144–154

Takagi M, Takada H, Imanaka T 1990 Nucleotide sequence and cloning in *Bacillus subtilis* of the *Bacillus stearothermophilus* pleiotropic regulatory gene *degT*. J Bacteriol 172:411–418

Vujaklija D, Ueda K, Hong S-K, Beppu T, Horinouchi S 1991 Identification of an A-factor-dependent promoter in the streptomycin biosynthetic cluster of *Streptomyces griseus*. Mol & Gen Genet 229:119–128

DISCUSSION

Demain: Is it known definitely that the spore pigments do *not* have antibiotic activity? This might be very important, from the point of view of screening for new antibiotics.

Chater: We have no information about their biological activity; we don't even know whether they are inside or outside the spore. We like to think of them as being outside. They could have structural roles; they could have an antibiotic function, to prevent the spores being eaten by myxobacteria or slime moulds, or they might be released as the spores germinate. Perhaps they help to slow down and prevent DNA expression, because these compounds might well be intercalating agents; if so, they could affect the organization of the chromosome in the spore as metabolism shuts down. Any of these things is possible.

Cavalier-Smith: What studies have been done on sporulation in Nature? In particular, what is known about the circumstances that induce it, and has anyone tried to measure any of the antibiotics, in natural environments?

Chater: People have tried to measure them, and in some circumstances they have been detected, so some are made, which is not altogether surprising; but a negative result doesn't tell you very much, because a small focus of *Streptomyces* organisms growing in a large soil particle will need to produce a large amount of antibiotic for it to be detected. As to whether there are environmental circumstances particularly favourable to sporulation, not very much is known, although it occurs quite rapidly when you inoculate soil; within 1–2 days the whole cycle has gone through to sporulation (Herron & Wellington 1990).

Nisbet: A fundamental question is whether all organisms make so-called secondary metabolites. This is related to an issue we face in industry, namely how to maximize the expression potential. It's one thing having a gene; it is another to get organisms to express those genes, especially if they have not been studied in detail. I am interested in what you and others think from a different standpoint, of the potential value of a genetic rather than a physiological approach to this problem and of introducing regulators (positive or negative effectors) that will operate at the gene level.

Chater: I don't think anything should or would prevent the traditional physiological and genetic approaches to strain improvement, but it would be helpful to supplement them with molecular approaches, especially at the very early stages of strain development. It could be useful to do very simple cloning experiments, where (if a strain is transformable) you introduce high copy numbers of a random library of DNA from the strain back into itself, and maybe also DNA from other strains, because there is now increasing evidence for cross-interactions. Those routes will not only provide you potentially with some way of dealing with yield improvement, but also, because they often specifically induce one pathway rather than another in the organism, they may help you in discrimination against other less desirable metabolites as well.

Davies: Is it known whether you can phenotypically suppress *bldA* mutants, in terms of overcoming the translation block occurring at the TTA or the TTT substitution that you showed, by adding alcohol or streptomycin to the medium?

Chater: We have attempted to use mistranslation-inducing circumstances to look for expression of TTA-containing genes in a *bldA* mutant, but we didn't see any evidence that the ability to read TTA codons was changed by streptomycin, for example.

Davies: This is interesting, because if you use a colour change (for example, using *lac* or *lux* expression) which is very sensitive, you ought to be able to detect very small amounts of phenotypic suppression.

Chater: At the same time as inducing mistranslation, you may be causing other problems, in association with antibiotic production. It is possible to suppress phenotypically (but not, so far as I know, by translational suppression) the effect of the *bldA* mutation on at least on one pathway, the *red* pathway, where, if we dramatically reduce the phosphate concentration in the medium, we do get red pigment production, even in a *bldA* mutant; so again there's some evidence for more complex regulation.

Cundliffe: Taking Julian Davies's suggestion one step further (and he was no doubt thinking about the experiments that he and Eva Kataja were doing at Harvard Medical School in the 1960s), where you have the production of the blue pigment (actinorhodin) in a *bldA* strain, in cases where you believe you are over-producing the mRNA for the *act*II-ORF4, maybe this involved misreading of the UUA messenger RNA codon by ribosomes. In a streptomycin-resistant strain you might eliminate it completely. I am thinking of the well-known translational restrictions of the ribosomes of *strA* strains.

Chater: A streptomycin resistance mutation is present in the strain in which the *act*II-ORF4 high copy number experiments were done, so this particular mutation doesn't prevent mistranslation.

Davies: But if you started with a streptomycin-resistant mutant, it may be *necessary* for the *bldA* phenotype.

Chater: We have exactly the same mutation, with the same phenotype, in streptomycin-sensitive strains as well.

Piepersberg: The DnrJ type of putative regulators also seem to occur in other pathways, apart from the streptomycin, daunorubicin and erythromycin pathways. For instance, it was shown recently that the puromycin gene cluster contains a homologue of this regulator class (Lacalle et al 1992), and the *rfb* gene cluster for lipopolysaccharide-O-chain formation in *Salmonella typhimurium* (Jiang et al 1991) seems to have one.

Secondly, the StrR regulator seems not to be only an activator; it also has repressor effects on other subclusters or operons of the *str* gene cluster. The TTA codon also seems to occur in production genes. We find a conserved TTA in a putative production gene in the streptomycin cluster which occurs in both *S. griseus* and *S. glaucescens*, the *strN* gene (Pissowotzki et al 1991).

However, this is at the beginning of an operon, which could mean that everywhere you need an operon control on the translational level, you insert a TTA codon in the first reading frame, at the N-terminal end. This is supported by the fact that in the streptomycin resistance genes we have only the TTA codon in *Streptomyces glaucescens*, where the *aphD* gene is at the beginning of the respective transcription unit, but not in *S. griseus*, where *aphD* is not localized in the same place and is not the first reading frame in the transcription unit (Distler et al 1987, Vögtli & Hütter 1987).

Chater: There is a rather high preponderance of TTA codons that occur within the first 30 codons. I imagine that ribosomes stuck at that position would interfere significantly with further ribosome binding to the initiation codon.

Beppu: I have a comment on the involvement of protein phosphorylation in the regulation of secondary metabolism in *Streptomyces*. There seem to be at least two different types of such phosphorylated protein in *Streptomyces*. One is a typical two-component regulatory system closely similar to OmpR-EnvZ and others that have been found from various prokaryotes. Another type is sensitive against the inhibitors of the eukaryotic protein kinases, such as staurosporin and its analogues. I would suggest that the similarity of the *Streptomyces* regulatory system to that of the eukaryotes is becoming clear.

Demain: I was wondering, from Wolfgang Piepersberg's comment, what function the TTA-carrying streptomycin biosynthetic gene might have. Do you know what enzyme it encodes?

Beppu: We recently found that the TTA codon is present in one of the regulatory genes which we cloned as a gene which restores spore formation in A-factor-negative *S. griseus*.

Piepersberg: The function of the StrN gene product is not known; it is encoded by a reading frame in front of the second amidinotransferase gene *strB2*. It has some similarity to aminoglucoside binding sites in its C-terminal half and we therefore think it is one of the genes in the streptidine pathway (Pissowotzki et al 1991).

Demain: Does its presence or absence affect resistance to streptomycin?

Piepersberg: That has not been tested yet, but it doesn't cause resistance by itself.

Demain: I gather that there are activators *and* repressors, or more than one regulatory molecule, in a particular synthetic pathway. Are they acting on the same genes, and on the same promoters, or are different promoters and different genes being controlled by these different regulatory loci?

Chater: The only information at the moment is from the *act* cluster, where the repressor is thought to autoregulate its own gene, and also to regulate the divergent transcription unit, which is for actinorhodin export (Caballero et al 1991). The positive regulatory element is perhaps regulating the transcription of all the other promoters, although that hasn't been looked at for many of them (Fernández-Moreno et al 1991). For the rest of the regulatory genes that I mentioned, there is almost no relevant information yet.

Davies: A-factor, and the γ-butyrolactones, are presumably also secondary metabolites, so one secondary metabolite controls the production of others within the producing organism and acts externally, on other organisms as well. In *Streptomyces coelicolor*, we are not quite sure what A-factor is doing; that's the problem. But in *S. griseus*, A-factor can act on a variety of different reactions. Is it believed that these molecules are real pheromones, in the sense that they work on other microbes, or is their main function considered to be effects on the organism that produces them?

Chater: For *Streptomyces griseus* we have to accept that A-factor is doing something to the producing organism. In *S. coelicolor*, we don't know its role, because you can lose production and there's no overt change in the phenotype, in relation to morphology or antibiotic production. In that case, one could argue that *S. coelicolor* is setting out to fool any nearby *S. griseus*-like organism into preparing itself for sporulation at the earliest possible stage, so that it no longer competes for growth. So one could imagine that organisms which use this A-factor type of process to stimulate their own more or less synchronous and rapid sporulation perhaps have effectively evolved a way of anticipating that times are going to get hard; they can sense incipient crowding. This sets them up to sporulate very rapidly. An organism which doesn't do that, like *S. coelicolor*, may have the benefit of not so readily becoming committed to sporulation, and being ready to reverse things if necessary, if circumstances change. Those two situations would lead to different ecological adaptations and, up to a point, to speciation.

Davies: What is known about the receptors for the butyrolactones, in terms of possible similarities to other receptors?

Beppu: We still have insufficient information about the possible involvement of autoregulatory factors in *S. coelicolor*, but I am quite confident that the butyrolactones such as A-factor and the virginiae butanolides are working as hormones or pheromones in at least several species of streptomycetes (Beppu

1992). Recently a family of variants of the butyrolactones, having slightly different structures, has been identified as autoregulatory factors for secondary metabolism and sometimes morphological differentiation in various *Streptomyces* species. The problem is the specificity of the receptors for these butyrolactones as the ligands. The binding protein found from *S. griseus* binds A-factor but not virginae butanolides at all, and vice versa for the binding protein from *S. virginiae*. This means that the two binding proteins possess very strict specificity, probably discriminating the keto (in A-factor) or hydroxy (in virginiae butanolides) group at the 6-position of the butyrolactones and thus preventing miscommunication between the different *Streptomyces* species using these compounds as signals.

However, I would also mention the fact that the specificity of these binding proteins is not so strict in relation to the acyl moieties of these butyrolactones. This may cause the situation that two different *Streptomyces* species use a common signal, which thus enables cross-talking to occur between the different species in the ecosystem.

Hopwood: The idea of one organism interfering with the 'radar' of another is interesting. There must be many examples in higher systems of this, which have evolved, perhaps in response to similar selection pressures.

Vining: In luminescent bacteria, the luciferase system is induced, and the induction can result from the production by external bacterial sources of specific secondary metabolites that can turn the system on. Once induced, the autoregulatory system operates to generate a large signal. Something similar might happen in streptomycetes; the autoregulators' role might ensure that sporulation or antibiotic production, when it occurs, is massive.

Chater: One part of your question, Dr Davies, was about homologies. There is an interesting finding from the Yamada group (Okamoto et al 1992) that the *Streptomyces virginiae* butyrolactone-binding protein is the homologue of a protein in *E. coli* called NusG, which has to do with the utilization of the lambda *N* gene product for transcriptional anti-termination. The binding-protein gene in *S. virginiae* is found in essentially the same position in relation to a nearby ribosomal protein gene as the *E. coli nusG* gene. It is interesting that NusG has developed this specificity in *S. virginiae*, and we would like to know whether the equivalent gene in other streptomycetes is in the same place and encodes a product with subtle differences, which might mean that it can bind different compounds.

Demain: Is it still considered correct that in some organisms the A-factor structural gene, or the regulatory genes, are plasmid-borne, and in others are chromosomal, or are they all thought to be chromosomal? For example, for *S. coelicolor*, *S. virginiae* and *S. griseus*, what is the location of the A-factor gene?

Davies: And are there any examples known of plasmid-encoded A-factor production?

Beppu: The genetic data indicate that the A-factor gene in *S. griseus* is carried on some genetically very unstable region which behaves like an unstable plasmid or transposon, but we have failed to prove that. There is no evidence that in *S. virginiae* and in *S. coelicolor* the gene is located on the chromosome.

Piepersberg: We think that in *S. griseus* the *afsA* gene is encoded most probably on the chromosome, and it can easily get lost by large deletions, which are longer than 200 kb, as we have recently shown (J. Distler & W. Piepersberg, unpublished). This occurs spontaneously at rather high rates.

Brandl: I would like to make a link to the possible practical applications of the A-factor. Do you think it's feasible to add the A-factor into a screening programme, in order to induce the production of secondary metabolites, or to increase or to stimulate their production?

Beppu: I suspect that this strategy may not work. As far as we have examined this, A-factor worked only for streptomycin production and sporulation in the A-factor-deficient mutants of *S. griseus*.

Rinehart: The question arises of how specific the individual A-factors are to the production of one antibiotic or another, or to one microbial species or another.

Beppu: The A-factor stimulates streptomycin production in the A-factor-deficient mutant cells. This does not exclude the possibility that A-factor works as a signal in other *Streptomyces*.

Rinehart: Is it only the original A-factor that stimulates streptomycin production or do analogues of A-factors also have this effect?

Demain: For example, analogues of A-factors, like those reported by Gräfe and co-workers (e.g. Gräfe & Sarfert 1985).

Beppu: In *S. griseus*, only A-factor is effective in inducing streptomycin biosynthesis and sporulation, and virginiae butanolide is not effective.

Davies: Will structural analogues of A-factor work?

Beppu: We have not identified such A-factor analogues in *S. griseus*.

Demain: None of the natural analogues work, but Yamada's group (Nihira et al 1988) has made many synthetic analogues of the virginiae butanolides, many of which had low activity and some of which had much higher activity than the natural compound.

Beppu: Yes, exactly. As I mentioned previously, the virginiae butanolides are a family of γ-butyrolactones possessing a hydroxy group at the 6-position with different acyl moieties. In the case of A-factor possessing the 6-keto group, we have not yet chemically identified such a family with different acyl side chains.

Piepersberg: You can modify the A-factor in the side chain, and it still works, in *S. griseus*. We did this by introducing a bromine residue into one of the two methyl groups at the end of the side chain (T. Kasberg & J. Distler, unpublished). This makes the molecule much more bulky in this region.

Beppu: With the ketone group at the 6-position?

Piepersberg: Yes; the stereochemistry of the synthetic A-factor derivatives is also retained.

Beppu: The situation is very similar to the virginiae butanolide family. So we have an A-factor family with the ketone at the 6-position, and the virginiae butanolide family with a hydroxy group at the 6-position.

References

Beppu T 1992 Secondary metabolites as chemical signals for cellular differentiation. Gene (Amst) 115:159–165

Caballero JL, Malpartida F, Hopwood DA 1991 Transcriptional organization and regulation of an antibiotic export complex in the producing *Streptomyces* culture. Mol & Gen Genet 228:372–380

Distler J, Ebert A, Mansouri K, Pissowotzki K, Stockmann M, Piepersberg W 1987 Gene cluster for streptomycin biosynthesis in *Streptomyces griseus*: nucleotide sequence of three genes and analysis of transcriptional activity. Nucleic Acids Res 15:8041–8056

Fernández-Moreno MA, Caballero JL, Hopwood DA, Malpartida F 1991 The *act* cluster contains regulatory and antibiotic export genes, direct targets for translational control by the *bldA* tRNA gene of Streptomyces. Cell 66:769–780

Gräfe U, Sarfert E 1985 Reconstitution by a butyrolactone autoregulator of the parental protein pattern in an asporogenous mutant of *Streptomyces griseus*. FEMS (Fed Eur Microbiol Soc) Microbiol Lett 28:249–253

Herron PR, Wellington EMH 1990 New method for extraction of streptomycete psores from soil and application to the study of lysogeny in sterile amended and non-sterile soil. Appl Environ Microbiol 56:1406–1412

Jiang X-M, Neal B, Santiago F, Lee SJ, Romana LK, Reeves PR 1991 Structure and sequence of the *rfb* (O antigen) gene cluster of *Salmonella* serovar typhimurium (strain LT2). Mol Microbiol 5:695–713

Lacalle RA, Tercero JA, Jimenez A 1992 Cloning of the complete biosynthetic gene cluster for an aminonucleoside antibiotic, puromycin, and its regulated expression in heterologous hosts. EMBO (Eur Mol Biol Organ) J 11:785–792

Nihira T, Shimizu Y, Kim HS, Yamada Y 1988 Structure–activity relationships of virginiae butanolide C, an inducer of virginiamycin production in *Streptomyces virginiae*. J Antibiot (Tokyo) 41:1828–1837

Okamoto S, Nihira T, Kataoka H, Suzuki A, Yamada Y 1992 Purification and molecular cloning of a butyrolactone autoregulator receptor from *Streptomyces virginiae*. J Biol Chem 267:1093–1098

Pissowotzki K, Mansouri K, Piepersberg W 1991 Genetics of streptomycin production in *Streptomyces griseus*: molecular strcuture and putative function of genes *strELMB2N*. Mol & Gen Genet 231:113–123

Vögtli M, Hütter R 1987 Characterization of the hydroxystreptomycin phosphotransferase gene (*sph*) of *Streptomyces glaucescens*: nucleotide sequencing and promoter analysis. Mol & Gen Genet 208:195–203

Terpenoid cyclases: design and function of electrophilic catalysts

David E. Cane

Department of Chemistry, Brown University, Providence, RI 02912, USA

Abstract. Terpenoid cyclases catalyse the cyclization of the universal acyclic precursors geranyl and farnesyl diphosphate to monoterpenes and sesquiterpenes, respectively. All such cyclases investigated to date are operationally soluble, moderately lipophilic proteins of relative molecular weight 40 000–100 000, requiring no cofactors other than a divalent metal, usually Mg^{2+} and occasionally Mn^{2+}. The focus of most work has been on the mechanisms of the cyclization reactions themselves. It is currently proposed that the cyclase binds the acyclic substrate in a suitable conformation and initiates the cyclization by ionization of the labile allylic diphosphate moiety. The use of stereospecifically labelled substrates and analysis of the sites of labelling in the derived cyclization products has allowed the proposal of detailed cyclization mechanisms. Further insight into the architecture and function of the cyclase active site has come from the study of substrate and intermediate analogues designed to act as potential inhibitors or anomalous substrates of the normal cyclization reaction. Progress has also been made on the cloning of the relevant structural genes for sesquiterpene cyclases. This has led to new insights into the basic requirements for cyclase catalysis and specificity.

1992 Secondary metabolites: their function and evolution. Wiley, Chichester (Ciba Foundation Symposium 171) p 163–183

As pointed out by J. W. Cornforth nearly 25 years ago in a review entitled 'Olefin alkylation in biosynthesis', while the majority of biologically formed carbon–carbon bonds are generated by condensations of the aldol and Claisen type, the biosynthesis of iosprenoids is distinguished by its nearly exclusive reliance on electrophilic reactions involving carbon–carbon double bonds (Cornforth 1968). The characteristic reactions of polyisoprenoid chain assembly and cyclization are initiated by one of two general reactions: (1) protonation or alkylation of a double bond or epoxide; or (2) ionization of an allylic diphosphate ester. Attack of the resulting electrophilic species on an olefinic bond, in an intermolecular or intramolecular fashion, generates one or more carbocationic intermediates, with positive charge ultimately being quenched by loss of a proton or capture of an external nucleophile such as water.

Although the biosynthesis of prenylated proteins, acyclic polyisoprenoids and cyclized terpenoids has been the subject of intensive study, the fascinating family of proteins which mediate the formation of this functionally diverse and structurally rich family of metabolites is still poorly understood from the point of view of structure and catalytic mechanism. Our own group has been particularly interested in the terpenoid cyclases, a family of enzymes which catalyse the cyclization of farnesyl diphosphate to any of 200 different sesquiterpene hydrocarbons or alcohols, as well as the related monoterpene cyclases mediating the cyclization of geranyl diphosphate (Cane 1990, Croteau 1987). Only a relative handful of such enzymes have so far been isolated and characterized, yet certain trends have already begun to emerge. All cyclases investigated to date are operationally soluble, moderately lipophilic, monomeric or homodimeric proteins of subunit relative molecular weight 40 000–65 000, requiring no cofactors other than a divalent metal, usually Mg^{2+} and occasionally Mn^{2+}. Experiments with inhibitors are consistent with the notion that all the events in a multistep cyclization take place at a single active site; there is no evidence supporting the intervention of any covalently bound intermediates. On the other hand, the few cyclases which have been cloned and sequenced show no homology at either the DNA or protein level to any other known protein, nor is there any substantive information about the nature and role of the amino acid residues which make up the cyclase active sites.

How, one might reasonably ask, can one claim to know something about the 'design and function of electrophilic catalysts' when so little is known about the structure and mode of action of the catalysts themselves? Fortunately, we now know enough about terpenoid cyclization reactions that we can at least begin to formulate appropriate questions and even to discern the shape, if not the details, of the answers we shall eventually obtain. The current state of affairs can best be put in perspective by considering what is perhaps the most thoroughly studied sesquiterpene cyclase, trichodiene synthase.

Trichodiene synthase has been isolated from a variety of *Fusarium* species (a group of plant pathogenic fungi which produce a number of potent metabolites toxic to livestock and humans). The synthase catalyses the cyclization of *trans,trans*-farnesyl diphosphate (FPP) (**1**) to trichodiene (**2**), the parent hydrocarbon of the trichothecene family of mycotoxins (see Fig. 1) (Hohn & VanMiddlesworth 1986, Hohn & Beremand 1989a). Extensive experiments with labelled substrates have led us to propose the cyclization mechanism illustrated in Fig. 1 in which the primary allylic substrate is initially rearranged to its tertiary allylic isomer, nerolidyl diphosphate (NPP) (**3**), which then undergoes ionization and cyclization, by way of a series of rearranged intermediates, with ultimate formation of trichodiene (Cane & Ha 1988, Cane et al 1985, 1990a). Although the isomerization and cyclization components of the reaction obviously differ in their gross consequences, they are believed to be very similar in

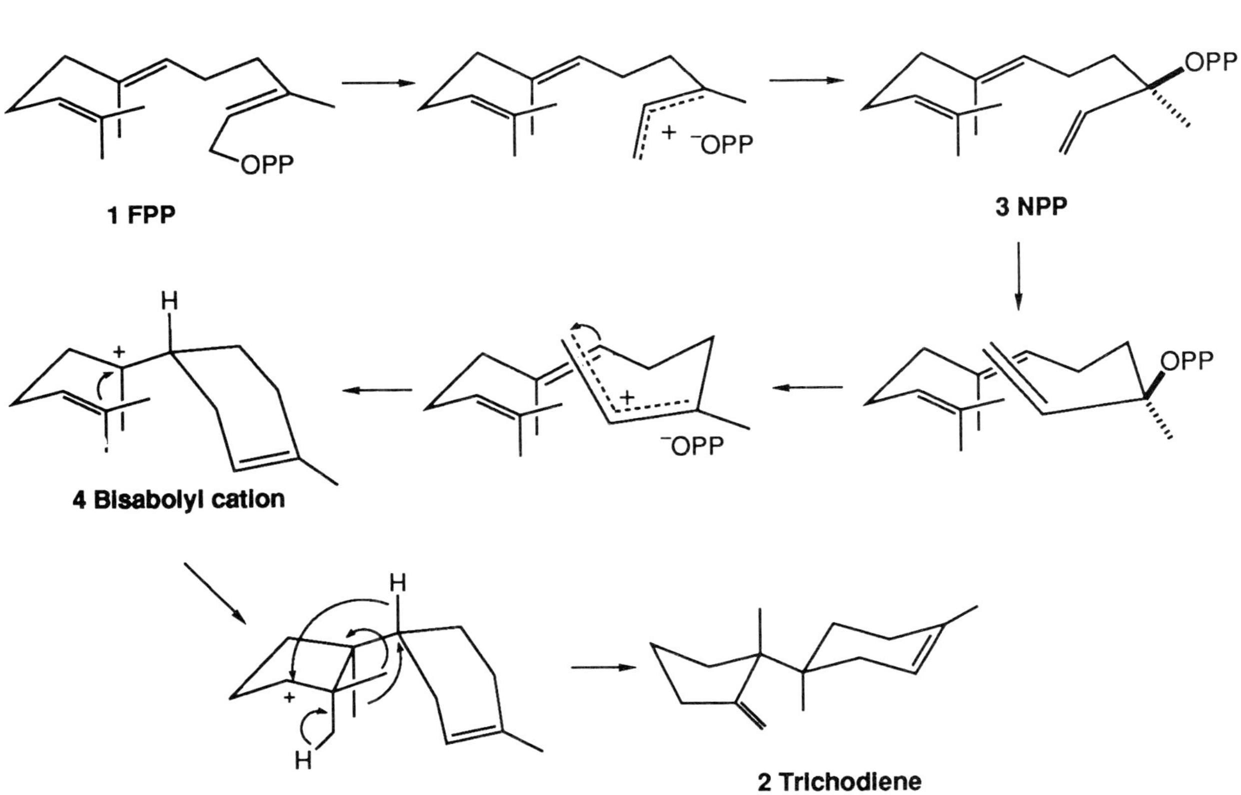

FIG. 1. Cyclization of farnesyl diphosphate (FPP) (**1**) to trichodiene (**2**).

microscopic mechanism, involving the ionization of an allylic diphosphate ester to an allylic cation–pyrophosphate anion pair, followed in each case by either recapture of the pyrophosphate anion or electrophilic attack on the neighbouring central double bond, respectively. The mechanism itself can be considered a prototype for all such cyclization reactions, in that the initial folding of the acyclic substrate, FPP, as well as that of the derived intermediate NPP, is believed to be a major determinant of the structure and stereochemistry of the eventually formed cyclic product (Cane 1985).

Allylic diphosphate ionization

Physical organic studies have suggested the special advantages of allylic diphosphate esters from the point of view of both acid lability and cation stability (Tidd 1971, Cane 1980). Moreover, studies of prenyl transferases, whose mode of action bears many mechanistic and catalytic similarities to those of the terpenoid cyclases, have established that the actual ionization substrate is the allylic diphosphate–divalent magnesium ion complex, the role of the metal presumably being to hold the diphosphate moiety in a defined conformation and to assist in its ionization by neutralization of the negative charge of the phosphoryl oxygen atoms (Poulter & Rilling 1981). How then is this ionization initiated at the cyclase active site? To date, none of the amino acids present at the active site of trichodiene synthase, nor any other sesquiterpene cyclase, has been identified, although the closely related monoterpene cyclases can be inhibited by sulphydryl-directed reagents. On the other hand, photoaffinity labelling of avian liver FPP synthase, which catalyses the mechanistically related intermolecular coupling of an allylic diphosphate substrate with the olefinic acceptor isopentenyl diphosphate, has implicated an arginine residue found within an active site peptide, EERYK, as possibly being involved in binding of the pyrophosphate moiety (Brems et al 1981). Indeed, closely related sequences have been found in the rat and human fetal liver FPP synthases, although the corresponding yeast gene shows only a weakly homologous sequence, with a basic lysine residue in place of the arginine (Cane 1992). Interestingly, cloning of the trichodiene synthase genes of both *Fusarium sporotrichioides* and *Gibberella pulicaris* has revealed the presence of similar basic amino acid-rich sequences (DRRYR in *F. sporotrichioides* and DHRYR in *G. pulicaris*) (Hohn & Beremand 1989b, Hohn & Desjardins 1991). The functional significance of these regions in trichodiene synthase remains to be demonstrated, although Hohn has found that replacement of the conserved arginine with lysine reduces the observed cyclase activity by 98%, while substitution by glutamate abolishes all but 0.1% of the trichodiene synthase activity (T. M. Hohn, personal communication 1991). In spite of these suggestive observations, a comparable basic amino acid-rich domain is absent from aristolochene synthase (T. M. Hohn, personal communication 1991) and preliminary sequence data on a third sesquiterpene cyclase, pentalenene synthase, have so far failed to reveal any such homologous sequences (D. E. Cane & C. R. Lamberson, unpublished results).

Edwards has noted the presence of two aspartate-rich domains in yeast hexaprenyl diphosphate synthase and has pointed out that the same (I,L, or V) XDDXXD consensus sequence is present in several FPP synthases as well (Ashby & Edwards 1990). This has led him to propose that the aspartate residues may somehow be involved in substrate binding, perhaps by chelation of the divalent metal. It is therefore of some interest that the *F. sporotrichioides* trichodiene synthase contains the sequence VLDDSKD beginning at amino acid 98, while the corresponding trichodiene synthase gene from *G. pulicaris* (*Tox5*) has the inferred peptide sequence VLDDSSD at the same site. Similarly, sequencing of aristolochene synthase has established the presence of a related heptapeptide, LIDDVLE, starting at amino acid 113 (T. M. Hohn, personal communication 1991). Homologous sequences are also present in pentalenene synthase, on the basis of preliminary sequence data (D. E. Cane & C. R. Lamberson, unpublished results). On the other hand, a scan of the PIR™ Protein Sequence database using the consensus sequence (I,L,V,H)X(D,N)DXX(D,E) provides some 1200 matches out of 27 000 sequences surveyed, close to the statistically expected frequency of one out of every 20 proteins, thereby raising interesting questions about the functional significance of this acidic peptide sequence. The identity of the active site residues involved in binding and ionization of the diphosphate moiety of allylic diphosphate esters therefore remains to be firmly established.

Substrate folding and stabilization of positive charge

In order to mediate the conversion of FPP to a single cyclic sesquiterpene, trichodiene synthase must not only control the folding of the substrate as well as that of several derived intermediates; it must also stabilize the various carbocationic species which have been generated at the active site, all the while avoiding accidental annihilation by these highly reactive electrophiles. Little if anything is known about the manner in which an enzyme can enforce a particular conformation on the lipophilic portion of a substrate such as FPP. Indeed, the problem is further complicated by the recognition that in the course of the formation of the eventual cyclized product, FPP must undergo substantial changes in bonding, hybridization and configuration. To illustrate the dimensions of the problem, Fig. 2 shows the presumed shapes of several key intermediates of the trichodiene synthase reaction, no two of which would be expected to be bound by an active site rigidly complementary to either one or the other. The single active site of trichodiene synthase must therefore somehow accommodate and control the reactivity of all of them. The necessity of allowing considerable freedom of movement in related squalene cyclizations was recognized long ago by Cornforth, leading him to postulate so-called X-group

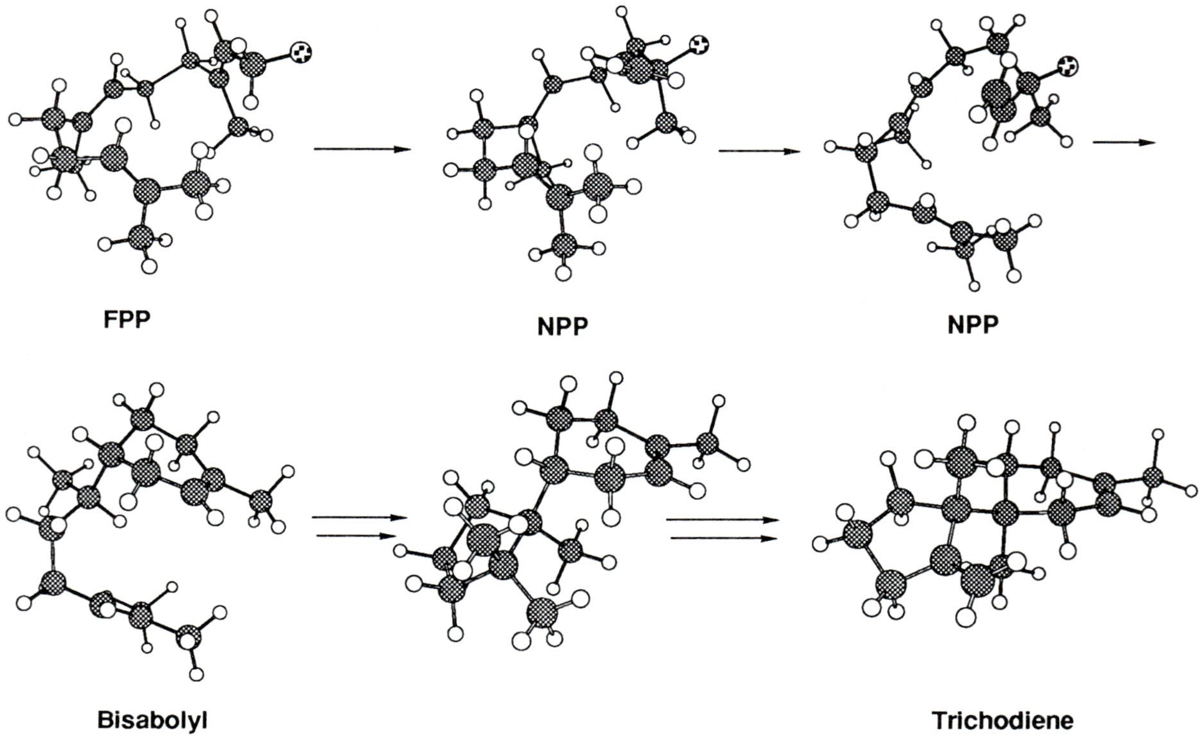

FIG. 2. Likely conformations of substrate FPP and intermediates in the formation of trichodiene. For simplicity, only the ester oxygen of the diphosphate moiety is shown. Carbocationic centres which undergo rearrangement are shown as tetrahedral centres carrying an extra hydrogen atom.

mechanisms to account for stereospecific rearrangements of carbonium ions at a single active site (Cornforth 1968). Unfortunately, there is still no firm evidence which explicitly supports the intervention of X-groups in terpenoid cyclizations. The problem of accounting for the ability of a cyclase active site to accommodate a variety of intermediates of varying shape and charge distribution remains unresolved.

In collaboration with Professor Robert Coates of the University of Illinois, we have probed the manner in which trichodiene synthase interacts with its various charged intermediates by testing a family of ammonium ion analogues of the proposed bisabolyl cation intermediate (**4**) (Fig. 1) for their relative ability to inhibit the normal cyclization reaction (Cane et al 1992). It was initially observed that the aza analogue (*R*)-**5** alone was at best a weak inhibitor of trichodiene synthase (Fig. 3). Interestingly, the protonated amine became a strong competitive inhibitor of the cyclization when incubated with inorganic pyrophosphate, itself a known competitive inhibitor of trichodiene synthase. Unexpectedly, the enantiomeric analogue, (*S*)-**5**, showed identical behaviour, indicating that the cyclase could not distinguish between the individual enantiomers of this inhibitor. Inhibition by the corresponding α-terpinyl cation analogues, (*R*)- and (*S*)-**6**, showed the same requirement for pyrophosphate ion but with a 4–8-fold weaker

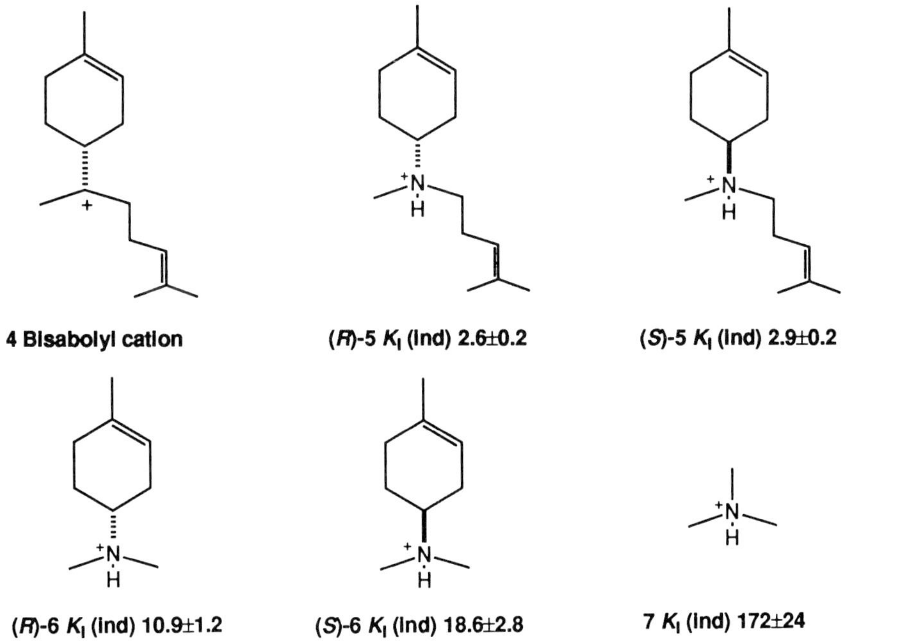

FIG. 3. Ammonium ion analogues of the bisabolyl cation as inhibitors of trichodiene synthase. Values for K_I (ind) are competitive inhibition constants in the presence of added inorganic pyrophosphate.

induced binding affinity. Once again, however, both enantiomers were essentially equally effective as inhibitors. Finally, the simplest analogue, trimethylammonium ion (7), while weaker by another factor of 10, also exhibited the same requirement for inorganic pyrophosphate. These results indicated that both electrostatic and hydrophobic interactions are important for the binding of each intermediate analogue. Moreover, the observation that inorganic pyrophosphate potentiates—indeed, that it is required for—inhibition suggests that during the normal course of the cyclization the pyrophosphate moiety is not simply a passive product of the initial ionization of the FPP substrate but may well play a role in the stabilization of the cascade of cationic intermediates which link FPP and trichodiene. Similar synergistic effects of inorganic pyrophosphate on terpenoid cyclase or chain elongation enzymes have previously been reported. For example, although neither pentalenene nor inorganic pyrophosphate alone at concentrations of 10 μM significantly inhibited the cyclization of FPP to the tricyclic sesquiterpene pentalenene, the combination of the two together at the same concentration increased the apparent K_m for FPP by a factor of seven (Cane & Pargellis 1987). Similarly, various cyclopropyl ammonium ion analogues of proposed intermediates in the conversion of presqualene diphosphate to squalene become potent inhibitors only in the presence of added inorganic pyrophosphate (Poulter et al 1989). The apparent inability of trichodiene synthase to discriminate between the enantiomeric ammonium ion analogues of the bisabolyl cation is itself reminiscent of our earlier observation that several monoterpene cyclases were unable to discriminate cleanly between the individual enantiomers of their natural tertiary allylic intermediate, linalyl diphosphate (Croteau et al 1986, 1988).

Permissive model of cyclase active sites

Taken together, these results are all supportive of what may be termed a *permissive* model of terpenoid cyclase active site structure. Rather than being rigidly complementary to some or all of the various cyclization intermediates, the cyclase appears to be considerably looser fitting, thereby accommodating a family of related structures derived from the initial folding and ionization of the precursor FPP. According to this model, the structure and stereochemistry of each intermediate is largely the consequence of the geometry of its immediate precursor. The cyclase need not be able to discriminate among all *possible* geometries of any given intermediate, since these substances are normally not encountered free in solution, but need only provide sufficient constraints so as to limit the conformations available to an intermediate which has been generated at the active site itself. In this sense, one can think of the active site of a cyclase such as trichodiene synthase as more closely resembling a mitten than a glove in its interaction with its natural substrates and intermediates. Whether this model is at all accurate will have to be determined by detailed structural analysis of the interaction of this and other cyclases with a range of substrate and intermediate analogues.

Cyclization of anomalous substrates

Further evidence for the ability of cyclase active sites to accommodate a range of anomalous substrate geometries has come from work carried out recently in our laboratory aimed at elucidating the intermediates of the reaction catalysed by the fungal cyclase, aristolochene synthase, isolated from *Aspergillus terreus*. On the basis of the cyclization of various labelled precursors, we have proposed that the formation of aristolochene (**8**) from FPP involves an initial cyclization to germacrene A (**9**) which in turn undergoes protonation of the 6,7-double bond and cyclization to aristolochene through the intermediacy of a eudesmane-type cation (Fig. 4) (Cane et al 1989, 1990b, Hohn & Plattner 1989b). In the course of the normal cyclization reaction, the postulated germacrene A intermediate never leaves the cyclase active site. Unfortunately, the thermal lability of germacrene A renders impractical any attempt to study the enzymic cyclization of synthetically prepared **9**. We therefore chose the alternative of incubating aristolochene synthase with (7*R*)-6,7-dihydroFPP (**10**). It was expected that cyclization of **10** would result in the formation of 6,7-dihydrogermacrene A (**11**). Since the latter intermediate lacks the double bond which would normally be the target of further protonation, **11** would be expected to be released from the active site, allowing expression of the otherwise cryptic initial cyclization step. Furthermore, since **11** is not subject to the facile Cope rearrangement characteristic of germacrene A, the abortive cyclization product would be able to accumulate in the incubation medium. In the event, **10** proved to be an effective competitive inhibitor of aristolochene synthase, with K_I 0.15 μM, compare to K_m 3.0 μM for FPP. Even more interestingly, aristolochene synthase converted **10** to a hydrocarbon of molecular mass 206 Da which proved to be identical by gas chromatography/mass spectrometry (GC/MS) comparison with one of two synthetically prepared epimers of dihydrogermacrene **11**, differing from one another only in the configuration of the isopropenyl side chain. Work is in progress to establish the full stereochemistry of the anomalous cyclization product (D. E. Cane & Y. S. Tsantrizos, unpublished).

In the meantime, we have also examined the role of **10** as a competitive inhibitor and anomalous substrate of *epi*-aristolochene synthase, a cyclase isolated and purified from tobacco (*Nicotiana tabacum*) tissue cultures by Professor Joseph Chappell of the University of Kentucky, which converts FPP to *epi*-aristolochene (**12**) (Vögeli et al 1990). The mechanism of the latter cyclization is believed to be closely related to the aristolochene synthase reaction, the major difference being the precise stereochemistry and folding of the FPP substrate as well as that of the derived germacrene A intermediate (Fig. 5).

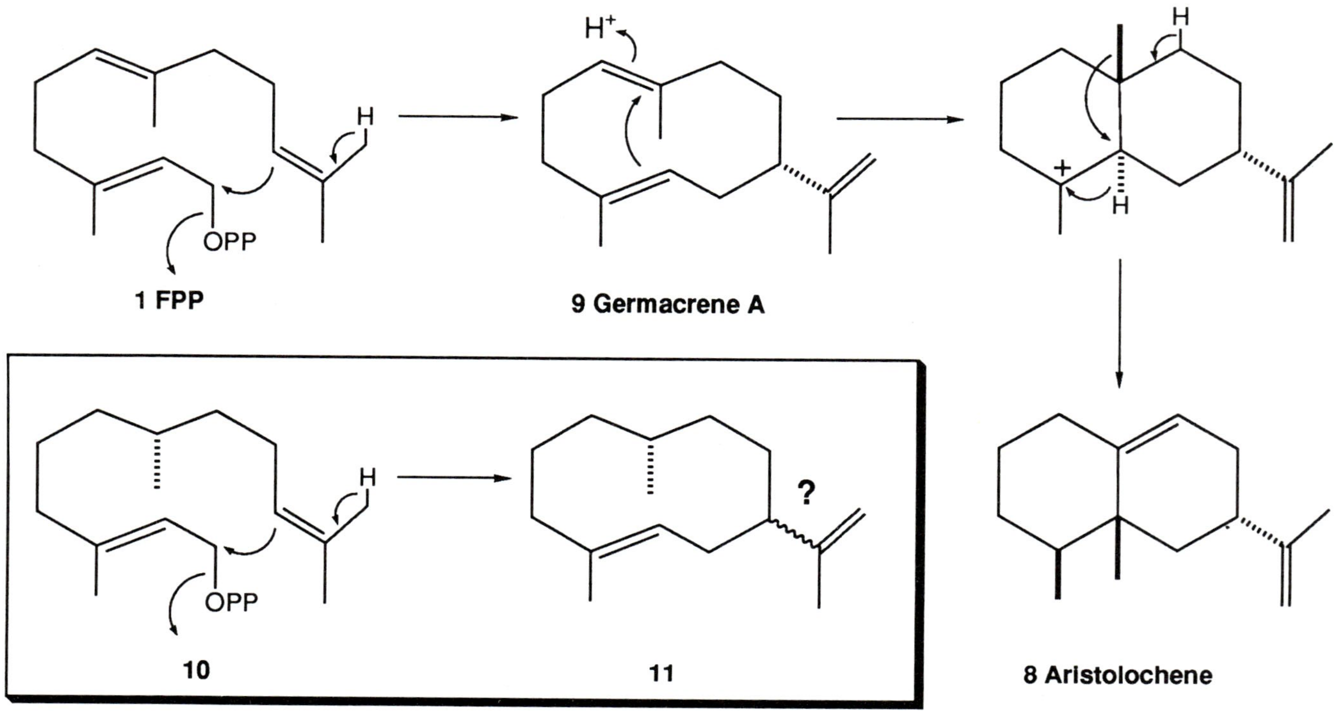

FIG. 4. Cyclization of FPP to aristolochene (**8**) through the intermediacy of germacrene A, and the cyclization of dihydroFPP (**10**) to dihydrogermacrene (**11**) by aristolochene synthase.

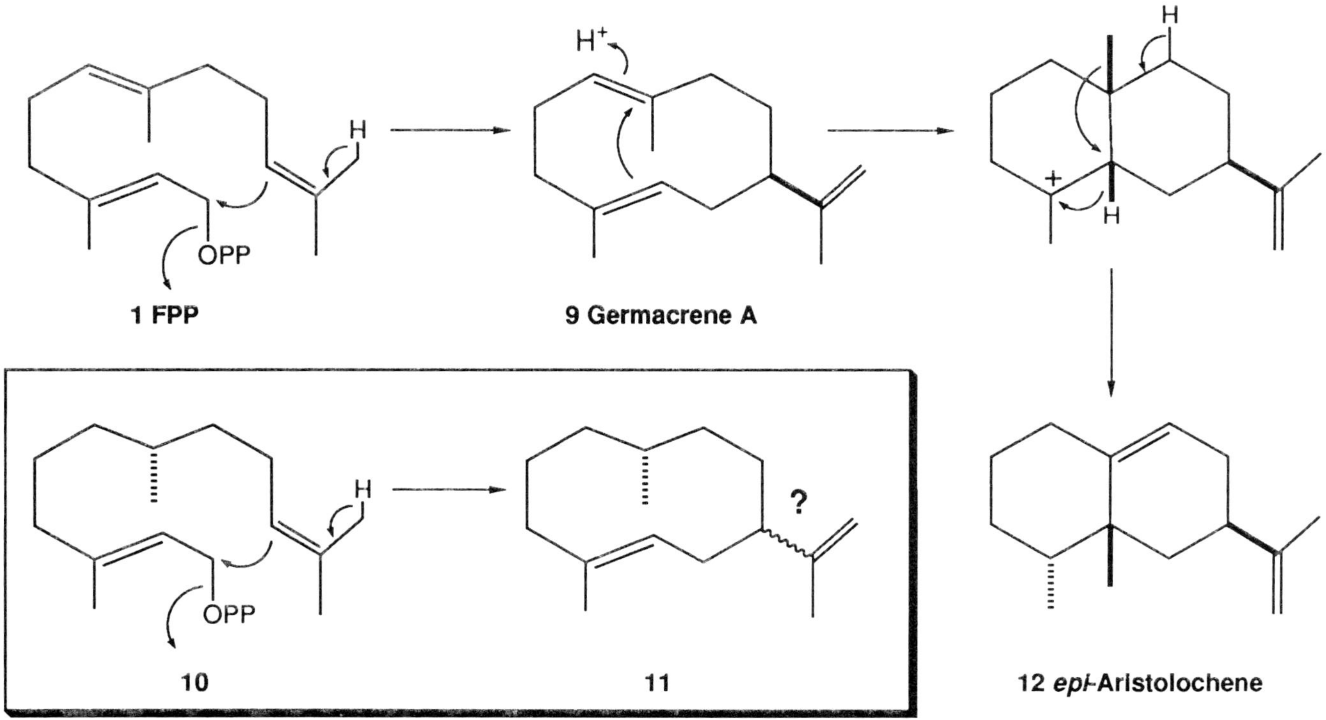

FIG. 5. Cyclization of FPP to *epi*-aristolochene (**12**) through the intermediacy of germacrene A and cyclization of dihydroFPP (**10**) to dihydrogermacrene (**11**) by *epi*-aristolochene synthase.

In fact, we have found that *epi*-aristolochene synthase converts **10** to a dihydrogermacrene A (**11**) isomer, but, unexpectedly, this isomer is *identical* by GC/MS comparison with the epimer of **11** generated by aristolochene synthase. Thus abortive cyclization of the anomalous substrate dihydroFPP (**10**) supports the proposed intermediacy of germacrene A in the normal cyclization catalysed by aristolochene and *epi*-aristolochene synthases. One of these two cyclases, however, is obviously converting dihydroFPP to the dihydrogermacrene product with the 2-propenyl moiety in an unnatural configuration. Presumably, the natural permissiveness of the cyclase active site is being reflected in binding of the anomalous substrate analogue in an unnatural conformation.

Future directions

Clearly, the understanding of terpenoid cyclase active site topology and catalytic mechanism is still at a rudimentary stage. Nonetheless, recent progress in the cloning and expression of terpenoid cyclase genes has opened up the possibility of further advances on a number of fronts:

1. The identification of substrate and intermediate analogues which act as effective competitive cyclization inhibitors, combined with the cloning and expression of cyclase genes (Hohn & Plattner 1989a), suggest that it may soon be possible to obtain X-ray crystallographic data on the precise geometry of binding of these inhibitors to the cyclase itself.
2. Experiments are in progress to identify effective analogues acting as irreversible inhibitors which can be used to identify active site residues by covalent modification.
3. Once sequences of additional cyclases become available, it should be possible to look for homologies at the DNA and amino acid sequence level, in order to identify conserved features of functional importance.
4. The availability of larger quantities of recombinant cyclases could allow detection of cyclization intermediates through the application of rapid quench and related kinetic techniques.

When these advances are combined with parallel advances in the understanding of terpenoid chain assembly (prenyl transferases, squalene synthase) and cyclization (oxidosqualene:lanosterol cyclase) enzymes, the way now seems open to an understanding of the way in which this important family of electrophilic catalysts carry out their complex and fascinating reactions.

Acknowledgements

Portions of the work described here which were carried out in the author's laboratory were supported by grants from the US Public Health Service, National Institutes of

Health GM30301 and GM22172. I would also like to thank Dr Thomas Hohn of the US Department of Agriculture in Peoria, Illinois for making available sequence data and other results in advance of publication, and Professor Joseph Chappell of the University of Kentucky for a gift of *epi*-aristolochene synthase.

References

Ashby MN, Edwards PA 1990 Elucidation of the deficiency in two yeast coenzyme Q mutants. Characterization of the structural gene encoding hexaprenyl pyrophosphate synthetase. J Biol Chem 265:13157–13164

Brems DN, Bruenger E, Rilling HC 1981 Isolation and characterization of a photoaffinity-labeled peptide from the catalytic site of prenyltransferase. Biochemistry 20:3711–3718

Cane DE 1980 The stereochemistry of allylic pyrophosphate metabolism. Tetrahedron 36:1109–1159

Cane DE 1985 Isoprenoid biosynthesis. Stereochemistry of the cyclization of allylic pyrophosphates. Acc Chem Res 18:220–226

Cane DE 1990 The enzymatic formation of sesquiterpenes. Chem Rev 9:1089–1103

Cane DE 1992 Biochemistry and genetics of antibiotic production. Isoprenoid antibiotics. In: Vining LC, Stuttard C (eds) Genetics and biochemistry of antibiotic production. Butterworth-Heinemann, Stoneham, MA, in press

Cane DE, Ha H 1988 Trichodiene biosynthesis and the role of nerolidyl pyrophosphate in the enzymatic cyclization of farnesyl pyrophosphate. J Am Chem Soc 110:6865–6870

Cane DE, Pargellis C 1987 Sesquiterpene biosynthesis. Partial purification and characterization of pentalenene synthetase. Arch Biochem Biophys 254:421–429

Cane DE, Ha H, Pargellis C, Waldmeier F, Swanson S, Murthy PPN 1985 Trichodiene biosynthesis and the stereochemistry of the enzymatic cyclization of farnesyl pyrophosphate. Bioorg Chem 13:246–265

Cane DE, Prabhakaran PC, Salaski EJ, Harrison PHM, Noguchi H, Rawlings BJ 1989 Aristolochene biosynthesis and enzymatic cyclization of farnesyl pyrophosphate. J Am Chem Soc 111:8914–8915

Cane DE, Pawlak JL, Horak RM, Hohn TM 1990a Trichodiene synthase. Studies of the cryptic allylic pyrophosphate isomerase activity using the anomalous substrate 6,7-dihydrofarnesyl pyrophosphate. Biochemistry 29:5476–5490

Cane DE, Prabhakaran PC, Oliver JS, McIlwaine DB 1990b Aristolochene biosynthesis. Stereochemistry of the deprotonation steps in the enzymatic cyclization of farnesyl pyrophosphate. J Am Chem Soc 112:3209–3210

Cane DE, Yang G, Coates RM, Pyun H, Hohn TM 1992 Trichodiene synthase. Synergistic inhibition by inorganic pyrophosphate and aza-analogs of the bisabolyl cation. J. Org Chem 57:3454–3462

Cornforth JW 1968 Olefin alkylation in biosynthesis. Angew Chem Int Ed Engl 7:903–964

Croteau R 1987 Biosynthesis and catabolism of monoterpenoids. Chem Rev 87:929–954

Croteau R, Satterwhite DM, Cane DE, Chang CC 1986 Biosynthesis of monoterpenes: enantioselectivity in the enzymatic cyclization of (+)- and (−)-linalyl pyrophosphate to (+)- and (−)-bornyl pyrophosphate. J Biol Chem 261:13438–13445

Croteau R, Satterwhite DM, Cane DE, Chang CC 1988 Enantioselectivity in the enzymatic cyclization of (+)- and (−)-linalyl pyrophosphate to (+)- and (−)-pinene and (+)- and (−)-camphene. J Biol Chem 263:10063–10071

Hohn TM, Beremand MN 1989a Regulation of trichodiene synthase in *Fusarium sporotrichioides* and *Gibberella pulicaris* (*Fusarium sambucinum*). Appl Environ Microbiol 55:1500–1503

Hohn TM, Beremand PD 1989b Isolation and nucleotide sequence of a sesquiterpene cyclase gene from the trichothecene-producing fungus *Fusarium sporotrichioides*. Gene (Amst) 79:131–138

Hohn TM, Desjardins AE 1991 Isolation and gene disruption of the *Tox5* gene encoding trichodiene synthase in *Gibberella pulicaris*. Mol Plant-Microbe Interact 5:249–256

Hohn TM, Plattner RD 1989a Expression of the trichodiene synthase gene of *Fusarium sporotrichioides* in *Escherichia coli* results in sesquiterpene production. Arch Biochem Biophys 275:92–97

Hohn TM, Plattner RD 1989b Purification and characterization of the sesquiterpene cyclase aristolochene synthase from *Penicillium roqueforti*. Arch Biochem Biophys 272:137–143

Hohn TM, VanMiddlesworth F 1986 Purification and characterization of the sesquiterpene cyclase trichodiene synthase from *Fusarium sporotrichioides*. Arch Biochem Biophys 251:756–761

Poulter CD, Rilling HC 1981 Prenyl transferases and isomerase. In: Porter JW, Spurgeon SL (eds) Biosynthesis of isoprenoid compounds. Wiley, New York, vol 1:161–224

Poulter CD, Capson TL, Thompson MD, Bard RS 1989 Squalene synthetase. Inhibition by ammonium analogues of carbocationic intermediates in the conversion of presqualene diphosphate to squalene. J Am Chem Soc 111:3734–3739

Tidd BK 1971 Some phosphate esters of biological importance. J Chem Soc Sect B 1168–1176

Vögeli U, Freeman JW, Chappell J 1990 Purification and characterization of an inducible sesquiterpene cyclase from elicitor-treated tobacco cell suspension cultures. Plant Physiol 93:182–187

DISCUSSION

Chater: Can one imagine a situation in which the substrate goes in through one 'door'and the product comes out through another, and everything possible happens inside, but the only thing that can get out is the product you want?

Cane: The experiments with the anomalous substrates indicate that abortive products can come out *earlier*. These cyclases show product inhibition and one can prevent the binding of the substrate with inorganic pyrophosphate. In fact, binding is potentiated for pentalenene synthase by the presence of pentalenene, so pentalenene plus inorganic pyrophosphate constitutes a more effective inhibitor than either alone.

There is a limitation to steady-state kinetics, but what those kinetics mean is that there is a single active site, so the inhibitors and substrate both compete for the same place. Steady-state kinetics do not tell you how you reach the active site or whether, when something leaves the enzyme, it flies off in the same or a different direction as when it approaches the enzyme. But there's no evidence that the various catalytic events happen in different places.

There are monoterpene cyclases which make product mixtures. In experiments in collaboration with Rod Croteau at Washington State University, we diverted the formation of the mixed products formed by pinene cyclase (Croteau et al

1987). The monoterpene cyclases tend to be more 'sloppy'; of course in any single turnover you form one product, but in a collection of turnovers, you tend to form different products, derived from quenching the common carbocations at different stages.

Orgel: What are the obstacles to obtaining enough of the enzyme to crystallize it and then proceed to determine the structure in the normal way?

Cane: As with many enzymes involved in this type of metabolism, where the products accumulate over a relatively long period, and are not themselves essential to the organism, there are very low titres of the enzyme. For example, from 20–30 litres of growth culture of *Streptomyces* we could isolate sufficient purified pentalenene synthase for sequencing, namely 100–200 pmol. In principle, there is no natural impediment to the over-expression of the cyclase gene.

Davies: What organisms are you using to produce the enzyme?

Cane: We are expressing trichodiene synthase in *E. coli*. Dr Hohn has shown that not only is the enzyme expressed, but it seems to appropriate endogenous farnesyl diphosphate, because the *E. coli* will produce trichodiene. This cyclase, as well as aristolochene synthase, does not seem to express well in the traditional vectors. This probably has nothing to do with it being lethal, because you can express the cyclase and the cells function perfectly well. It is probably more to do with secondary structure and codon usage. But I think it can be done. [Note added in proof: With trichodiene synthase, we have recently achieved substantial levels of over-expression, 15–25% of total protein (D. E. Cane & Z. Wu, unpublished).]

Rinehart: I gather that aristolochene synthases are different from *Penicillium roqueforti* and from *Aspergillus terreus*. Is that based on sequencing?

Cane: There are three pieces of evidence. One is the gross molecular size: the *Penicillium* cyclase is just under 40 kDa, whereas the *Aspergillus* enzyme is close to 60 kDa. Secondly, antibodies to the aristolochene synthase from *Penicillium roqueforti* do not cross-react with the *Aspergillus terreus* enzyme. And, thirdly, the cloned *P. roqueforti* structural gene for aristolochene synthase, when used as a probe against *A. terreus* DNA, shows no stringent hybridization; so there appear to be sequence differences.

Trichodiene synthase is present in a large number of *Fusarium* species that are closely related. You can use DNA from *F. sporotrichioides* as a probe for *Gibberella pulicaris* DNA. But *Myrrothecium verrucaria* also produces other trichothecanes and there's no hybridization to *M. verrucaria* DNA. So there is already a strong divergence, even within closely related fungal species.

Rinehart: Do different *Penicillium* species give the same aristolochene synthase, and do different *Aspergillus* species all give the same synthase?

Cane: I can't say, because the aristolochene synthase activity has been detected only in the two fungi. In *P. roqueforti*, aristolochene is the hydrocarbon precursor of a toxin, called PR toxin; this is presumably present in roquefort cheese! There is a family of structurally related compounds, such as the eremofortins, which occur as mycotoxins in various other *Aspergillus* species. Some of these are also plant pathogens. Nobody knows the oxidation product of aristolochene in *Aspergillus terreus*. We found aristolochene completely by accident; we were looking for a cyclase that would make the precursor of quadrone. We still have no idea what the quadrone cyclase is, but we found three other sesquiterpene hydrocarbons, and the cyclase for one of them.

Davies: Have you ever used specific antibodies to look at the presence of cyclases in other strains? In general, does the anti-cyclase or anti-synthase antibody pick up this kind of activity?

Cane: We are doing experiments using polyclonal antibodies that people have given us; some experiments are being done in other labs. An antibody raised by Tom Hohn of the US Department of Agriculture against aristolochene synthase from *Penicillium roqueforti* does not cross-react with the *epi*-aristolochene synthase from the tobacco plant (T.M. Hohn, personal communication); antibodies to trichodiene synthase do not cross-react with any other cyclases. The idea of an essentially universal DNA probe with which to look for cyclases in a range of organisms doesn't seem to be on.

Brückner: Do you know whether there are different enzymes for the first steps in steroid biosynthesis, for the construction of membranes, and for secondary metabolite synthesis?

Cane: This is not known. We have not looked at the prenyl transferases of the fungal systems; in the streptomycetes, as far as one knows, there are no steroids; so the cyclic sesquiterpene pathway is in that sense unusual. The *Streptomyces* which produces pentalenene is not the only streptomycete with the ability to make cyclic terpenes (Gerber 1971). In the fungi we don't know whether there is any specialization in terms of the source of acyclic prenyl units.

Brückner: In yeast there are two different HMG-CoA reductases, but in higher fungi, I don't know.

Haslam: It's my experience in the area of plant phenolic chemistry that far from things being totally promiscuous, they are only partially promiscuous, in the sense that of the number of pathways of phenol oxidation which are possible, only a certain number of these are found. I wonder whether you agree, Dr Cane, with the proposition that once a certain chemical type of intermediate has been set up, part of the whole process has already been determined, and then follows a purely chemically determined pathway.

Jeremy Knowles has worked on the cyclization of a seven-carbon sugar (DAHP) to give 3-dehydro-quinate, at the beginning of the shikimate pathway (Knowles 1989, Bender et al 1989a,b). The enzyme 3-dehydroquinate synthase catalyses in principle four distinct chemical steps: an oxidation, an elimination, a reduction, and then a cyclization. The thrust of his work is that only the oxidation and reduction are enzymically controlled. The elimination and cyclization reactions are determined purely by the intrinsic chemistry of the system, and do not need enzyme catalysis. I wonder whether, once you have set up a particular type of carbocation intermediate, the rest of the reaction then follows as a result of the inherent chemistry of the intermediates.

Cane: I would think so. The notion is that even though the chemistry is not mediated by the enzyme, it is still taking place within the cavity of the enzyme, and therefore one is preventing side reactions. For the cyclases, one is preventing interception of intermediates by water. The idea that one would have to make a carbocationic rearrangement go faster doesn't work out, from what is known about the rates, in a super acid solution or gas phase. One is looking for control, basically, and this has to come from (among other things) holding groups in proximity to one another, or holding them in the right stereo-electronic orientation, in the same way, for example, that many pyridoxal-dependent processes depend on the conformation of the amino acid relative to the pyridoxine ring. The chemistry needs little catalysis, but the product-forming event needs channelling, to prevent all possible reactions from happening. By retarding the processes that you don't want, you are not necessarily making the reactions that you do want go substantially faster, because the turnover rates are very small compared to what are probably the inherent chemical rates. Isn't that what you think about IPNS, Dr Baldwin?

Baldwin: In reactions that proceed through potentially very reactive intermediates, which can be intercepted or diverted, it must be a critical role of the enzyme protein to channel the pathway.

It seems to me that in all these terpene cation cyclizations, once you have established the first carbonium ion, the process of the reaction is the translation in space of a positive charge, through (in the case of the squalene cyclization) an enormous distance of about 12Å. Control of the pathway may be better achieved, not by conformational interactions with the surrounding environment at the active site, but by some means of stabilizing the motion of a positive charge, because the electrostatic effect is a much more potent effect than non-bonding interactions. This would fit in with what you seem to find, namely a very sloppy active site that can adjust quite readily even to different enantiomers.

Cane: The squalene case is a good comparison. When you are separating positive charge, after a separation of about 3 Å it doesn't cost much to get it further, because the attraction is falling off as the square of the distance. But with squalene, the charge goes out to one end of the molecule, after which there is a series of rearrangements by which the charge comes back quite close to where it started. With the squalene oxide to lanosterol cyclization, the whole molecule turns inside out, so there is a gross change in shape. The eventually formed sterol is not going to look anything like the initial folding of the substrate.

Bu'Lock: To what extent is this due not so much to the conformation of the cyclizing molecule, but to the location of the ultimate acceptor site, which in many cases will be a potential proton acceptor? In the alternative cyclizations of squalene oxide in mammalian systems versus plant systems, in the plant systems, cycloartenol, a sterol with a cyclopropane ring, is generated, where the animal systems generate lanosterol. This seems a *priori* to depend on whether you have the proton acceptor on the α or β side of the system.

Cane: The position of counter bases may be critical. The only piece of relevant information that I have, because no one knows anything about what's present at the active sites, is in the formation of pentalenene. We have shown that a proton is removed in an initial cyclization, but that this proton is re-added to the adjacent carbon without any 'cross-talk' with the medium (Cane et al 1984). There is presumably a single base mediating deprotonation–reprotonation. We have looked at the stereochemistry of the final deprotonation which quenches this system (Cane et al 1991). We can't prove that this deprotonation involves the same base, but we know that the proton which is removed is very close to where the original base would be present. It is conceivable that a single base does triple duty: it pulls off a proton; it puts the proton back on an adjacent carbon; and then, after a further cyclization and rearrangement, it removes a proton from a site immediately adjacent to the one where it removed a proton initially. In principle, the reaction could have stopped after the first deprotonation, but that is not sufficient, because it is capable of going on further. We want to find out what is present around the substrate; this will address some of these questions. My own view is that once one knows the full structure of any enzyme, all understanding of how it works immediately disappears; that is either a psychological or a conceptual principle!

Leadlay: Are you saying you don't *need* many of these counter groups on the enzyme? You seem to be saying that there is quite a sloppy interaction; is it that you just have one base, or very few, and the intermediates can actually rotate around, and in this way you can get deprotonation and quench at different points?

Cane: I think you need a few counter groups, because of the way the charge moves; if you track it in any given cyclization, it moves from one side of the active site to the other. Some people have speculated that a major tactic for controlling the charge is that the enzyme uses the pyrophosphate moiety to steer things. Our own inhibition studies, and also some nice work by Dale Poulter on squalene synthase, have shown that the pyrophosphate group is critical to the effectiveness of certain ammonium ion analogues (Poulter et al 1989). He has also made bi-substrate analogues of prenyl transferase and shown that these can be cyclized. His argument is that the pyrophosphate is acting as a base.

Williams: On the question of the tolerances that the enzymes will allow, it was striking that in your ammonium ion analogue of the cation intermediate, you could remove very large hydrocarbon side chains; the compound was then just a weaker inhibitor, by a factor of four to eight. If in fact removal of the hydrocarbon side chain left a cavity, you would expect an enormously greater effect. For example, in selecting a given amino acid in protein synthesis, you select very nicely between the subtle differences of glycine and alanine. Here, we are removing a very large side chain and only getting a small effect. This would seem, from the evidence, to preclude a cavity being left when that large side chain is removed; the enzyme may somehow rearrange itself to accommodate the structural change so that a cavity does not remain. Does that seem reasonable?

Cane: It's reasonable, but I can't say whether it's true or not! You certainly would not fill up the cavity with water, or you would have a big problem of maintaining free carbocations without quenching. The trouble with the steady-state experiment is that I can't even claim, for example, that any two of these inhibitors bind in the same way, or that any single one binds in a unique way. Obtaining insight into what else is happening is impossible from a steady-state experiment. All one can say is that there is some hydrophobic component. How the enzyme compensates for the absence of a component which would normally be bound, I don't know. One can conceive of ways of binding the ammonium ion inhibitors in which the inhibitor goes into the active site completely backwards, and all the enzyme is really seeing is the ammonium ion analogue and some extra pieces of hydrocarbon.

Williams: I agree that you wouldn't expect the channel to be filled by water, but it's conceivable that hydrophobic side chains that were originally apart move in to fill the cavity.

Cane: Yes, phenylalanine rings, say, might just move closer; that is perfectly conceivable.

Cavalier-Smith: I would like to ask about the evolutionary implications of the enzyme not speeding-up reactions, but selecting from various potential pathways. When the enzyme initially evolved, did it just reduce the number of possibilities, and was therefore rather sloppy, yielding not one specific product, but a mixture? Then, later in evolution, it would be improved, further reducing the number of products. Or do you think that initially it would be rather precise, perhaps having evolved from some other enzyme that was already precise, and therefore be capable of selecting just one of the products, right from the start?

Cane: I would imagine that sloppiness would precede precision, because in the natural cyclization, the first thing it has had to learn is to isomerize the 2,3 double bond. Once it has evolved to do that, controlling the conformation of the rest of the cyclizing molecule, as well as the points of deprotonation, in principle follows. I don't know whether there is any selective advantage for an enzyme in remaining sloppy; that is, if an organism at some point derives an advantage from having not one but five different monoterpenes, then there would be no more pressure to get better. But although sloppiness might precede precision, it's not clear to me, at this stage, that one is inherently more primitive than the other.

Cavalier-Smith: There are different chlorophylls in different groups of algae, and there is some evolutionary indication that their chloroplasts were 'experimenting' with diversity, to start with, and then, as they diverged, stabilized in one direction or another. Perhaps the sort of side chain modifications that are seen in the chlorophylls are analogous to what you are discussing? The enzymes could initially have been sloppy as to what they were adding, and then became more precise.

Cane: I don't know. You can propose that this is an adaptation to particular types of membrane interactions, with multiple ways of solving the same problem; I would tend to favour that view.

Orgel: In relation to the metal ion dependence, where the enzyme works better with manganese than with magnesium, is the suggestion that manganese is the naturally occurring metal?

Cane: It was a *quantitative* effect, in that one can measure the rate of reaction as a function of ion concentration, and, as with many kinases, manganese is often more effective than magnesium at low concentrations; but in many cases manganese becomes inhibitory at higher concentration, presumably because of non-specific interactions with thiols. There are a small number of cyclase enzymes for which manganese is simply superior. I don't think anyone has a way of sorting out which particular complex of geranyl diphosphate is the physiological substrate.

Orgel: Is it clear that the function of the metal ion is to complex the pyrophosphate group?

Cane: Yes; studies of prenyl transferases have shown that the metal ion is essential; the stoichiometry of binding has also been worked out. It has also been shown by Rod Croteau that monoterpene cyclases can be inhibited by sulphydryl-directed reagents. In at least one case, you cannot protect the active site with geranyl diphosphate alone, but you can protect it against inactivation with geranyl diphosphate–magnesium complex (R. B. Croteau, personal communication).

Baldwin: On the question of channelling, and how this could have evolved towards more precision, it seems to me that there is, underlying this discussion, the idea that these kinds of enzymes may be acting not as classical Pauling catalysts but as devices that restrict reaction pathways. The implication is that those pathways existed before the enzymes did, and the proteins came later, to provide selectivity.

Orgel: Particularly if they have something to do with eliminating water as a potential competitor, rather than accelerating a reaction.

Baldwin: That would be another version of the same idea of restricting the selection of the pathway from all of those reactions that are open to the system, once the reaction began to proceed.

Cavalier-Smith: Did Dr Cane say that the reaction is actually a lot slower than if it didn't have the enzyme there?

Cane: No; without the enzyme, there is a magnesium ion-catalysed solvolysis, but in a typical enzymic incubation, the cyclization rates are usually a factor of a hundred or a thousand more than the background solvolysis catalysed solely by magnesium. I was saying that the rate of the allowed reaction must be substantially faster than the rate of the competing reactions which would give off other possible products. So, what has been catalysed, essentially, has been the ionization of the substrate. Upon ionization, the formation of one or another product should follow quite rapidly. You now have to channel things so that only one or a few of the many conceivable products are generated. But without ionization there is no generation of the product.

Cavalier-Smith: So there is catalysis—an effective speeding up of the reaction—and selection, and the two are equally important.

Leadlay: Are these enzymes under thermodynamic control, or kinetic control? That is to say, is the final quenching of the cation essentially irreversible?

Cane: It is irreversible; there's no case known in which one can take a product and convert it back to the starting material.

References

Bender SL, Mehdi S, Knowles JR 1989a Dehydroquinate synthase: the role of divalent metal cations and of nicotinamide adenine dinucleotide in catalysis. Biochemistry 28:7555–7560

Bender SL, Widlanski I, Knowles JR 1989b Dehydroquinate synthase: the use of substrate analogues to probe the early steps of the catalyzed reaction. Biochemistry 28:7560–7572

Cane DE, Abell C, Tillman AM 1984 Pentalenene biosynthesis and the enzymatic cyclization of farnesyl pyrophosphate. Proof that the cyclization is catalyzed by a single enzyme. Bioorg Chem 12:312–328

Cane DE, Abell C, Harrison PHM et al 1991 Terpenoid biosynthesis and the stereochemistry of enzyme-catalyzed allylic addition–elimination reactions. Phil Trans R Soc B Biol Sci 332:123–129

Croteau RB, Wheeler CJ, Cane DE, Ebert R, Ha H 1987 Isotopically sensitive branching in the formation of cyclic monoterpenes: proof that (−)-(α)-pinene and (−)-(β)-pinene are synthesized by sage cyclase II by deprotonation of a common intermediate. Biochemistry 26:5383–5389

Gerber NN 1971 Sesquiterpenoids from actinomycetes: cadin-4-ene-1-ol. Phytochemistry 10:185–189

Knowles JR 1989 Aldrichimica Acta 22(3):59–66

Poulter CD, Capson TL, Thompson MD, Bard RS 1989 Squalene synthetase. Inhibition by ammonium analogues of carbocationic intermediates in the conversion of presqualene diphosphate to squalene. J Am Chem Soc 111:3734–3739

Roles of secondary metabolites from microbes

Leo C. Vining

Biology Department, Dalhousie University, Halifax, Nova Scotia, Canada B3H 4J1

Abstract. The common feature of the seemingly diverse array of biological activities exhibited by microbial secondary metabolites is their survival value for the producing organism. The propensity to form these compounds is unevenly distributed in microbial taxa and seems more closely associated with existence in a competitive environment than with phylogeny. The characteristic multibranched elaboration of secondary biosynthetic pathways and the marked species specificity of the end products are consistent with their evolution by an 'inventive' mechanism. The species specificity suggests that distinctive terminal reactions may be of recent origin. However, comparisons of the nucleotide sequence of genes involved in the biosynthesis of phenazine and polyketide metabolites with related genes of primary pathways indicate that the secondary pathways have not evolved exclusively within the organisms in which they are now found. Sequence similarities with related primary pathway genes in phylogenetically distant organisms suggest that gene transfer has played an important part in the evolution of secondary metabolism. The diversity of products may reflect the many roles for which secondary metabolites have been selected after the genes for their biosynthesis have transferred to organisms with different physiologies and different environmental challenges.

1992 Secondary metabolites: their function and evolution. Wiley, Chichester (Ciba Foundation Symposium 171) p 184–198

The evidence that secondary metabolism benefits the organism, and that this benefit comes predominantly from the functions of the metabolites, is now so convincing that arguments to the contrary have only historical interest. However, accepting that these products have roles does not mean that we can readily agree upon, or sometimes even perceive, what these roles may be. Attempts to discover an underlying benefit common to the production of all secondary metabolites have not been profitable, and it seems likely that diversity of function as well as of chemical structure is a characteristic of the activity (Demain 1989). The diversity can be plausibly accounted for by the selective influence of environments, both physiological and ecological, associated with the organism in which the secondary metabolic pathway has evolved. Contemporary pathways of secondary metabolism have their starting points within the primary metabolic network. There are a limited number of initiation sites from which primary intermediates or precursors are diverted into idiosyncratic syntheses. The diversity of the products generated stems from the multiplicity of biochemical choices that each pathway has afforded. Whether the options were a consequence of gene mutation, relaxed enzyme specificity or even spontaneous chemical change, choice has been exercised through selection by the environment of the phenotype conferring a competitive advantage. The roles of secondary metabolites reflect the variety of ways in which such advantages can be gained.

Roles of secondary metabolites

From a broad perspective, secondary metabolic products might be expected to benefit the producing organism in either of two ways: they might improve its ability to grow, reproduce or disperse in appropriate conditions, or they might afford protection from competition or predation. The majority of compounds fit into these categories (Vining 1990). Among those conferring a growth advantage are siderophores (ferric iron-chelating compounds) which, in association with the sophisticated mechanisms used to recover them, have strong selective value in iron-deficient environments (Crosa 1989). Autoregulators provide the chemical signals that activate light production in luminescent bacteria and probably also coordinate sporulation in streptomycetes. Other secondary metabolites serve as chemical signals promoting differentiation of vegetative cells to form fruiting bodies, or enhancing sexual reproduction through pheromone activity. Although many such compounds have been discovered, the preponderance of known secondary metabolites have a more aggressive role. The production of antibiotics, targeted through a variety of biochemical mechanisms at other microorganisms, is an important factor in the competition for resources. Since microorganisms compete with animals for the same food resources, and themselves are a food source for protozoa and detritivores, it is not surprising that many secondary metabolites are potently toxic to large organisms (Janzen 1977). Whereas the mechanism of antibiotic action is normally directed at intermediary metabolism or macromolecular synthesis, toxicity against animals can be aimed at higher-level functions such as hormonal activity or nerve transmission.

The properties of some microbial secondary metabolites equip them for a role as deterrents. Along with potent mycotoxins, fungi competing with insects and rodents for stored grain produce pigments and other less dangerous products that advertise the presence of a microorganism. In providing a warning signal, this may be the closest that microbial secondary metabolites come to benevolence in their interactions with animals. However, in plant–microbial interactions the production of growth-promoting agents or nodulation factors by rhizosphere microorganisms directly benefits the plant. Plants can also benefit indirectly, as in the production by microorganisms of secondary metabolites that suppress plant pathogens (Lynch 1987). In such three-way interactions the active agent is not necessarily an antibiotic; the siderophore pseudobactin 358, excreted by a rhizosphere-colonizing *Pseudomonas putida* strain, prevents growth of harmful bacteria and fungi by sequestering all available iron in the root environment (Schroth & Hancock 1982). On the whole, however, microbial secondary metabolism has severely negative consequences for plants, and the properties of some toxic substances leave little doubt about their target. Phaseolotoxin biosynthesized by *Pseudomonas syringae* has an oligopeptide transport component that efficiently delivers the active antimetabolite to the plant cell interior (Mitchell et al 1981). The genetic systems of *Agrobacterium* species are structured to ensure plant-directed auxin biosynthesis in deleterious amounts. The concomitant directed production in the tumorous plant tissue of opines, usable only by the invading bacterium, emphasizes the value of the secondary metabolic activity to the microorganism.

Though the roles of many microbial secondary metabolites can be deduced with reasonable confidence, firm supporting evidence is not usually available. There are also some substances for which a role can only be guessed at. Products discovered because of their anticancer activity or their ability to inhibit enzymes functioning in the physiology or immune systems of animals may owe their activity to an evolutionary linkage to the biochemistry of common ancestral organisms; however, their continued production in modern microorganisms suggests that they also have a useful role in the producing species.

Distribution of secondary metabolism

Conclusions about the range of microorganisms in which secondary metabolism occurs are hazardous because of the inevitable bias in the search for products. However, the results from many kinds of screening procedures suggest that some microbial groups are more prolific than others. In general, these are organisms inhabiting resource-rich environments. Existence in harsh or resource-poor conditions might be expected to select narrowly for characteristics that cope with the environment and demand metabolic economy. On the other hand, survival in environments with more abundant resources requires that a species compete successfully with numerous others, each of which may be similarly or better endowed (Case & Gilpin 1974). Consistent with this view, microorganisms included in the newly defined 'domain' Archaea (formerly Archaebacteria) are not reported to commonly produce toxic secondary metabolites. In the 'domain' Bacteria (formerly Eubacteria), the ability is unevenly distributed and, also in accordance with expectations, it seems to be absent from groups such as anaerobes and chemolithotrophs that inhabit the more exacting environments. It has been most frequently reported in the Gram-positive filamentous actinomycetes, but it also occurs in other Gram-positive as well as in Gram-negative bacterial families.

Within the Bacteria, the distribution of secondary metabolism in phyla established by 16S RNA sequencing (Woese 1987, Woese et al 1990) is distinctly uneven. The Proteobacteria (a group that includes the purple photosynthetic bacteria and many non-photosynthetic genera) contain some species that produce secondary metabolites abundantly. However, other species in the group do not, and even within the phylogenetically more homogeneous γ subdivision of Proteobacteria, the fluorescent pseudomonads are frequently producers whereas the enterics are not noted for secondary metabolism. The most prolific bacterial groups known so far are all soil organisms, living in a strongly competitive environment. In the domain Eucarya (which includes all eukaryotic organisms), secondary metabolism is widespread in two kingdoms, the Fungi and the Plantae. In a third, the Animalia, it is clearly present in some members of the class Insecta, where the products have been adapted into communication and defence systems. It also is present in some algae and marine invertebrates (see chapter by K. L. Rinehart: this volume 1992), where the postulated functions are broadly similar to those of terrestrial counterparts. Secondary metabolism in the Eucarya thus seems to have its primary role in chemical signalling and chemical weaponry; the producing organisms are those in which such uses are well integrated into the pattern of existence. Notwithstanding the bias in screening procedures, it is difficult to avoid the conclusion that, overall, secondary metabolism is more closely associated with habitat and lifestyle than with genealogy.

Origin of diversity in secondary metabolism

Secondary metabolic pathways are most readily accounted for as the products of 'inventive' evolution, a process in which the haphazard generation of a catalytic sequence yields a product that sometimes fortuitously fulfils a useful function (Wong 1981). Secondary pathways represent adaptive extensions of the pre-existing metabolic apparatus, presumed to be the essential primary pathways that supported normal life functions in ancestral organisms. The mechanisms by which they have arisen are probably similar to those by which the globins have evolved (Czelusniak et al 1982). Duplication of primary metabolic genes and mutation of the copies furnished altered proteins; these modified primary pathway enzymes catalysed reactions that converted primary metabolic intermediates to new products that were acted upon in turn by enzymes of relaxed specificity recruited from other primary roles (Jensen 1976). By a series of such random steps the diverted intermediates yielded new metabolites that inadvertently fulfilled a useful function. The success of the phenotype ensured that the genetic system responsible for the diversionary pathway became fixed into the genome (Zähner et al 1983, Vining 1992). Continuation of this process has led to the gradual elaboration of secondary metabolic pathways.

One noteworthy feature of secondary metabolism is the diversity of products; another is their dispersion. Few compounds are present in more than one species and often their biosynthesis is strain specific. *A priori* this lack of universality suggests that the pathways are of recent evolutionary origin, having appeared since the separation of closely related species. However, the course of secondary metabolic evolution as an inventive, fortuitous and continuing process may have led from a relatively small number of successful early reactions to multibranched pathways in which the terminal steps have been added in more recent times. The diversification of products would have created a range of potentially bioactive molecules, thus providing the variation needed for natural selection. The most recent pathway elaborations may have occurred in a wide variety of organisms, each of which has selected for a different end product, or group of end products.

The substantial progress made lately in cloning and sequencing secondary metabolic pathway genes makes possible the testing of speculative schemes, such as that outlined above, by comparing appropriate genes of primary and secondary metabolism. In the remainder of this chapter some aspects of the role and evolution of secondary metabolites will be examined by focusing on two secondary metabolic systems, one branching from the shikimate pathway of primary aromatic biosynthesis and the other representing a diversion via polyketide intermediates from the primary pathway for fatty acid biosynthesis.

Shikimate pathway metabolites

Some recent information on a secondary pathway that diverts shikimate intermediates from primary metabolism bears on the question of pathway evolution by an 'inventive' mechanism. By drawing upon chorismate as a precursor (Fig. 1), the branching reaction gives rise to a variety of phenazine secondary metabolites in some coryneform bacteria, pseudomonads and actinomycetes. Although the early steps in the pathway were previously uncertain, it is now clear that anthranilate is an intermediate. *Pseudomonas aeruginosa*, which produces the phenazine antibiotic pyocyanine, contains two forms of anthranilate synthase, the enzyme complex converting chorismate to anthranilate (Essar et al 1990). One form (TrpEG) has the characteristics expected of the initial enzyme for tryptophan biosynthesis; the other (PhnAB) appears predominantly during late growth when pyocyanine is produced, and has features more in keeping with a role in secondary metabolism. PhnAB forms part of a dedicated complex that does not make free anthranilate available for tryptophan biosynthesis, but channels it exclusively to phenazines.

The nucleotide sequences of the pairs of genes (*trpEG* and *phnAB*) encoding the two enzymes possess sufficient similarity (26–39% amino acid sequence identity) to indicate a common ancestry (Essar et al 1990). However, *trpEG* of

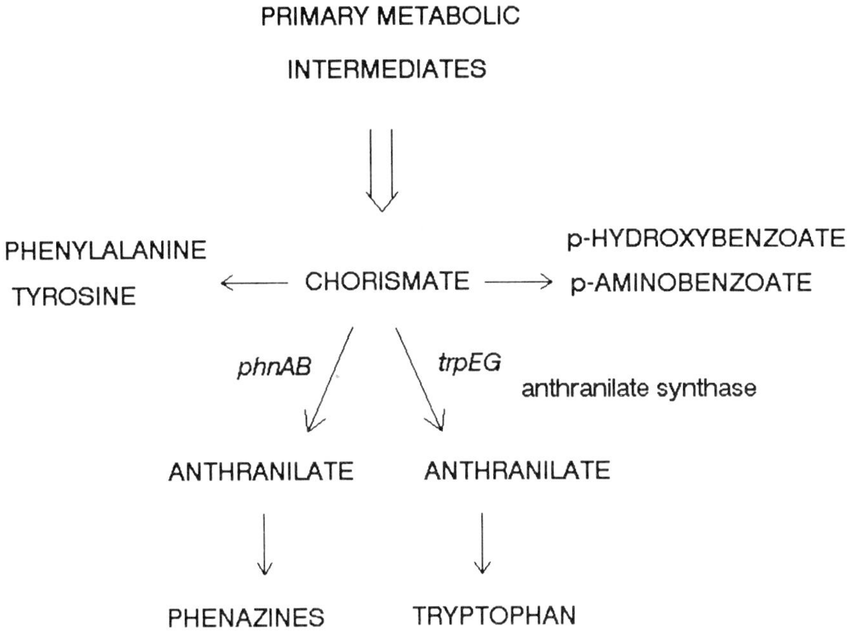

FIG. 1. Biosynthesis of phenazine secondary metabolites by withdrawal of chorismate from the shikimate pathway of primary metabolism.

P. aeruginosa is more closely related (79–87% amino acid sequence identity) to the anthranilate synthase genes (i.e. *trpEG*) in a separate species, *Pseudomonas putida*, than to the cohabiting *phnAB* pair. In fact, *phnAB* of *P. aeruginosa* is closer to *trpEG* in *Escherichia coli* (46–48% amino acid sequence identity) than to *trpEG* in either of the *Pseudomonas* species. Since phenazines are not produced in *E. coli* or *P. putida*, the Phn pathway may have originated in a bacterial lineage before the separation of enterics and the fluorescent pseudomonads and been lost in most species. An alternative interpretation is that the *phn* genes were transferred horizontally into *P. aeruginosa* after its separation from *P. putida* (Fig. 2). Irrespective of whether a recent gene transfer or evolution within the phenazine-producing lineage is responsible for the presence of the secondary pathway in *P. aeruginosa*, the gene duplication that initiated it was not a recent event. However, the product of the progenitor enzyme complex was probably not pyocyanine. More likely the initial secondary metabolite was phenazine-1,6-dicarboxylic acid, a substance from which pyocyanine could have been formed in *P. aeruginosa* by late-stage reactions (Turner & Messenger 1986).

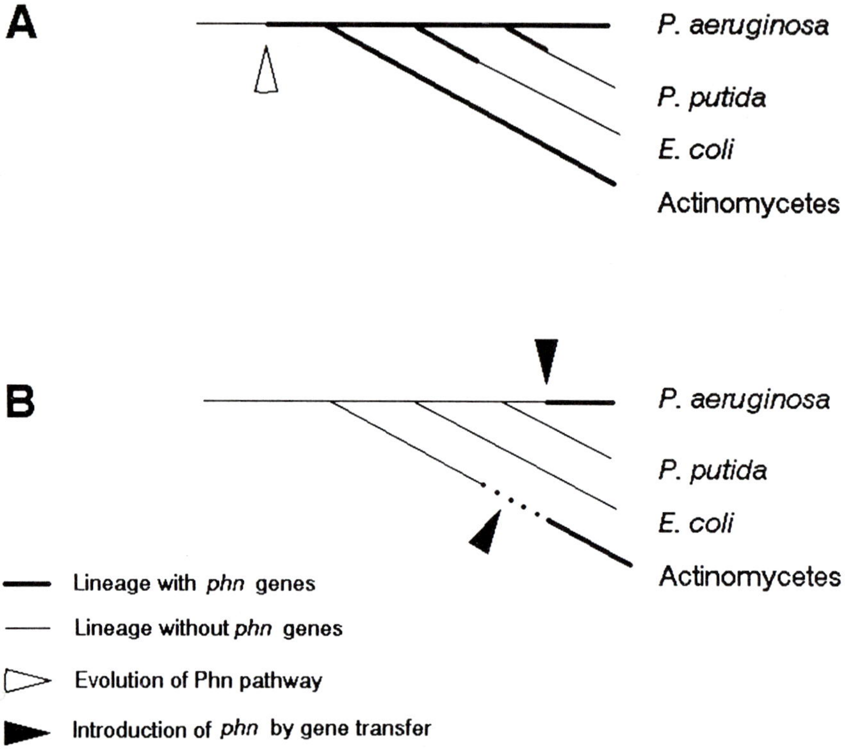

FIG. 2. Alternative explanations for the dispersion of phenazine biosynthesis: A, early evolution and selective loss of the pathway in divergent lineages; B, late introduction of the pathway by gene transfer from a 'reservoir' organism.

Polyketide-derived metabolites

The biosynthetic sequences that lead to polyketide intermediates are related to those of fatty acid biosynthesis in linking activated acyl precursors head-to-tail by enzyme-catalysed reactions (see also the chapter by D. A. Hopwood & C. Khosla: this volume 1992). In the fatty acid synthase (FAS) system, the intervention of ketoreductase, dehydratase and dehydrogenase activities after each acyl condensation leads to a predominantly reduced carbon chain, whereas the polyketide synthase (PKS) system lacks some or all of the steps between the condensation reactions. Although all polyketides are made by a similar sequence of reactions, the product can vary in the number and type of acyl units introduced, and to some degree in the extent of reduction or substitution. In addition, a variety of post-PKS processing reactions that bring about cyclizations, chain scissions and further substitutions can lead to many different end products; these exhibit a wide range of biochemical properties and biological activities. It is noteworthy that these differences in structure and function are associated with polyketides biosynthesized in different microorganisms. In plants, and in various lower animal and single-celled eukaryotes, polyketide-derived metabolites exhibit additional types of structure and biological function.

Recent investigations of the molecular genetics of fatty acid and polyketide biosynthesis have provided considerable information on the two systems (Hopwood & Sherman 1990), making it possible to explore some of the questions about the evolution of secondary metabolism discussed earlier. In particular, it is possible to establish in a few examples whether the PKS genes (or derived polypeptides) have strong sequence identity with FAS genes or enzymes in the same organism. In several streptomycetes where genes directing the biosynthesis of aromatic polyketide-derived metabolites have been sequenced, the PKS has proven to be a multienzyme complex comparable to the Type II FAS present in many bacteria and in plants (Bibb et al 1989, Sherman et al 1989). Although published data on the FAS of streptomycetes possessing the Type II PKS are so far lacking, some information on the closely related actinomycete *Saccharopolyspora erythraea* suggests that its FAS belongs in Type II (Revill & Leadlay 1991); thus it is likely that in streptomycetes there are both FAS and PKS of this type. However, the deduced sequence of amino acids in the acyl carrier proteins of the putative *S. erythraea* FAS and the streptomycete PKSs showed only about 25% identity. The sequence identity between the acyl carrier proteins in the FASs of *S. erythraea* and *E. coli* or *Rhizobium meliloti* was over 40%. Although these values indicate that the PKS and Type II FAS shared common ancestry, they do not suggest a close evolutionary relationship.

In *S. erythraea* the indigenous polyketide-derived metabolite is erythromycin, a macrocyclic lactone with an aglycone formed from propionyl precursors with partial reduction. It is assembled by a PKS consisting of six synthase units that act sequentially to introduce the propionyl components. The synthase units are linked to form a multifunctional enzyme; each unit is encoded by a DNA module that contains a set of genes directing the appropriate functions of that synthase unit. In the arrangement of its genes and in the enzyme multifunctionality of the encoded protein, each module bears similarities to the Type I FAS of vertebrate animals (Cortes et al 1990, Donadio et al 1991). Moreover, the acyl carrier protein domains in two synthase units of erythromycin PKS do not have strong similarity (25% amino acid sequence identity) to the acyl carrier protein of the putative *S. erythraea* FAS (Revill & Leadlay 1990). Sequence data on the PKS for avermectin biosynthesis in *Streptomyces avermitilis* have shown that it resembles the PKS for erythromycin biosynthesis, and in addition that it bears marked similarities to the animal FAS system (MacNeil et al 1991). Dichotomous evolutionary affinities of PKS and FAS have also been observed in *Penicillium urticae*, which produces the mycotoxin patulin from a polyketide intermediate via 6-methylsalicylate. The PKS for 6-methylsalicylate in this fungus is a multifunctional enzyme, again with marked similarity to the Type I vertebrate FAS (Beck et al 1990, Wang et al 1991). In the organization of functional domains, in the subunit structure of the native enzyme and in the release of the product as a free acid rather than as a coenzyme A ester, the resemblance to vertebrate FAS is closer than to the FAS of *P. urticae*, which possesses typical fungal characteristics (Wiesner et al 1988). Moreover, the derived amino acid sequence in the active site region of the acyl carrier protein domain of the PKS has only 13% identity with the sequence in the fungal FAS, compared to 67% identity with the acyl carrier protein sequence in vertebrate FAS (Wang et al 1991).

While sequence similarities between the FAS and PKS genes in each of these systems again indicate common ancestry, they do not support the concept that secondary metabolic pathway genes have arisen by duplication and modification of similar primary genes in the organism where the pathway is now found. In this respect the evidence is consistent with that on the origin of anthranilate synthases in *P. aeruginosa*. These secondary metabolic pathways may have arisen elsewhere and been acquired through gene transfer. Much of the diversity of structure and function in secondary metabolites, and their species specificity, can then be explained by post-transfer refinement of pathways to optimize the selective advantage they confer. The more recent metabolic elaborations have evidently been shaped by selection in the wide variety of physiological and environmental circumstances of the organisms that acquired the genes.

This scenario raises some intriguing questions. Have certain core secondary metabolic pathways arisen early in evolution and been maintained in reservoir organisms that served as donors for gene transfer? If so, widely distributed pathways with a common biochemistry may prove to have a relatively ancient core component, on to which the more recent processing reactions responsible for the species specificity of the product have been grafted. Some major adaptations may have taken place before transfer. The thiotemplate mechanism by which most secondary metabolic peptides are made has functional similarity to the Type I mechanisms for fatty acid and polyketide synthesis (Kleinkauf & von Döhren 1990). An evolutionary relationship is supported by similarities in derived amino acid sequences obtained by nucleotide sequencing of the genes (Gutierrez et al 1991). In these and other pathways, the early steps associated with intermediates that are enzyme-bound, unstable or of doubtful selective value may have evolved only infrequently, but gene transfer has made the benefits of occasional success widely available.

The distribution of PKS genetic systems related to Type I and Type II FAS has some unexpected features. Recent molecular evidence (Pazirandeh et al 1989) is consistent with the evolution of Type I multifunctional FAS from Type II multienzyme complexes by gene fusion—probably on more than one occasion to account for the Type I variations that have been found in bacteria, fungi, algae and vertebrates (McCarthy & Hardie 1984). However, the persistence of Type II complexes in many bacteria and in plants, where fatty acid synthesis is a chloroplast activity and the genes may have been introduced by endosymbiotic transfer, suggests that the multienzyme complex is advantageous in some prokaryotic or prokaryote-like hosts. Assuming that FAS gene fusion has been limited to certain phylogenetic lineages, the presence of PKS with the characteristics of vertebrate Type I FAS in actinomycetes and in a fungus suggests that these secondary metabolic systems evolved in a lineage within which the vertebrate-type FAS gene fusion had occurred, and were subsequently transferred to other organisms.

Acknowledgement

Research on the biosynthesis of secondary metabolites has been supported by the Natural Sciences and Engineering Research Council of Canada.

References

Beck J, Ripka S, Siegner A, Schiltz E, Schweizer E 1990 The multifunctional 6-methylsalicylic acid synthase gene of *Penicillium patulum*: its gene structure relative to that of other polyketide synthases. Eur J Biochem 192:487–498

Bibb MJ, Biro S, Motamedi H, Collins JF, Hutchinson CR 1989 Analysis of the nucleotide sequence of the *Streptomyces glaucescens tcmI* genes provides key information about the enzymology of polyketide antibiotic biosynthesis. EMBO (Eur Mol Biol Organ) J 8:2727–2736

Case TJ, Gilpin ME 1974 Interference competition and niche theory. Proc Natl Acad Sci USA 71:3073–3077

Cortes J, Haydock SF, Roberts GA, Bevitt DJ, Leadlay PF 1990 An unusually large multifunctional polypeptide in the erythromycin-producing polyketide synthase of *Saccharopolyspora erythraea*. Nature (Lond) 348:176–178

Crosa JH 1989 Genetics and molecular biology of siderophore-mediated iron transport in bacteria. Microbiol Rev 53:517–530

Czelusniak J, Goodman M, Hewett-Emmett D, Weiss ML, Venta PJ, Tashian RE 1982 Phylogenetic origins and adaptive evolution of avian and mammalian haemoglobin genes. Nature (Lond) 298:297–300

Demain AL 1989 Functions of secondary metabolites. In: Hershberger CL, Queener SW, Hegeman G (eds) Genetics and molecular biology of industrial microorganisms. American Society for Microbiology, Washington, DC, p 1–11

Donadio S, Staver MJ, McAlpine JB, Swanson SJ, Katz L 1991 Modular organization of genes required for complex polyketide biosynthesis. Science (Wash DC) 252:675–679

Essar DW, Eberly L, Hadero A, Crawford IP 1990 Identification and characterization of genes for a second anthranilate synthetase in *Pseudomonas aeruginosa*: interchangeability of the two anthranilate synthetases and evolutionary implications. J Bacteriol 172:884–900

Gutierrez S, Diez B, Montenegro E, Martin J-F 1991 Characterization of the *Cephalosporium acremonium pcbAB* gene encoding α-aminoadipyl-cysteinyl-valine synthetase, a large multidomain peptide synthetase: linkage to the *pcbC* gene as a cluster of early cephalosporin biosynthetic genes and evidence of multiple functional domains. J Bacteriol 173:2354–2365

Hopwood DA, Khosla C 1992 Genes for polyketide secondary metabolic pathways in microorganisms and plants. In: Secondary metabolites: their function and evolution. Wiley, Chichester (Ciba Found Symp 171) p 88–112

Hopwood DA, Sherman DH 1990 Molecular genetics of polyketides and its comparison to fatty acid biosynthesis. Annu Rev Genet 24:37–66

Janzen DH 1977 Why fruit rots, seeds mold and meat spoils. Am Nat 111:691–713

Jensen RA 1976 Enzyme recruitment in evolution of new function. Annu Rev Microbiol 30:409–425

Kleinkauf H, von Döhren H 1990 Nonribosomal biosynthesis of peptide antibiotics. Eur J Biochem 192:1–15

Lynch JM 1987 Biological control within microbial communities of the rhizosphere. Symp Soc Gen Microbiol 41:55–82

McNeil DJ, Occi T, MacNeil T et al 1991 Complex organization of the avermectin polyketide synthase genes. Abstracts of the Eighth International Symposium on the Biology of Actinomycetes, Madison WI (abstr PI-053)

McCarthy AD, Hardie DG 1984 Fatty acid synthase—an example of protein evolution by gene fusion. Trends Biochem Sci 9:60–63

Mitchell RE, Johnston JS, Ferguson AR 1981 Phaseolotoxin and other phosphosulfamyl compounds: biological effects. Physiol Plant Pathol 19:227–235

Pazirandeh M, Chirala SS, Huang W-Y, Wakil SJ 1989 Characterization of recombinant thioesterase and acyl carrier protein domains of chicken fatty acid synthase expressed in *Escherichia coli*. J Biol Chem 264:18195–18201

Revill WP, Leadlay PF 1991 Cloning, characterization and high-level expression in *Escherichia coli* of the *Saccharopolyspora erythraea* gene encoding an acyl carrier protein potentially involved in fatty acid biosynthesis. J Bacteriol 173:4379–4385

Rinehart KL 1992 Secondary metabolites from marine organisms. In: Secondary metabolites: their function and evolution. Wiley, Chichester (Ciba Found Symp 171) p 236–254

Schroth MN, Hancock JG 1982 Disease suppressive soil and root-colonizing bacteria. Science (Wash DC) 216:1337–1381

Sherman DH, Malpartida F, Bibb MJ, Kieser HM, Bibb MJ, Hopwood DH 1989 Structure and deduced function of the granaticin-producing polyketide synthase gene cluster of *Streptomyces violaceoruber* Tü22. EMBO (Eur Mol Biol Organ) J 8:2717–2725

Turner JM, Messenger AJ 1986 Occurrence, biochemistry and physiology of phenazine pigment production. Adv Microb Physiol 27:211–275

Vining LC 1990 Functions of secondary metabolites. Annu Rev Microbiol 44:395–427

Vining LC 1992 Secondary metabolism, inventive evolution and biochemical diversity—a review. Gene (Amst) 115:135–140

Wang I-K, Reeves C, Gaucher M 1991 Isolation and sequencing of a genomic DNA clone containing the 3′-terminus of the 6-methylsalicylic acid polyketide synthetase gene of *Penicillium urticae*. Can J Microbiol 37:86–95

Wiesner P, Beck K-F, Ripka S, Muller G, Lucke S, Schweizer E 1988 Isolation and sequence analysis of the fatty acid synthetase FAS2 gene from *Penicillium patulum*. Eur J Biochem 177:69–79

Woese CR 1987 Bacterial evolution. Microbiol Rev 51:221–271

Woese CR, Kandler O, Wheelis MK 1990 Towards a natural system of organisms: proposal for the domains Archaea, Bacteria and Eucarya. Proc Natl Acad Sci USA 87:4576–4579

Wong JT-F 1981 Coevolution of genetic code and amino acid biosynthesis. Trends Biochem Sci 6:33–36

Zähner H, Anke H, Anke T 1983 Evolution and secondary pathways. In: Bennett JW, Ciegler A (eds) Secondary metabolism and differentiation in fungi. Marcel Dekker, New York, p 153–171

DISCUSSION

Hopwood: I tend to agree with your view of evolution, but an extra factor which might have led to the polyketide synthase and the fatty acid synthase in the same species being now more different than those in different species could have been selection that favoured making them distinct, so that transfer of substrates and intermediates between the pathways was avoided. Could that explain some of these sequence relationships?

Vining: You are suggesting that different functions for homologous genes in the same organism might have led to divergence?

Hopwood: Yes. There might have been selection for the genes to diverge more sharply in the same organism, in order to keep the pathways separate. An analogy to this could be the regulatory genes that Keith Chater discussed. There is an activator gene (*act*II-ORF4) in the actinorhodin pathway, and a corresponding gene (*redD*) in the undecylprodigiosin pathway in the same organism, *Streptomyces coelicolor* (Fernández-Moreno et al 1991, Narva & Feitelson 1990). At the protein level, their sequences look very similar, but one will not substitute for the other. On the other hand, if you bring in a member of the same gene family from a different species (*Streptomyces peucetius*), it will substitute for the function of the actinorhodin regulator (C. R. Hutchinson & F. Malpartida, personal communication). You could argue that in the same species there has been selection for them to be distinct, whereas that hasn't applied between species.

Vining: I think this is a valid point. The presence of identical sequences would lead to genetic instability that would tend to eliminate these strains.

Cane: I had a comment along the same lines, but invoking a different means of selection, in terms of quaternary structure, or protein organization. It may be that the range of variability allowable in a multifunctional protein is much narrower than the range for a protein that is organized into subunits. A fatal mistake in just one part of the reading frame basically means the rest of the protein is useless. The similarities between a microbial polyketide synthase and a multifunctional gene for an animal (Type I) fatty acid synthase may reflect the constraints on its organization, rather than its origins. Because essentially you are implying that there is a constant rate of somatic mutation, and that if two things are different, it simply means that even if they have started at the same place, they will spread apart at some rate. If you say that some changes will be more costly than others, in different environments, whether it's within the same organism or just because of the type of catalysis, then you may see different rates of changes. This point was raised earlier by Jack Baldwin in relation to genes in *Penicillium*, where some have diverged more than others (see p 125).

Cavalier-Smith: David Hopwood's suggestion about divergent selection within a species is plausible for the bacterial examples, but the examples of the resemblances between the *Penicillium* PKS and animal FAS genes seem to me to very strongly favour the lateral transfer of one of the genes. It would be hard to explain that by divergence.

Cane: If the similarity is not an expression of who got what from whom, but is due to the fact that the present-day organisms had a common ancestor, and the constraint on them was for the enzyme to stay multifunctional, for whatever reasons are required in terms of what's being synthesized, then they might have diverged more slowly. This was my point.

Cavalier-Smith: The products of both the fatty acid synthase (FAS) and the 6-methylsalicylic acid synthase (6-MSAS) PKS genes of *Penicillium patulum* are multifunctional. They resemble each other by only 13%, yet the 6-MSAS protein resembles the vertebrate FAS by 67%. This is not explained by your hypothesis, Dr Cane.

Vining: McCarthy & Hardie (1984) postulated that the reason why the Type II FAS system has persisted in prokaryotes and the Type I has evolved in eukaryotic cells is that the multienzyme (Type II) complex becomes inefficient in the large cells, through dilution. The trade-off between greater efficiency and the risks from mutation described by Dr Cane favours putting together a multifunctional enzyme in eukaryotes, where fatty acid synthesis takes place in the cytosol, but keeping the ancestral multienzyme organization in the small, compact prokaryotes.

Cane: But this is not required, because *Mycobacterium smegmatis* has a large (Type I) fatty acid synthase.

Vining: Yes; so does *Corynebacterium*, so the distinction between prokaryotes and eukaryotes is not absolute. However, chloroplasts in plants have the Type II FAS system, and fatty acid synthesis in plants is confined to the chloroplasts.

Hopwood: The problem that Tom Cavalier-Smith is highlighting is that it is not just that all these synthases are Type I systems, but that the order of the domains along the length of the proteins is similar.

Cavalier-Smith: Yes, but there is also the incredibly impressive sequence similarity (67%) between the PKS of *Penicillium* and the mammalian Type I FAS, and to explain that in terms of divergence is difficult. I don't see how you could build a plausible phylogenetic tree.

Hopwood: It depends whether you are emphasizing similarities or differences; I was suggesting in my first remark that there may have been a *selection* for divergence.

Vining: The need to diverge should alter the primary as well as the secondary metabolic pathway copies of a gene. It would be useful to compare nucleotide sequences for a primary gene in organisms with and without a related secondary pathway. If divergence is due to selection within an organism, the presence of a secondary pathway should increase the rate of change.

Wächtershäuser: Dr Vining, you made the interesting observation that heterotrophic bacteria are more prone to develop secondary metabolites than autotrophs, and you explained this in terms of competition: when the going gets tough, organisms have to become more inventive. There may be an alternative explanation. If you consider function and biosynthesis, and if you consider that both the functional aspect and the biosynthetic aspect involve cellular machineries, the question arises: which came first? Simple logic suggests that biosynthesis precedes function. However, this would have required something like 'foresight'. If you turn this around and say that the function of a secondary metabolite came before the biosynthesis, there is one explanation that could satisfy such an approach. The original secondary metabolite must have come from outside the organism, which would then have had to be a heterotroph, which took in all kinds of organic compounds; then the heterotroph finds that it has, inside itself, a function, and only later does it find a synthesis for the same or a similar structure.

Vining: I don't think there's any theoretical advantage to the product coming from outside; after all, it would have to be made somewhere else. If we think of the synthesis of this product as a fairly ancient but postribosomal process, then it can be plausibly explained by gene duplication within an organism. Mutations in one copy of the gene created a new metabolic reaction that initiated a pathway to a new product.

Davies: But amino acids were surely present on earth before their biosynthesis.

Vining: Some; but not all.

Davies: Many have been found in primordial soups.

Vining: Amino acids such as glycine, alanine and aspartic acid are formed in appreciable yield by synthesis under simulated abiotic conditions, and may well have been available from the 'soup' for incorporation into primitive proteins. They are also formed by fairly simple biochemical pathways, which could have arisen by retrograde evolution. However, a majority of the 20 protein amino acids (e.g. arginine, methionine, tryptophan) are formed in negligible amounts under simulated abiotic conditions. They are biosynthesized by relatively complex pathways, often in families that originate with one of the 'prebiotic' group of amino acids. The evidence that amino acids in this second group are of postbiotic origin and were formed by inventive evolution, in much the same way that secondary metabolites have probably arisen, is rather persuasive (Parsot et al 1987). For example, various features of the genetic code can be plausibly explained by its expansion to take advantage of the protein amino acids made available by postbiotic inventive evolution (Wong 1981).

Wächtershäuser: Many of the pheromones of insects are derived from the food they take in, from terpenes and so forth. Many of the animal secondary products are also food-derived, such as retinal.

Vining: To extend this idea to microbial secondary metabolites, you would have to say that such compounds were around in the prebiotic era, at the chemical stage of organic synthesis which provided all the substances in the organic soup. Otherwise you are obliged to invoke biosynthesis in some other organism to supply the compounds, and the question about how these compounds arise has not been answered. The difficulty I see in postulating prebiotic synthesis is the need to explain some fairly complicated structures.

It's conceivable that some of the early intermediates in a synthetic pathway might have entered the pathway in that way, but I don't see why they could not equally have arisen by the kind of process that Dr Zähner has proposed, by a kind of biochemical exploration.

Wächtershäuser: Consider ethylene, the simplest secondary product; it is structurally simple, yet its biosynthesis is extremely complicated. Its hormonal action is undoubtedly also complex. The easiest explanation of the emergence of this kind of situation is that the ethylene comes from outside at first; it establishes some sort of a function. Then you get a kind of functional improvement; much later, its biosynthesis evolves.

Vining: I don't disagree with what you are saying, but I don't think you can explain all secondary metabolites in this way. I suspect that there are many reasons for their present-day existence; both types of synthesis (prebiotic and postbiotic) are possible, but only the chemically simple substances are likely to be of prebiotic origin.

References

Fernández-Moreno MA, Caballero JL, Hopwood DA, Malpartida F 1991 The *act* cluster contains regulatory and antibiotic export genes, direct targets for translational control by the *bldA* tRNA gene of Streptomyces. Cell 66:769–780

McCarthy AD, Hardie DG 1984 Fatty acid synthase—an example of protein evolution by gene fusion. Trends Biochem Sci 9:60–63

Narva KE, Feitelson JS 1990 Nucleotide sequence and transcriptional analysis of the *redD* locus of *Streptomyces coelicolor* A3(2). J Bacteriol 172:326–333

Parsot C, Saint-Girons I, Cohen GN 1987 Enzyme specialization during the evolution of amino acid biosynthetic pathways. Microbiol Sci 4:260–262

Wong JT-F 1981 Coevolution of genetic code and amino acid biosynthesis. Trends Biochem Sci 6:33–36

Self-protection mechanisms in antibiotic producers

Eric Cundliffe

Department of Biochemistry, University of Leicester, Leicester, LE1 7RH, UK

Abstract. Various ways in which antibiotic-producing organisms are able to resist the actions of their products are discussed. Examples are given of antibiotic inactivation and also the modification of antibiotic target sites (most notably, ribosomes) to which drugs would otherwise bind and thereby exert their usual inhibitory effects. An interesting variation on the latter theme involves the duplication of target enzymes so that both sensitive and resistant versions are produced, the latter inducibly. Speculative discussion of antibiotic efflux leads to examples of cloned resistance determinants that probably encode components of efflux systems. Although of interest in their own right, resistance mechanisms should not be viewed narrowly when the physiology of antibiotic producers is considered. Thus, chemical modification of drug molecules may not only fulfil a protective role within the cell but may also provide substrates for efflux. Recent evidence that such considerations apply to macrolide antibiotics is presented. The control of resistance in producing organisms is also discussed with particular reference to the induction of novobiocin resistance in *Streptomyces sphaeroides*. This involves the interplay of novobiocin-sensitive and -resistant forms of DNA gyrase and features a promoter that displays a dramatic response to changes in DNA topology.

1992 Secondary metabolites: their function and evolution. Wiley, Chichester (Ciba Foundation Symposium 171) p 199–214

This symposium deals with the function and evolution of secondary metabolites, including antibiotics. Discussion will also centre upon the organization and regulation of the genes involved in antibiotic production, and the origins and rationale of secondary metabolism in general. In one sense, antibiotics are typical secondary metabolites and certain features of their production and its regulation may also apply in a broader context. Even so, the biological activities of antibiotics make them special, because organisms cannot indulge in this particular aspect of secondary metabolism without making adequate provision for their own protection. In reviewing the ways in which this is achieved, particularly in the context of actinomycetes and their production of antibacterial drugs, the aim is for breadth of coverage rather than encyclopaedic depth.

Self-protection mechanisms and their expression

Antibiotic-producing organisms have available a range of options to ensure their own well-being. These include modification or replacement of the targets (enzymes or cellular components) to which the toxic product would normally bind, inactivation or sequestration of intracellular drug molecules, and the utilization of membrane permeability barriers to separate the antibiotics from their targets, perhaps in conjunction with efflux mechanisms. In 'the choice' of any such option(s), the biosynthetic route to a given antibiotic must be considered, because some are synthesized as biologically active molecules inside the cell, whereas others are produced and secreted as inert derivatives that are activated outside the cell. For example, *Streptomyces griseus* produces and secretes 6-phosphorylstreptomycin, which is dephosphorylated (and thereby activated) outside the mycelium by a specific enzyme, also produced by *S. griseus*. Inside the mycelium, streptomycin 6-phosphotransferase, SPH(6), acts upon the 6-OH group of streptidine and subsequent intermediates in the antibiotic biosynthetic pathway, thereby ensuring that they are maintained in the inactive 6-phosphoryl form (Walker & Skorvaga 1973). Streptomycin itself

TABLE 1 Enzymes inactivating autogenous antibiotics in producing organisms (*Streptomyces* species)

Species	*Produced antibiotic*	*Enzyme(s)*[a]
S. griseus	Streptomycin	SPH(6), SPH(3′′)
S. fradiae	Neomycin	APH(3′), AAC(3)
S. ribosifidicus	Ribostamycin	APH(3′), AAC(3)
S. rimosus forma *paromomycinus*	Paromomycin	APH(3′), AAC(3)
S. tenebrarius	Nebramycin	AAC(6′), AAC(2′)
S. hygroscopicus NRRL 2387	Hygromycin B	HPH
S. vinaceus	Viomycin	VPH
S. capreolus	Capreomycin	CPH, CAC
S. alboniger	Puromycin	PAC
S. hygroscopicus ATCC 21705	Bialaphos	DPAT (PAT)
S. antibioticus	Oleandomycin	MGT

[a]SPH(6), streptomycin 6-phosphotransferase (EC 2.7.1.72); SPH(3′′), streptomycin 3′′-phosphotransferase (EC 2.7.1.87); APH(3′), aminoglycoside 3′-phosphotransferase (EC 2.7.1.95); AAC(3), aminoglycoside N^3-acetyltransferase (EC 2.3.1.81); AAC(6′), aminoglycoside $N^{6'}$-acetyltransferase (EC 2.3.1.82); AAC(2′), aminoglycoside $N^{2'}$-acetyltransferase (EC 2.3.1.59); HPH, hygromycin phosphotransferase; VPH, viomycin phosphotransferase (EC 2.7.1.103); CPH, capreomycin phosphotransferase; CAC, capreomycin acetyltransferase; PAC, puromycin acetyltransferase; DPAT (PAT), demethylphosphinothricin (or phosphinothricin) acetyltransferase; MGT, macrolide glycosyltransferase.

For references and additional examples see Cundliffe 1989.

is also a substrate for SPH(6), which can therefore deal with any active form of the antibiotic that makes its way back into the mycelium. This, however, is unlikely to be the whole story. One supposes that streptomycin does not enjoy unimpeded access back into *S. griseus*, for, if it did so, a futile and costly cycle of re- and dephosphorylation would ensue. Antibiotic-inactivating enzymes in other producing organisms (Table 1; see also Cundliffe 1989) may fulfil similar roles to that enacted by SPH(6) in *S. griseus*, but the presence of multiple inactivating enzymes in individual strains remains an enigma.

Streptomycin is an inhibitor of protein synthesis that acts by binding directly to ribosomes which, in *S. griseus*, remain fully sensitive to the drug during antibiotic production (idiophase). However, other organisms that produce anti-ribosomal drugs may not be able to afford such complacency, particularly if active drugs are produced intracellularly. Thus, *Streptomyces azureus* renders its ribosomes (and thereby itself) insensitive to the thiostrepton that it produces by methylating a specific nucleotide within one of the RNA components of the ribosome. In consequence, thiostrepton no longer binds to the ribosome (Thompson et al 1982). Quite apart from the obvious benefit to *S. azureus*, the thiostrepton resistance gene (*tsr*) is also extremely useful to *Homo sapiens* as the selectable marker on many *Streptomyces* cloning vectors.

Ribosomal modification is employed widely but not universally (as noted above) in the protection of organisms that produce inhibitors of protein synthesis (Table 2; see also Cundliffe 1989). Thus far, all documented examples involve methylation of ribosomal RNA at single sites, each characteristic of a given resistance phenotype. Certain other essential cellular enzymes that are normally the targets for antibiotics or toxins are also resistant to the effects of such drugs in the respective producing organisms (Table 3 and Cundliffe 1989), although

TABLE 2 Resistance to autogenous antibiotics resulting from the methylation of ribosomal RNA in producing organisms

Organism	*Antibiotic produced*	*Resistance gene(s)*	*Site of RNA methylation*[a]
Streptomyces azureus	Thiostrepton	*tsr*	23S; A-1067
Saccharopolyspora erythraea	Erythromycin	*ermE*	23S; A-2058
Streptomyces fradiae	Tylosin	*tlrA*	23S; A-2058
S. thermotolerans	Carbomycin	*carB*	23S; A-2058
S. pactum	Pactamycin	*pct*	16S; A-964
S. kanamyceticus	Kanamycin	*kan*	16S; G-1405
S. tenjimariensis	Istamycin	*kamA*	16S; A-1408
S. tenebrarius	Nebramycin	*kgmB* and *kamB*	16S; G-1405 and A-1408

[a]Equivalent site within *Escherichia coli* rRNA.
For references and additional information, see Cundliffe 1989, Ballesta & Cundliffe 1991.

TABLE 3 Resistance to autogenous antibiotics exerted at the level of non-ribosomal target sites

Organism	*Produced antibiotic*	*Resistant target*
Streptomyces sphaeroides	Novobiocin	DNA gyrase
Nocardia mediterranei[a]	Rifamycin	RNA polymerase
Amanita sp.[b]	α-Amanitin	RNA polymerase II
Streptomyces cinnamomeus	Kirrothricin	EFTu
Cephalosporium caerulens[b]	Cerulenin	Fatty acyl synthase
Streptomyces arenae	Pentalenolactone	GPDH
Pseudomonas fluorescens	Pseudomonic acid	Ile-tRNA synthase
Pseudomonas syringae subsp. "*tabaci*"	Tabtoxinine-β-lactam	GS
Pseudomonas syringae pv. *phaseolicola*	Phaseolotoxin	OCT

EFTu, protein synthesis elongation factor Tu; GS, glutamine synthase; GPDH, glyceraldehyde-3-phosphate dehydrogenase; OCT, ornithine carbamoyltransferase.
[a]Now, *Amycolatopsis mediterranei.*
[b]Eukaryotic organism.
For further details and references, see Cundliffe 1989, also Ferguson et al 1980.

in most cases further details are not available. However, glutamine synthase is known to be adenylylated in *Pseudomonas syringae* subsp. "*tabaci*" and thereby rendered resistant to tabtoxinine-β-lactam (Knight et al 1986). Other strains achieve resistance by duplicating the targets of the antibiotic, by producing resistant as well as sensitive versions of the enzymes. For example, *Streptomyces arenae* synthesizes a pentalenolactone-sensitive form of glyceraldehyde-3-phosphate dehydrogenase (GPDH) when grown in complex media but, when pentalenolactone production is induced, a drug-resistant GPDH activity replaces the sensitive one. This change in drug response does not involve modification of pre-existing GPDH. Two distinct genes encoding different isoforms of the enzyme are present in *S. arenae* and their expression is differentially regulated (Fröhlich et al 1989). A conceptually similar situation is encountered in the novobiocin producer, *Streptomyces sphaeroides*, where the mode of induction of resistance has been studied in detail (Thiara & Cundliffe 1988, 1989).

Novobiocin is a coumarin antibiotic that acts upon the B subunit of DNA gyrase, a tetrameric enzyme of the A_2B_2 genre, inhibiting the binding of ATP to the B subunit. Gyrase, also known as DNA topoisomerase II, is a ubiquitous and essential bacterial enzyme that uses the energy of ATP hydrolysis to introduce negative supercoils into relaxed, closed circular duplex DNA *in vitro* (for a review, see Reece & Maxwell 1991). In contrast, DNA topoisomerase I acts in the opposite sense and removes such supercoils. In intact bacteria,

DNA superhelicity is subject to perturbation at each site of transcription (and is also affected by events occurring at replication forks) and is dynamically controlled by the complementary activities of the DNA topoisomerases.

Two *gyrB* genes (encoding DNA gyrase B protein) are present in *S. sphaeroides*, whereas other bacteria, including actinomycetes, normally possess only one (Thiara & Cundliffe 1989). When *S. sphaeroides* is grown in the absence of antibiotic, only novobiocin-sensitive DNA gyrase activity is detectable in mycelial extracts. However, when novobiocin is added to the growth medium (and, presumably, during a novobiocin-production fermentation), drug-resistant DNA gyrase appears. The mechanism of induction of $gyrB^R$ (which encodes the novobiocin-resistant gyrase B protein) was investigated by transplanting the promoter into a plasmid where it controlled the expression of a gene conferring resistance to neomycin. Strains of *Streptomyces lividans* containing this plasmid grew around novobiocin discs on plates containing normally inhibitory concentrations of neomycin. Crucially, however, a similar pattern of growth was seen around ciprofloxacin discs.

Ciprofloxacin is a synthetic fluoroquinolone that is chemically unrelated to novobiocin and would not be expected, *a priori*, to mimic novobiocin in binding to any putative receptor. In fact, DNA gyrase A protein appears to be the principal target of ciprofloxacin (for a review, see Reece & Maxwell 1991), suggesting that induction of the $gyrB^R$ promoter stems from the inhibition of gyrase activity and does not involve classical positive or negative control (Thiara & Cundliffe 1989). The model put forward to account for this effect was derived from an earlier one (Menzel & Gellert 1983) which invoked homeostatic control of the relative levels of DNA topoisomerase I and II in *Escherichia coli*. It is proposed that the addition of novobiocin to a culture of *S. sphaeroides*, already expressing $gyrB^S$ (which encodes the novobiocin-sensitive gyrase B protein), leads to a transient shift in the relative activities of gyrase and DNA topoisomerase I, with a consequent reduction in DNA superhelicity. Normally, in *E. coli* and other bacteria, this would stimulate expression of *gyrA* (which encodes the A subunit of gyrase) and *gyrB*, while reducing the expression of *topA* (which encodes DNA topoisomerase I). In *S. sphaeroides*, this stimulus also turns on (*sic*) the $gyrB^R$ promoter (Thiara & Cundliffe 1989). Many promoters probably respond, one way or the other, to a given change in DNA topology, and those regulating the production of DNA topoisomerases may be more sensitive than most. Apparently, the $gyrB^R$ promoter (especially in its native state in the *S. sphaeroides* genome) is extreme even by such standards and elicits a hair trigger response to changes in DNA topology.

One could, of course, wonder why all this is necessary. Why resort to such interplay between the two DNA gyrase activities? Why does *S. sphaeroides* not simply dispense with $gyrB^S$ and express $gyrB^R$ constitutively? A plausible rationale is suggested by the close proximity of $gyrB^S$ and *gyrA* in the *S. sphaeroides* genome, where they are separated by only 50 bp (A.S. Thiara, personal communication 1991), and may be inextricably linked in the regulatory sense. The archetypal *S. sphaeroides* strain presumably did not possess *gyrB*^{R}, which is not obviously near *gyrB*^{S} in the extant genome although the amino acid sequences of their deduced products are closely similar. Perhaps the resistance gene evolved following the duplication of *gyrB*^{S} and translocation of one homologue within the genome. How such putative events may have been related to evolution of the capacity to produce novobiocin is not clear, but it would be interesting to know whether *gyrB*^{R} is presently clustered with novobiocin-production genes in *S. sphaeroides*.

Induction of resistance genes by antibiotics or their precursors may be fairly common in producing organisms and may not be confined to any one type of resistance mechanism. Alternatively, resistance genes may be co-regulated with those for antibiotic biosynthesis and other genes of secondary metabolism. Such appears to be the case in *S. griseus*, where *strR* (encoding a positive regulator of the streptomycin-biosynthetic gene cluster) and *aphD*, encoding SPH(6), are co-transcribed (Distler et al 1987) as part of a regulatory cascade initiated by A-factor, a low molecular weight γ-butyrolactone 'hormone' that is indispensable for drug-production in *S. griseus* (for review, see Horinouchi et al 1991).

Not all antibiotic resistance genes are inducible, despite the widespread notion that producers are typically sensitive to their products if challenged during the active growth phase. For specific organisms that may well be so, but as a generalization it simply will not do. Two lines of reasoning demand that some resistance genes may be expressed constitutively—perhaps necessarily so. For example, not all antibiotics are produced exclusively during a distinct idiophase that is separable in time from normal growth and replication, even though maximal production may usually occur as cultures enter stationary phase. Therefore, some organisms may need to be resistant to their toxic products on a constitutive basis. Also, the mode of expression of a given resistance gene may be influenced by peculiarities of the encoded mechanism. Thus, some of the methylase enzymes that act on ribosomal RNA, and render producing organisms resistant to anti-ribosomal drugs, do not modify RNA within mature ribosomes. Rather, they act *in vitro* only upon isolated ribosomal RNA and, assuming similar behaviour *in vivo*, presumably modify newly synthesized rRNA prior to its assembly into particles. Given the longevity of ribosomes, this means that effective resistance levels might be attainable with such enzymes only when they are produced constitutively (as in the thiostrepton producer, *S. azureus*) or, at the very least, when they appear well ahead of drug production. In organisms where such enzymes are produced inducibly (e.g. the *tlrA* product in *Streptomyces fradiae*, producer of tylosin) one expects, and finds, additional resistance mechanisms (Baltz & Seno 1988, Zalacain & Cundliffe 1991).

Antibiotic efflux

The manner in which antibiotics cross cell surfaces has received almost unilateral attention. That is, uptake mechanisms have been studied in some detail, although not to the extent that they deserve, but the fact that many antibiotics can also get out of bacterial cells has attracted comparatively less attention. Yet that is implicit in the operational definition of a 'bacteriostatic' agent. (Those agents that cannot get out of cells in drug-free media are bactericidal unless, of course, they get detoxified.) Antibiotics penetrate into cells either by direct diffusion or by hitching a ride on transport systems that normally serve other needs. The essence of the argument here is that similar considerations may also apply to the way(s) in which antibiotics get out of cells, and that efflux systems utilized by some producing organisms may involve components whose homologues may be quite widespread among bacteria, although in producers they may be supplemented with additional components to establish active, rather than passive, processes.

Passive drug efflux mechanisms, perhaps mediated by membrane-associated components and/or transmembrane proteins, could serve the needs of some antibiotic-producing organisms, especially in the wild where production levels may be low and the secreted product may be free to diffuse away. However, the performance of industrial strains in closed fermenters implies the existence of active mechanisms in order to drive drug export against concentration gradients. Most simply, this could involve the coupling of metabolic energy to otherwise passive systems through ATP-binding proteins, membrane potentials or the like. And, as an added refinement, specific antibiotic-binding proteins might serve the dual function of helping to protect the cell and more efficiently delivering drug molecules to the transporter. Recently, as a result of sequence analysis, genes that probably contribute to such modular efflux systems have been identified in various antibiotic producers. Thus, *drrA* from *Streptomyces peucetius* (producer of daunorubicin) and *tlrC* from the tylosin producer, *S. fradiae*, apparently encode ATP-binding proteins similar to those involved in ribose uptake into, or haemolysin secretion from, *E. coli* (Guilfoile & Hutchinson 1991, Rosteck et al 1991). There is also similarity to putative ATP-binding domains in the mammalian P-glycoprotein (Mdr1) that is responsible for the efflux of cytotoxic compounds (including daunorubicin) from drug-resistant tumour cells. Significantly, *drrA* confers resistance to daunorubicin only in the presence of *drrB*, the product of which is extremely hydrophobic and probably membrane associated (Guilfoile & Hutchinson 1991). Similarly in *S. fradiae* the products of *tlrB* and *tlrC* may also act in concert, and one possible function for the former (apparently a cytoplasmic protein) would be that of a soluble drug receptor (Rosteck et al 1991).

Some antibiotic efflux systems may not be modular. Several producing organisms have yielded genes whose deduced products resemble that of *E. coli tet A* and appear to be large proteins with multiple transmembrane domains.

Such genes include *mmr*, the methylenomycin resistance gene from plasmid SCP1 of *Streptomyces coelicolor* (Neal & Chater 1987), *otrB* from the oxytetracycline producer *Streptomyces rimosus* (Reynes et al 1988), *act*II-ORF2 from the actinorhodin gene cluster of *Streptomyces coelicolor* (Fernández-Moreno et al 1991) and *tcmA* from the tetracenomycin C producer, *Streptomyces glaucescens* (P.G. Guilfoile & C.R. Hutchinson, personal communication 1991). Like the TetA protein, the products of these respective genes may utilize transmembrane electrochemical gradients (e.g. proton-motive forces) and act as drug efflux pumps. Accessory proteins may also be involved. Thus, the product of *act*II-ORF3 appears to complement that of *act*II-ORF2 in the secretion of actinorhodin from *S. coelicolor* (Fernández-Moreno et al 1991).

Before we leave the subject of antibiotic efflux, it is germane to consider the nature of the transported substrates. No doubt some antibiotics are secreted as the active molecular species, but others clearly are not. The efflux of streptomycin from *S. griseus* in an inert phosphorylated form has already been mentioned, and it is likely that other antibiotics listed in Table 1, particularly aminoglycosides of the neomycin family (such as ribostamycin and paromomycin), are also secreted from the respective producers in derivatized form. Similar considerations probably apply to macrolides. Extracts of the oleandomycin producer, *Streptomyces antibioticus*, inactivate the drug if supplemented with UDP-glucose, and an extracellular glucosidase (also produced by *S. antibioticus*) can re-potentiate the inactivated drug (Vilches et al 1992). These observations strongly suggest that oleandomycin is exported from the producer as an inert glucosylated derivative. Earlier (Kuo et al 1989), erythromycin had been shown to be inactivated by 2′-*O*-glucosylation in *Streptomyces vendargensis* (an organism not known to produce any macrolide), and the macrolide glycosyltransferase (MGT) in *S. antibioticus* probably has a similar mode of action. There is also an inducible MGT in *S. lividans*, again active on the 2′-OH of its substrates (Cundliffe 1992) and the gene encoding it (*mgt*) has been cloned and sequenced (Jenkins & Cundliffe 1991). An obvious homologue, displaying over 80% sequence identity to *S. lividans mgt*, is present in the spiramycin producer, *Streptomyces ambofaciens* (J.-L. Pernodet, personal communication 1991) and the latter strain also contains MGT activity (L.A. Merson-Davies & E. Cundliffe, unpublished data 1992). Interestingly, however, this enzyme differs in substrate preference from that in *S. lividans*. It is therefore tempting to speculate that macrolide producers collectively utilize a family of MGT enzymes with substrate specificities that reflect the nature of the respective autogenous drugs. If so, one would expect that oleandomycin is not the only macrolide to be secreted in glycosylated form.

Summary

In order to avoid self-intoxication, antibiotic-producing organisms deploy a range of defence mechanisms that, broadly, illustrate two strategies. On the one hand, cells may be (or become) de-sensitized to the potentially toxic effects of their products; alternatively, antibiotic levels within the cell may be controlled within tolerable limits. In the latter context, efficient mechanisms of drug efflux (the subject of much current interest) may be complemented by the inactivation or sequestration of intracellular drug molecules. Presumably also, in order to avoid energetically wasteful futile cycles of repeated uptake and efflux, permeability barriers to drug uptake may be quite common among antibiotic producers. The conceptual alternative to the inactivation and/or exclusion of antibiotics involves modification or replacement of the cellular target sites to which drugs normally bind and through which they exert their characteristic inhibitory effects. Commonly, however, multiple resistance mechanisms are employed by antibiotic producers, even by those that elaborate only a single toxic product, and no one type of mechanism should be considered, *a priori*, to preclude the complementary use of another.

Expression of resistance mechanisms in antibiotic-producing organisms may be constitutive or inducible and, again, there are two conceptually different models. In some instances, it is the first appearance of the antibiotic product (or one of its precursors) that triggers the expression of resistance. Otherwise, resistance genes may be co-regulated with those for antibiotic production, with which they are commonly clustered in the genome.

Acknowledgements

Work from the author's laboratory cited in this article was supported by the Science and Engineering Research Council, The Wellcome Trust and Eli Lilly and Co.

References

Ballesta JPG, Cundliffe E 1991 Site-specific methylation of 16S rRNA caused by *pct*, a pactamycin resistance determinant from the producing organism, *Streptomyces pactum*. J Bacteriol 173:7213–7218

Baltz RH, Seno ET 1988 Genetics of *Streptomyces fradiae* and tylosin biosynthesis. Annu Rev Microbiol 42:547–574

Cundliffe E 1989 How antibiotic-producing organisms avoid suicide. Annu Rev Microbiol 43:207–233

Cundliffe E 1992 Glycosylation of macrolide antibiotics in extracts of *Streptomyces lividans*. Antimicrob Agents Chemother 36:348–352

Distler J, Ebert A, Mansouri K, Pissowotzki K, Stockmann M, Piepersberg W 1987 Gene cluster for streptomycin biosynthesis in *Streptomyces griseus*: nucleotide sequence of three genes and analysis of transcriptional activity. Nucleic Acids Res 15:8041–8056

Ferguson AR, Johnston JS, Mitchell RE 1980 Resistance of *Pseudomonas syringae* pv. *phaseolicola* to its own toxin, phaseolotoxin. FEMS (Fed Eur Microbiol Soc) Microbiol Lett 7:123–125

Fernández-Moreno MA, Caballero JL, Hopwood DA, Malpartida F 1991 The *act* cluster contains regulatory and antibiotic export genes, direct targets for translational control by the *bldA* tRNA gene of *Streptomyces*. Cell 66:769–780

Fröhlich K-U, Wiedmann M, Lottspeich F, Mecke D 1989 Substitution of a pentalenolactone-sensitive glyceraldehyde-3-phosphate dehydrogenase by a genetically distinct resistant isoform accompanies pentalenolactone production in *Streptomyces arenae*. J Bacteriol 171:6696–6702

Guilfoile PG, Hutchinson CR 1991 A bacterial analog of the *mdr* gene of mammalian tumor cells is present in *Streptomyces peucetius*, the producer of daunorubicin and doxorubicin. Proc Natl Acad Sci USA 88:8553–8557

Horinouchi S, Miyake K, Hong S-K, Vujaklija D, Ueda K, Beppu T 1991 Regulation by A-factor and *afsR* of secondary metabolism and morphogenesis in *Streptomyces*. Actinomycetologica 5:119–125

Jenkins G, Cundliffe E 1991 Cloning and characterisation of two inducible genes from *Streptomyces lividans*, responsible for resistance to lincomycin and macrolide antibiotics. Gene (Amst) 108:55–62

Knight TJ, Durbin RD, Langston-Unkefer PJ 1986 Role of glutamine synthase adenylylation in the self-protection of *Pseudomonas syringae* subsp. "tabaci" from its toxin, tabtoxinine-β-lactam. J Bacteriol 166:224–229

Kuo M-S, Chirby DG, Argoudelis AD, Cialdella JI, Coats JH, Marshall VP 1989 Microbial glycosylation of erythromycin A. Antimicrob Agents Chemother 33:2089–2091

Menzel R, Gellert M 1983 Regulation of the genes for *E. coli* DNA gyrase: homeostatic control of DNA supercoiling. Cell 34:105–113

Neal RJ, Chater KF 1987 Nucleotide sequence analysis reveals similarities between proteins determining methylenomycin A resistance in *Streptomyces* and tetracycline resistance in eubacteria. Gene (Amst) 58:229–241

Reece RJ, Maxwell A 1991 DNA gyrase: structure and function. Crit Rev Biochem Mol Biol 26:335–375

Reynes JP, Calmels T, Drocourt D, Tiraby G 1988 Cloning, expression in *Escherichia coli*, and nucleotide sequence of a tetracycline-resistance gene from *Streptomyces rimosus*. J Gen Microbiol 134:585–598

Rosteck PR Jr, Reynolds PA, Hershberger CL 1991 Homology between proteins controlling *Streptomyces fradiae* tylosin resistance and ATP-binding transport. Gene (Amst) 102:27–32

Thiara AS, Cundliffe E 1988 Cloning and characterization of a DNA gyrase B gene from *Streptomyces sphaeroides* that confers resistance to novobiocin. EMBO (Eur Mol Biol Organ) J 7:2255–2259

Thiara AS, Cundliffe E 1989 Interplay of novobiocin-resistant and -sensitive DNA gyrase activities in self-protection of the novobiocin producer, *Streptomyces sphaeroides*. Gene (Amst) 81:65–72

Thompson J, Schmidt F, Cundliffe E 1982 Site of action of a ribosomal RNA methylase conferring resistance to thiostrepton. J Biol Chem 257:7915–7917

Vilches C, Hernandez C, Mendez C, Salas JA 1992 Role of glycosylation and deglycosylation in biosynthesis of and resistance to oleandomycin in the producer organism, *Streptomyces antibioticus*. J Bacteriol 174:161–165

Walker JB, Skorvaga M 1973 Phosphorylation of streptomycin and dihydrostreptomycin by *Streptomyces*. J Biol Chem 248:2435–2440

Zalacain M, Cundliffe E 1991 Cloning of *tlrD*, a fourth resistance gene, from the tylosin producer, *Streptomyces fradiae*. Gene (Amst) 97:137–142

DISCUSSION

Turner: You discussed streptomycin phosphotransferases being part of the biosynthetic gene cluster. Are you suggesting that streptomycin is exported as the phosphorylated derivative and dephosphorylated outside the cell?

Davies: Surely it's not exported as the phosphorylated form; as I understand it, the phosphate is removed *during* the export process.

Turner: If so, how should one envisage the process happening?

Cundliffe: The important point is whether the activating enzyme is extracellular or not. The literature on streptomycin clearly states that there are extracellular phosphatases that can re-activate 6-phosphorylstreptomycin—and Wolfgang Piepersberg is agreeing with me as I speak!

Turner: How do you envisage the situation in a β-lactam-producing streptomycete such as *S. clavuligerus*? Japanese workers (Kimura et al 1990) identified a β-lactamase gene next to the cluster of the β-lactam biosynthetic genes. It seems a 'sledgehammer' approach to make a β-lactamase right next to the antibiotic. How can it handle that and keep the antibiotic and enzyme apart, without destroying the antibiotic that it makes?

Cundliffe: The answer is that your guess is as good as mine!

Vining: Streptomyces clavuligerus also excretes a protein that binds β-lactamase (Doran et al 1990); so the β-lactamase is taken out of the picture when cephamycin is being made.

Demain: Also, clavulanic acid is a low molecular weight inhibitor of β-lactamase. There's a real interplay of enzymes and small molecules and proteins in *S. clavuligerus*, and how they are regulated is a very interesting question.

Nisbet: Who is to say that these antibiotic-inactivating enzymes aren't just important regulatory proteins of some sort, the function of which is regulated by the antibiotics? One should perhaps think of β-lactamases in this context. It's rather strange that when many of the β-lactamases and the β-lactamase inhibitors interact, the latter actually inhibit the hydrolytic enzyme. Something that's just a processing enzyme would presumably take the substrate, split it and spew it out, and not be inhibited by the substrate that it has interacted with, unless these proteins have some other functions and the β-lactamase inhibitors provide a way of regulating that function.

Cundliffe: That's an interesting suggestion. However, what might be true for β-lactamase in the context in which you raise it may not be true for the macrolide glycosyltransferase, or for aminoglycoside phosphotransferases. I would hesitate to generalize too much in that direction, although it seems plausible.

Davies: But isn't it true that β-lactamases are made by almost every microbial species and may have nothing to do with β-lactam synthesis?

Turner: Perhaps, but if the β-lactamase gene is located next to all the other genes on the antibiotic cluster, it suggests a self-protection mechanism.

Cane: What is known about the mechanisms of polyether self-resistance?

Leadlay: Iain Hunter and his colleagues (Linton & Hunter 1988) have devised a way to test for the conferral of resistance to polyethers in *Streptomyces lividans*, and have managed to clone two separate pieces of DNA from the tetronasin producer, *Streptomyces longisporoflavus*, which each confer resistance to the polyether in the presence of high sodium levels. Together, we have now shown that one of these resistance determinants encodes a system with an ATP-binding component and a hydrophobic component, very similar to other ATP-dependent transport systems (K.J. Linton, H.N. Cooper, I.S. Hunter & P.F. Leadlay, unpublished work). Self-resistance is obviously a particularly sharp question for an ionophore-producing organism, because of the mode of action of the antibiotic. My own view is that any resistance mechanism has to be very specific, because in the end the ionophore has to kill other bacteria. If an ionophore can enter any streptomycete and promptly be pumped out, it will be ineffective. But there is a second resistance determinant in *S. longisporoflavus*, about which we don't know anything.

Cane: John Robinson's lab. and my lab. have also tried to feed very late-stage presumptive intermediates in monensin formation (which have a hydroxyl instead of a methoxyl, or a methyl instead of a hydroxymethyl: Ashworth et al 1989). Both intermediates are produced either by mutants or in a normal culture, and yet neither can serve as an exogenous precursor. This may mean of course that they are not precursors; it's hard to see any grounds why one or the other would not be. One possibility is that they are simply seen as polyethers and are prevented from getting anywhere near the biosynthetic machinery, and are just kept out of the cell.

Cundliffe: Clearly, these organisms don't want to indulge in futile cycles, whereby they make a drug, put it out, then let it back in, and spend energy in putting it out again. One supposes that there will be, at least in some producing organisms, barriers to the readmission of the drug. It's easy to see how that can be arranged if the uptake of such a drug is normally carrier mediated. It's less easy to see (except in the trivial context of putting an extra membrane around the cell and making it Gram negative) how this is done for any antibiotic that might have the capacity to get in by simple diffusion through a lipid bilayer, not obviously carrier mediated. How many antibiotics do that, I don't know. It's a possibility that one has to consider.

Piepersberg: I am wondering how strongly you would support the suggestion that one part of resistance mechanisms might be derived from the biosynthetic machinery for the toxic compound? For example, with the antibiotic bialaphos, acetylation is one of the intermediate steps, which is in fact designed for protection against the formation of a toxic intermediate, which afterwards is inactivated by adding the two alanine residues. These, in turn, are needed to sharpen up the product as a weapon, because otherwise it wouldn't be taken up by the putative target cells. Which do you think, in terms of the evolution of this particular pathway, was the stronger selective advantage—the protective effect, or the biosynthetic effect?

Cundliffe: I can't answer that, but I am very happy with what you say. It fits the model. Consider the biosynthesis of the compound bialaphos as an acetylated compound. Part way through the biosynthetic pathway, in order for the producing organism to be shielded metabolically, a toxic precursor of bialaphos is detoxified by acetylation. This is entirely analogous with what happens with streptomycin (although, as you point out, there is the additional feature that the subsequent elaboration of the bialaphos molecule involves adding a couple of alanines, whereas, for streptomycin, it involves adding a couple of sugars).

Davies: Another interesting point about bialaphos is that an acetyl residue is removed before the antibiotic is secreted. So the producing organism may use two different modifications to permit protection and de-protection of the antibiotic molecule during synthesis.

Cavalier-Smith: If one presumes that at the time of the origin of a new antibiotic, the producer cell must be less sensitive to it than the target, one can imagine a scenario involving lateral gene transfer in which a fungus first evolved the antibiotic, for example a β-lactam, directed against competitive bacteria; then some bacteria evolved defence mechanisms of various sorts and as a result became able to take up the antibiotic biosynthesis genes of the fungi, by lateral transfer, and then perhaps even send them on to other bacteria. This could get over some of the initial problems more easily than if a cell that already had a peptidoglycan wall started making the antibiotic directed against its own wall. What do you feel about that sort of scenario?

Cundliffe: We have commented in the past on the chicken-and-egg situation that one finds whenever there is a resistance mechanism for which the gene does not seem to be part of the biosynthetic pathway. The example we usually give is thiostrepton resistance, which involves methylation of a ribose residue in a specific nucleotide within ribosomal RNA. There is no evidence that the enzyme catalysing such methylation binds thiostrepton, or that the reaction plays any role in the biosynthesis of thiostrepton, although we know very little about this. Nor is there evidence that *Streptomyces azureus* gains any advantage other than thiostrepton resistance from possession of the methylase gene; so did the future thiostrepton producer 'decide' to have the resistance gene before it decided to make the antibiotic? You can't talk about the simultaneous acquisition of the resistance gene and the biosynthetic genes from somewhere else; this just pushes the argument one step further away, unless you suppose that the evolution of thiostrepton production occurred originally in an organism which was of such a type that it wouldn't have been sensitive to the antibiotic anyway, like a eukaryote.

Cavalier-Smith: This could happen, in theory, for the ribosome-directed antibiotics, but in most cases there isn't evidence that eukaryotes make the antibiotic. That's why I mentioned β-lactam, because this is an example where the antibiotic is made both by sensitive organisms and by ones, namely the fungi, that couldn't be sensitive because they do not have peptidoglycan walls.

Cundliffe: Lateral transfer is a much better proposition there than in other cases. We had the idea at one stage that because thiostrepton resistance involves pentose methylation (and this is relatively rare as a modification of ribosomal RNA in bacteria, but extremely common in eukaryotes), perhaps the thiostrepton resistance enzyme was originally a eukaryotic enzyme, and is therefore responsible for the uniform resistance of eukaryotic ribosomes to this agent. I should emphasize that we have no evidence to support this idea!

Orgel: All the mechanisms that you discussed involve the initial active development of a resistance mechanism. Is there a possibility of the completely fortuitous appearance of resistance mechanisms? There are, for example, nucleotide analogues that are accepted by one DNA polymerase but not by another. Could you start off by creating a molecule to which the producer's enzyme is not sensitive, but to which the enzyme of a competitor happens to be sensitive? Then, before the competitor organism has time to become resistant, the inventer would have a reason for improving its offensive strategy and inventing a defence mechanism. Could resistance evolve in a two-step way like that?

Cundliffe: I would imagine so. Then, to transpose the argument to a different type of drug receptor, I suppose one could imagine that some organisms elaborated drugs which are now revealed to us as ribosome inhibitors, but they initially elaborated them as compounds with very poor activity against ribosomes; then there would be, from some point, co-evolution of the resistance mechanism *and* the biosynthetic pathway, to give a more toxic drug.

Beppu: Professor Cundliffe, you showed several examples of efflux mechanisms, via transmembrane proteins. Do they show distinct homology with haemolysin B, or the P-glycoprotein?

Cundliffe: No; those that most clearly resemble P-glycoprotein are the ones that we know as ATP-binding proteins (because the P-glycoprotein is also an ATP-binding protein). There are some antibiotic targets where it is difficult to see how you could change them to make them insensitive to drugs, and DNA is one of them, because most of the drugs that intercalate into DNA are not very site specific. Therefore it was knowing of the existence of things like P-glycoprotein, where there is a rapid efflux of anti-tumour agents (cytotoxic drugs) from mammalian tumour cells, that suggested that something like this could happen in the actinomycetes that make those cytotoxic agents—for example, in bacteria like the daunorubicin producer *Streptomyces peucetius*. So it wasn't such a surprise when they were found.

Beppu: We recently obtained a gene showing clear homology (about 30% identity) with *E. coli* haemolysin B protein. This gene was obtained as a positive regulatory gene for sporulation in *S. griseus*. This kind of gene was also identified as being involved in sporulation in *Bacillus subtilis*. This type of protein, known as the ATP-dependent efflux pump, could have another such physiological role.

Cundliffe: Exactly.

Zähner: I have two questions in relation to the efflux systems. How similar are these systems that you discussed, and how specific are they?

Cundliffe: They are all fairly similar. The *tlrC* gene of *Streptomyces fradiae* is closely similar to the *carA* gene of *Streptomyces thermotolerans*, and so on.

What is the specificity of these effluxes? This is a very good question. Very few people have reported on that. It has usually been a case of trying the obvious substrate and saying that we get resistance to this antibiotic in clones. But in fact, the *carA* gene, which comes from the carbomycin producer and gives carbomycin resistance in *Streptomyces lividans*, gives hypersensitivity to erythromycin and tylosin in *S. lividans*. This was the first indication that the *carA* resistance mechanism had something to do with the way drugs cross cell surfaces. In one case (e.g. carbomycin) the drug may either be kept out or be driven out; in another case (e.g. tylosin) the drug may be let in.

I would like to know, for example, how broad a resistance phenotype is conferred by the gene (*drrA*) encoding the putative ATP-binding protein involved in daunomycin production and resistance. I don't know whether anyone has tried compounds that are not of the daunomycin–adriamycin type to see whether this gene can confer resistance to anything else. So we do not yet have the answer to your question.

Hunter: Work done by my group may provide you with a partial answer. We tested the tetronasin resistance gene (see p 209) for cross-resistance with monensin. There was no cross-resistance.

Piepersberg: The other group (besides the ATP-dependent transporters) of transmembrane protein exporters shows quite a wide range of substrate specificities. We have a system of low substrate specificity in staphylococci, the QacA protein, which exports various drugs (toxic dyes and antiseptic compounds) and is very similar in its protein primary structure to the Mmr type of resistance determinants (Rouch et al 1990). But in other cases, very high substrate specificity is observed within that same family of proteins. We have recently described a resistance determinant (*LmrA*) from the lincomycin producer (*S. lincolnensis*) which exports lincomycin; this putatively proton gradient-driven exporter is very specific for this antibiotic and doesn't export even very similar compounds, such as clindamycin or celesticetin (Zhang et al 1992).

Escoubas: Professor Cundliffe, you said that often self-protection mechanisms are multiple in one organism. Is it possible to speculate that they would also represent natural defence mechanisms against a competitor's compounds? This goes back to the question of how flexible the system needs to be to accommodate different substrates, in order to detoxify them, or to pump them out of the cell, or prevent them entering cells.

Cundliffe: If an organism possessed more than one gene which, when cloned in *S. lividans*, revealed itself as a resistance determinant, then by examining the resistance character of those clones, you could speculate that the original organism might, fortuitously perhaps, be resistant to a range of antibiotics. I am thinking of cases like the nebramycin producer, *S. tenebrarius*, which has two different rRNA methylases and two aminoglycoside acetyltransferases. This organism is therefore broadly resistant to aminoglycosides, but is perfectly sensitive to macrolides and other inhibitors of protein synthesis; so it doesn't have resistance beyond that group of compounds. So the answer to your question is that within certain limits, this seems very possible.

Chater: When I discussed the complexity of biosynthetic genes in terms of transcription units, earlier (p 144–156), I wanted to stress that this would be compatible with the idea of a modular assembly of the gene clusters. And if the modular assembly and reassembly of clusters goes on during the development of differences in antibiotic pathways, there is an adaptive benefit for each individual module to carry around with it relevant resistance genes, because when the module goes into a new context, it may be causing production of a new compound to which the host would otherwise be sensitive.

Demain: I am wondering why uridine and cytidine are not methylated. Do all the examples of resistance concern only adenosine and guanosine?

Cundliffe: In the very highly conserved sequences in ribosomal RNA, purines are represented more frequently than pyrimidines and every one of the resistance sites that we have found in rRNA is in a highly conserved sequence: because, if it wasn't, it wouldn't be stopping the drug from binding into one of the active sites of the ribosomal enzyme.

References

Ashworth DM, Holmes DS, Robinson JA, Oikawa H, Cane DE 1989 Selection of a specifically blocked mutant of *Streptomyces cinnamonensis*. Isolation and synthesis of 26-deshydroxymonensin A. J Antibiot (Tokyo) 42:1088–1099

Doran JL, Leskiw BK, Aippersbach S, Jensen SE 1990 Isolation and characterization of a β-lactamase-inhibitory protein from *Streptomyces clavuligerus* and cloning and analysis of the corresponding gene. J Bacteriol 172:4909–4918

Kimura H, Miyanoshita H, Sumino Y 1990 DNA and applications thereof. Japan Patent Application No. 2-3762

Linton KJ, Hunter IS 1988 Cloning of DNA sequences from *Streptomyces longisporoflavus* which confer increased resistance to the ionophore antibiotic tetronasin in *Streptomyces lividans*. Heredity 61:304–305

Rouch DA, Cram DS, DiBerardino D, Littlejohn TG, Skurray RA 1990 Efflux-mediated antiseptic resistance gene *qacA* from *Staphylococcus aureus*: common ancestry with tetracycline- and sugar-transport proteins. Mol Microbiol 4:2051–2062

Zhang H-Z, Schmidt H, Piepersberg W 1992 Molecular cloning and characterization of two lincomycin resistance genes, *lmrA*, and *lmrB*, from *Streptomyces lincolnensis* 78-11. Mol Microbiol 6:2147–2158

Useful functions of microbial metabolites

Louis J. Nisbet

Xenova Limited, 240 Bath Road, Slough, Berks SL1 4EF, UK

Abstract. The mood-enhancing effects of fungi and their medicinal properties have been recognized for centuries. Ergot was initially used by midwives to speed childbirth in the Middle Ages. More recently their pharmacological action on dopamine receptors has been exploited to treat post-partum bleeding, migraine, Parkinson's disease and senile dementia. Further indications of the potential value of microbial metabolites are exemplified by the discovery and development of cyclosporin, to treat organ rejection, and mevinolin, a cholesterol-lowering drug. Such discoveries are not unexpected because we have known for some time that fungi regulate morphogenesis, differentiation and sexuality via hormonal molecules, ranging from peptides through to steroidal molecules similar in structure to human sex hormones. A combination of the power of molecular biology to design screens based on isolated disease mechanisms with the chemical inventiveness of microorganisms is providing numerous new pharmacophores for drug development.

1992 Secondary metabolites: their function and evolution. Wiley, Chichester (Ciba Foundation Symposium 171) p 215–235

From the outset, I wish to dispel two myths: the first is that microbes make secondary metabolites to kill off other competing organisms and are therefore only of value in searching for new antibiotics; the second is that it is not possible to find molecules that selectively interfere with protein–protein interactions such as those between growth factors and their receptors.

Martin & Demain (1978) provided one of the most erudite definitions of secondary metabolites, although this is really a set of observations about secondary metabolites and does not address their origins or their functions: 'Secondary metabolites are those metabolites which are often produced in a phase subsequent to growth, have no function in growth (although they may have a survival function), are produced by certain taxonomic groups of microorganisms, have unusual chemical structure, and are often formed as mixtures of closely related members of a chemical family. Whereas primary metabolism is basically the same for all living systems, secondary metabolism is restricted to plants and microorganisms and is often species or strain specific.'

Zähner et al (1982) described secondary metabolism as a games room for molecular roulette, and others have proposed hypotheses that attempt to

TABLE 1 Rationale for secondary metabolism

Derailment of metabolism
Biochemical dead-end
By-product of primary metabolism
Detoxification mechanism
Result of limited growth
Result of weak regulation of metabolism

explain how secondary metabolites are produced (Table 1). These proposals do not address the evolutionary significance of secondary metabolites or the advantages that they confer on the producing microorganisms or plant.

Believers in theistic evolution might consider that God experimented first in simpler organisms and, building on success, eventually put together the more complex biological systems that make up the human being! Whatever one's theological disposition, it is a matter of experimental observation that simple microorganisms do have receptors and enzymes and many of them have structures and functions in common with those in mammalian systems. It is therefore no surprise that regulatory molecules are found in simple organisms. Indeed, the regulatory properties of microbial metabolites are the *raison d'être* for secondary metabolites.

Roth et al (1986) and other workers have provided a tremendous body of evidence linking human receptor structure and function with those of unicellular organisms. The ligands for a number of human cell receptors, expressed either on the surface or intracellularly, interact with homologous receptors in simple prokaryotes and eukaryotes, including fungi (Table 2). The application of

TABLE 2 Binding of radiolabelled vertebrate ligands to receptors in microbes

Radioactive ligand	*Prokaryote*	*Eukaryote*
Human chorionic gonadotropin (HCG)	*Pseudomonas maltophilia*	—
Thyroid-stimulating hormone (TSH)	*Escherichia coli* *Yersinia enterocolitica*	—
Opioids	—	*Amoeba proteus*
Corticosterone	—	*Candida albicans*
Oestrogen	—	*Saccharomyces cerevisiae* *Paracoccidioides brasiliensis*
Antheridiol	—	*Achlya ambisexualis*

typical Scatchard analysis, exemplified by the binding of corticosterone to *Candida albicans*, yields a maximum binding coefficient (518 fmol/mg) and a dissociation constant (5.5 nM) which are typical for molecules that bind to receptors. In retrospect, it should probably have been recognized long ago that molecules from microbes would have a potential for the regulation of mammalian systems, because small molecules regulate microbial growth, differentiation and sexuality.

Fungi have different mating types, male and female, and the interaction of these mating types is regulated by small hormonal molecules, ranging from simple peptides to steroids structurally analogous to mammalian steroid sex hormones. The dissociation constants of these different molecules lie within the nanomolar (10^{-9}) to femtomolar (10^{-15}) range, characteristic of receptor-mediated processes (Table 3).

The cellular responses to these events are highly analogous to our own, so that, for example, regulators of mammalian cell signal transduction and second messenger production can be expected to have closely related, if not identical, counterparts in fungi. In the light of the foregoing comments it would be logical to view secondary metabolites in terms of a pharmacophore [Gr. *pharmakon*—medicine + Gr. *phore*—bearing]: 'the group of atoms in a drug molecule which is responsible for the action of the compound' (Dorlands Illustrated Medical Dictionary 1981). In the case of the penicillins, cephalosporins and other β-lactam antibiotics, then, the pharmacophore is the β-lactam nucleus that is common to all of these structures. Other rings and substitutions affect the potency, spectrum of action, β-lactamase stability and β-lactamase inhibition properties, but the fundamental activity is driven by the β-lactam nucleus.

In the case of the ryegrass endophyte *Acremonium loliae*, the microorganism synthesizes diverse molecules that are not commonly found in *Acremonium* species: tryptophol, a plant growth hormone analogue which increases biomass and increases drought resistance; lolitrem B, the tremorgenic compound that causes neurological disorders in sheep and cattle; as well as peramine, an insect

TABLE 3 Microbial pharmacophores

Hormone	*Producer*	*Chemical type*	*Activity level (M)*
Antheridiol	*Achlya bisexualis*	Steroid	1×10^{-11}
Oogonial	*Achlya bisexualis*	Steroid	1×10^{-10}
Sirenian	*Allomyces arbuscula*	Alicyclic	1×10^{-11}
Trisporin acid C	*Blakeslea trispora*	Alicyclic	1×10^{-16}
α-Factor	*Saccharomyces cerevisiae*	Peptide	1×10^{-11}
a-Factor	*Saccharomyces cerevisiae*	Peptide	4×10^{-9}
Tremerogen A-10	*Tremella mesenterica*	Lipo-peptide	2×10^{-9}

antifeedant. It is notable that ryegrass infected with the endophyte is much healthier, taller and more disease resistant, than uninfected rye grass (Clay 1988). Some of these small molecule effectors can be generated in culture, provided that the fungus is grown under appropriate conditions. Plowman et al (1990) reported the ethnobotanical use of the Amazonian sedges *Cyperus articulatus* and *C. prolixus*, in birth control, in the induction of labour, and in producing hallucinogenic effects. However, herbarium specimens from South America and live specimens collected in Ecuador revealed the presence of the systemic ascomycete *Balansia cyperi*. Pure cultures of *B. cyperi* were shown to produce unidentified indole alkaloids which could account for the medicinal properties of the plant extracts.

Other natural examples of small regulatory molecules are to be found in fungal pathogens of insects. There are very good examples of both South American and African species of insects whose behaviour is altered before they are ultimately killed by an invading pathogenic fungus. The ponerine ant, when infected by *Cordyceps* spp., is driven to the top of the leaf canopy as a result of a behavioural modification induced by the fungus. The advantage for the fungus is that it requires light for sporulation, and the phenomenon illustrates a further means by which the microorganism has secured an environmental advantage for its own competitive ends (Evans 1987).

Secondary metabolites thus confer competitive advantages on the producing organisms and may markedly influence growth or development. As a stimulus to our thinking, rather than as a serious proposal, we should consider such metabolites as 'metabolophores' [Gr. *metabole*—change + Gr. *phore*—bearing], meaning 'a molecule containing a specific group of atoms the function of which is to change the growth, morphology or differentiation of an organism; its mating type; its host; or its parasite, saprophyte or predator'.

Microbial biodiversity: a source of chemical diversity

'Biodiversity' is an ecological term used to describe the different macro- and micro-ecosystems within a geographical area, the diversity of different species within an ecosystem, and the genetic diversity within species. Organisms can be systematically isolated and screened according to their taxonomy, but although this may have some value for the actinomycetes, the higher taxonomic groupings of the fungi are generally artificial and bear little relation to the ecology and physiology of the groups.

To date, natural product research has tended to concentrate on the isolation of molecules from actinomycetes, and this is particularly the case with the discovery of novel antibiotics. The actinomycetes are Gram-positive prokaryotes comprising 63 genera and are relatively well documented because of their value in drug discovery. They are generally regarded as sporulating, soil-borne organisms, but show considerable ecological and biochemical diversity. Biodiversity, ecological niche exploitation, association with other organisms, and sheer abundance reach their greatest complexity with the fungi, although, unlike the actinomycetes, they have scarcely been explored for bioactivity. The fungal kingdom is composed of groups that have evolved along separate evolutionary paths, and they are an extraordinarily diverse group. There are approximately 69 000 species of fungi known, but well over 1000 new species are being discovered each year, and a conservative world total may be in the region of 1.5 million (Table 4). A considerable proportion of the new records are from the tropics, and the lack of available data on the biodiversity of fungi in such complex and unexplored ecosystems is primarily due to the lack of mycologists working in the area. During an ecological study of tropical forest in the El Verde forest reserve in Puerto Rico, Lodge (1988) estimated that 20–30% of the agarics (the large-fruiting basidiomycetes) encountered were new species. Probably fewer than 1% of existing fungal species are represented in culture collections. The total untapped microbial resource potentially available for new drug screening is enormous.

So what is the evidence that microbes harbour a rich source of drug leads in addition to antibiotics? We should remember that it was the Deuteromycotina that first provided β-lactam antibiotics from *Penicillium, Acremonium* and *Aspergillus* species. Also mevinolin (Mevacor®), the new cholesterol-lowering drug, was first isolated from *Aspergillus terreus*, and cyclosporin A, an immunosuppressant that has revolutionized the management of rejection in organ transplants, comes from another member of the Fungi Imperfecti, *Tolypocladium inflatum*. Another fungal group, the Ascomycotina, includes the sclerotium-forming genus *Claviceps* which provide ergotamine and other dopaminergic alkaloids that are used in treating migraine, post-partum bleeding and Parkinson's disease. A number of fungal and actinomycetes metabolites have now been demonstrated to have agonistic or antagonistic properties against a wide range of receptors, particularly neuroreceptors (Table 5).

TABLE 4 Microbial diversity

	Number of species		*Species in culture collections*	
Group	*Described*	*Estimate of total*	*Number*	*% of total*
Algae	40 000	60 000	1 600	2.5
Bacteria	3 000	30 000	2 300	7.0
Fungi	69 000	1 500 000	11 500	0.8
Viruses	5 000	130 000	2 200	2.0
TOTAL	117 000	1 720 000	17 600	1.0

The table includes data from Hawksworth 1991.

TABLE 5 Pharmacological action of microbial metabolites

Ergot alkaloids	Dopamine agonists
Muscarine	Acetylcholine agonist
Muscimol	GABA agonist
Avermectin	GABA antagonist
Psilocybin	5-HT antagonist
Asperlicin Tetronothiodin Viginiamycin M_1	Cholecystokinin antagonists
BE-18257A WS-7338 FR-139317	Endothelin antagonists
ES-242-1	*N*-Methyl-D-aspartate (NMDA) antagonist
L156 602	C5a antagonist
Anantin	Atrial natriuretic factor antagonist
2-Methoxy-5-methyl-benzoquinone	Thromboxane A_2 antagonist

It is significant that some microbial products first discovered as antimicrobial agents have subsequently been shown to possess highly potent activity in mammalian systems (Table 6). These include glycopeptide and macrolide antibiotics, the antihelminthic invermectin, and the antifungal agents cyclosporin A and mevinolin. One indisputable observation is the diversity of microbially derived chemical structures. Also, microbes often provide drug leads (Fig. 1) that are beyond the imagination and deductive reasoning that are possible from our existing knowledge of the molecular and cellular processes that drive diseases.

Designing screens to detect useful microbial metabolites

Of the fifteen top-selling pharmaceutical products of 1991, four were microbially derived, with aggregate sales of \$3.8 billion. Indeed, microbially derived

TABLE 6 Pharmacological functions of antibiotics

Antibiotic	*Antibiotic activity*	*Pharmacological function*
Vancomycin	Gram-positive bacteria	Hypotensive agent
Erythromycin	Gram-positive bacteria	Motilin agonist
Ivermectin	Antihelminthic	GABA antagonist
Cyclosporin	Antifungal	Inhibition of interleukin 2 (IL-2) synthesis
Mevinolin	Antifungal	Cholesterol lowering

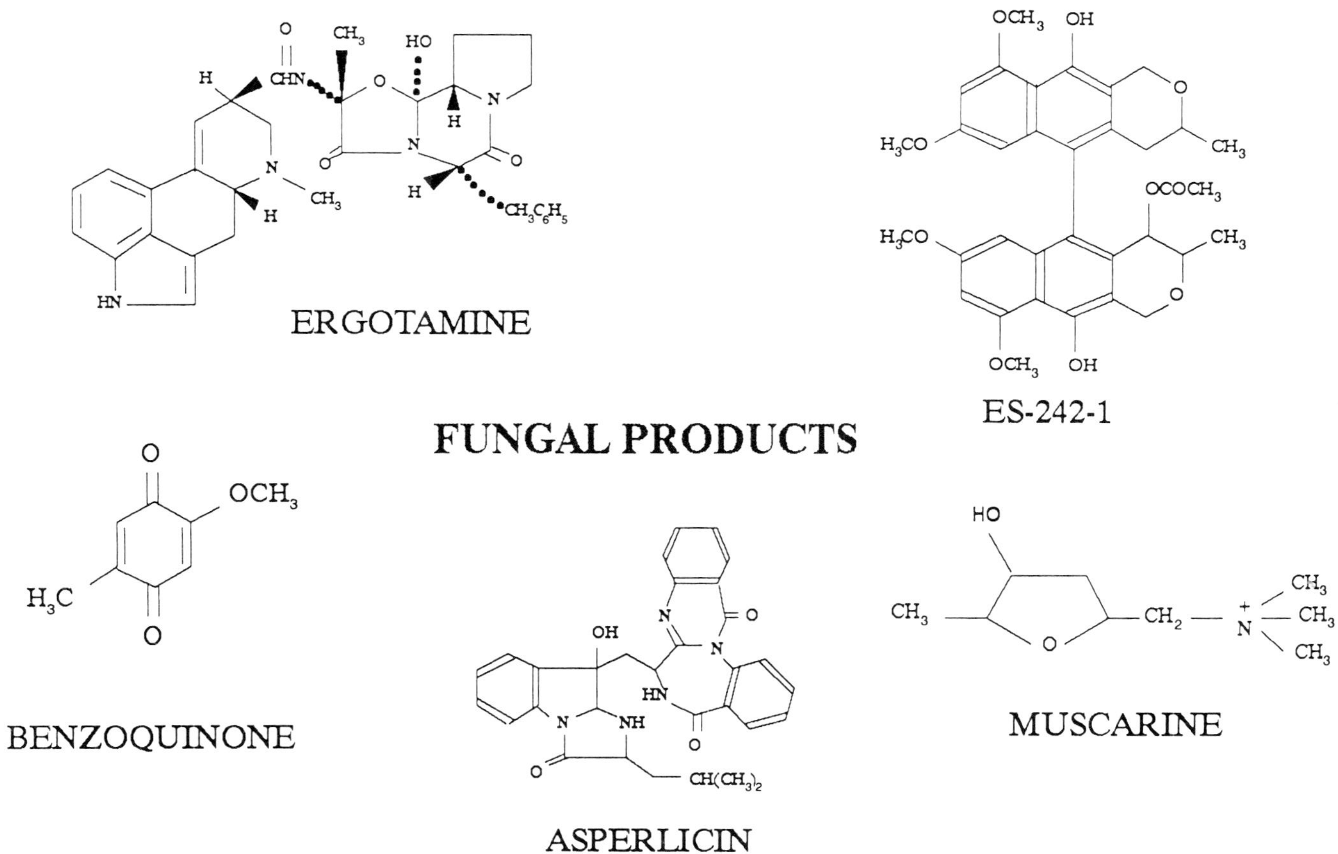

FIG. 1. Structures of fungal products demonstrating pharmacological activity.

TABLE 7 Role of molecular biology in defining and isolating targets for microbial screening

Disease mechanisms	Infection and replication process of HIV Role of cytokines in inflammation Structure and function of complex receptors
Recombinant DNA assay tools	Cloned enzymes: HIV protease expressed in *E. coli* Cloned receptors: IL-1 and IL-6 receptors Polypeptide ligands: cytokines such as IL-2 and TNF
Monoclonal antibodies	Epitope mapping of receptors: CD4 receptor
Functional assays	Cell clones, growth factor-dependent proliferation of cells, and signal transduction and second messenger pathways Disease models based on chimeric and transgenic animals

antibiotics make up most of the $16.5 billion annual sales in the anti-infectives market. For some time it has been possible to readily screen for new and improved antibiotics by screening against intact organisms or targeting a cellular process uniquely present in bacteria or other microbes (Jeffs & Nisbet 1988).

Molecular biology techniques have enabled the identification of enzymes encoded by the human immunodeficiency virus (HIV) and have provided the quantities of such enzymes required for high capacity screens within a short time of the discovery of the causative agent (Table 7). Similarly, the role of the CD4 receptor in viral infectivity and the assay for this interaction were made facile by the cloning, expression and isolation of the receptor and by the preparation of epitope-specific antibodies to the receptor. Other areas of molecular biology

TABLE 8 Technological advances that have enhanced screening efficiency

Recent developments	
Chromogenic substrates	Protease, phosphatase
Radioligands	Hunter–Bolton iodination of cytokines
Immunoassays	ELISA, RIA, biotin–streptavidin assay
Microtitre systems	Automation of assay preparation, washing, harvesting, reading
Robotic work stations	Zymark®, Biomek®
Future prospects	
Biosensors	Enzyme inhibition and receptor–ligand assays
Molecular sizing	Selection of small molecules/removal of interfering substances
Novel assays	Amerlex® Magnetic Particle Separation Scintillation Proximity Assay (SPA)

have enhanced the validity of cellular and animal models of disease by the creation of chimeric and transgenic species in which the relevant human protein(s) replace the murine or other species form.

New technologies developed during the biotechnology era have increased the efficiency, quality control and cost control of high through-put screening (Table 8). Particular advances include the development of immunoassay techniques that have facilitated the measurement of protein–protein interactions; robotic work stations; the newer microtitre plate assay systems and stacking plate systems for reading chromogenic substrate assays; the 96-well scintillation counters for radioligand assays; and the Scintillation Proximity Assay (SPA®, Amersham International, UK) that allows radioligand binding assays to be performed without the need to separate the bound ligand from unbound ligand. Furthermore, the screening of microbial samples for drug leads has traditionally been complicated by the presence of hydrolytic enzymes that destroy receptors or protein-based ligands, resulting in high rates of false-positive results. High through-put membrane filtration or automated chemical fractionation systems can be used to enrich samples in the desired small molecule fractions.

Screening microbial metabolites for inhibition of protein–protein interactions

One of the great assets derived from the applications of molecular biology is our vastly improved understanding of the roles of protein effectors in human disease. The cytokines interleukin 1 and tumour necrosis factor drive a number of immune-inflammatory diseases such as rheumatoid arthritis and septic shock; the growth factors transforming growth factor-α and basic fibroblast growth factor promote the proliferation of cancer cells via their respective receptors. Because of the complexity of the protein ligand–receptor interactions, and the lack of information on the molecular events in the binding and activation processes, these are not amenable to traditional drug design approaches such as those that led to the development of beta-blockers and H_2 receptor antagonists. Also, a few years ago, there were sceptics who questioned the value of small molecules in relation to the inhibition of multi-site binding interactions. One way to uncover a small molecule drug lead in these areas is to screen for one.

One of the protein–protein binding processes that Xenova targeted at an early stage of the company's development is the interaction between plasminogen activator (PA) and its endogenous inhibitor, plasminogen activator inhibitor (PAI). Plasminogen activators are proteins that regulate the production of plasmin, which in turn dissolves blood clots. PAI, which is produced by endothelial cells proximal to clots, is over-produced in certain conditions and hinders clot lysis. Clearly, an inhibitor of PAI

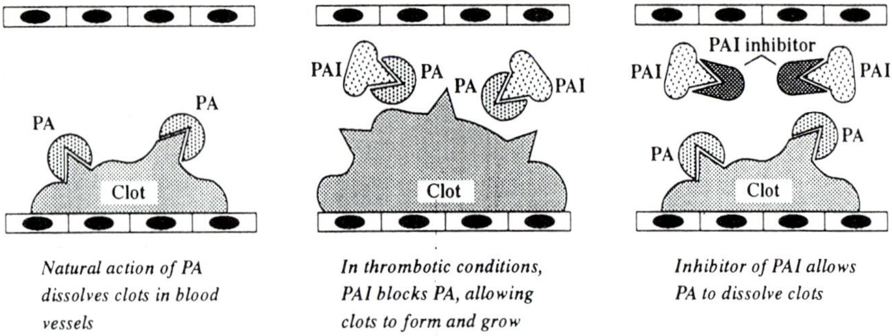

FIG. 2. Role of plasminogen activator inhibitor (PAI) in thrombosis.

would protect the catalytic activity of PA and result in natural dissolution of the clot (Fig. 2).

A total of 20 000 microbial samples were screened and 120 (0.6%) were initially found to be active in the PAI–PA binding assay. As a result of specificity and relative potency studies, eleven of these were selected for chemical analysis and two have now been characterized as different small molecule inhibitors of PAI–PA binding. The precise mode of action of the molecules is under investigation, although they do appear to bind to PAI in a manner which interferes with PAI–PA binding and preserves the clot-dissolving activity of PA.

Conclusions

Despite its serendipitous nature, natural product screening has had a profound effect on the pharmaceutical and agrochemical industries. The era of antibiotic discovery based on the random screening of secondary metabolites from microbes is being replaced by sophisticated, target-directed, mode-of-action screens. This new approach relies on the technological tools of modern molecular biology and involves the search for low molecular mass molecules with specific activities. In conjunction with this new approach, a knowledge of ecology and an appreciation of biodiversity is used to direct microbial isolation programmes in order to uncover new regulatory molecules. Understanding the role of secondary metabolites in Nature can provide valuable insight into their regulatory properties. Natural product screening has traditionally been regarded as a random 'hit or miss' exercise, but recent research suggests that the search among microbes for molecules with receptor activity in mammalian systems has a previously unrecognized value and a strong rationale behind it.

Acknowledgements

I am grateful to F. Fox for advice on the possible role of ecology and biodiversity in the discovery of bioactive metabolites, and to M. Moore for critical review of the manuscript.

References

Clay K 1988 Fungal endophytes of grasses: a defensive mutualism between plants and fungi. Ecology 69:10–16
Dorlands Illustrated Medical Dictionary 1981 26th edition. WB Saunders, Philadelphia, PA, USA
Evans HC 1987 Mycopathogens of insects of epigeal and aerial habitats. In: Wilding N, Collins NM, Hammond PM (eds) Insect–fungus interactions. Academic Press, London, p 205–238
Hawksworth DL 1991 The fungal dimensions of biodiversity: magnitude, significance and conservation. Mycol Res 95:641–655
Jeffs PW, Nisbet LJ 1988 Glycopeptide antibiotics: a comprehensive approach to discovery, isolation, and structure determination. In: Actor P, Daneo-Moore L, Higgins ML (eds) Antibiotic inhibition of bacterial cell surface assembly and function. American Society for Microbiology, Washington, DC, chapter 53:509–530
Lodge J 1988 Three new *Mycena* species (Basidiomycotina, Tricholomataceae) from Puerto Rico. Trans Br Mycol Soc 91:109–116
Martin JF, Demain AL 1978 Fungal development and metabolite formation. In: Smith JE, Berry DR (eds) The filamentous fungi, vol 3: Development mycology. Edward Arnold, London, p 426–450
Plowman TC, Leuchtmann A, Blaney C 1990 Significance of the fungus *Balansia cyperi* infecting medicinal species of *Cyperus* (Cyperaceae) from Amazonia. Econ Bot 44 (4):452–462
Roth J, Leroith D, Collier ES 1986 The evolutionary origins of intercellular communication and the Maginot Lines of the mind. Ann NY Acad Sci 463:1–11
Zähner H, Drautz H, Weber W 1982 Novel approaches to metabolite screening. In: Bu'Lock JD, Nisbet LJ, Winstanley DJ (eds) Bioactive microbial products, vol 1: Search and discovery. Academic Press, London & New York, p 51–70

DISCUSSION

Rinehart: Are you using mainly enzymic assays or whole-cell assays to look for new compounds, and how does the through-put compare for the two?

Nisbet: On cellular assays versus enzyme- or receptor-based assays, I would say that the rate at which you can reasonably screen against cellular assays is probably one-quarter to one-fifth of the rate using a receptor- or enzyme-based assay. Cellular assays cannot be automated quite so well; you have to filter sterilize the supernatant before you put it back in the cell culture. You not only may have to separate large and small molecules; you need to avoid bacterial contamination, because the cell systems may need 2–4 days to develop; so it does affect through-put. We would be thinking of 12 000–20 000 samples at the outside, per year, through a cell-based assay, but four to six times that number through an enzyme-based or receptor-based assay. The advantage of a cell assay is that instead of, for example, just looking at the interaction of interleukin 1 (IL-1) with its receptor, if you want to look at the functional properties of IL-1, you are really interested in seeing how it interacts with the cell and then elicits the production of another molecule, such as IL-2. If you use a cell assay and measure the production of IL-2, you are looking at many potential points of interference in the process; technically, you have many more assays running and therefore a higher potential for success.

Rinehart: Would you normally go through a secondary whole-cell screen to complement the enzyme-based screening?

Nisbet: Yes. If we start with an enzyme- or receptor-based assay, we inevitably go through a cellular system; because we are not just interested in binders; compounds must have some functional activity.

Zähner: Problems arise with proteases that degrade the receptors or the enzymes in the assays. How have you solved these problems?

Nisbet: In different ways! Historically, we have used membranes to exclude proteins. But there are a variety of membranes available, with different charges, or other characteristics in relation to lipophilic or polar molecules, and different pore sizes. It's quite easy to exclude antibiotics using membranes with a nominal 10 kDa cut-off, so it's not straightforward, but membrane technology is the way that we have done this.

However, we should be prepared to look more broadly at initial crude separation systems based on the chromatographic properties of compounds. One should be able to set up automated chemistry laboratories to absorb and desorb materials from different columns, to pull out polar compounds or to separate compounds on the basis of charge. This would be something to investigate, especially with the robotic systems that are now available.

Hunter: Your assays are critically dependent on having enough of a metabolite present to detect it. I imagine that you grow organisms under different conditions, to cover different nutritional regimes. Can you give us some feeling for how often the genetic potential of a microorganism is actually expressed and fulfilled?

Nisbet: Our feeling is that we should try to learn from the precise work that people like you are doing with individual organisms and apply that on a more generic basis to groups of organisms. We tend to know which organism we are dealing with; we are going for microbial diversity, and that means knowing what we have got, so that we don't keep testing the same thing. Beyond that, we try to group organisms in a given period of time, so that we can gain knowledge from your precise studies on a particular microbial species. We then grow the organisms in media demonstrated to be useful for stimulating the production of what we commonly call antibiotics. What we want is chemical diversity, and antibiotics have a rich diversity of structure and function. We are looking to see if they have other functions, and are also looking for new molecules that may not have been detected as antibiotics, perhaps because that's not their primary mode of action, or their primary role in the producing organism. Our physiologists design media that are carbon limited, phosphate limited, or nitrogen limited, if that's relevant. Largely they use complex nutrients; it's one thing to work with *Aspergillus* or a streptomycete in a fully defined medium and describe whether there is a particular type of repression, say; but it is very different to engineer a culture system that is applicable to a whole range of organisms. Complex nutrients are still very important in that respect. They may contain lots of carbon, nitrogen and phosphorus, but it's not readily available; the organism must work to get at it. It will give you biomass, but then you growth limit the organism with the complex nutrients, which it has to work to obtain; this introduces a slow feed, and may overcome some of the repressive effects discussed earlier.

Davies: If you *had* to choose a few very simple media, would you try to avoid using glucose as a carbon source when growing microbes for secondary metabolite production?

Nisbet: We have done that, by and large, but avoidance of glucose is not a total disaster, if you have other carbon sources which will kick in later in the cell growth cycle. Glycerol and amino acids such as proline can be very useful sources of C and N respectively.

Demain: Since you have such unique screens, why turn to Nature at all? Why not just arrange to study all of the compound collections of pharmaceutical companies, especially the old antibiotics, and find something new that way?

Nisbet: The imagination of chemists and biologists in the pharmaceutical industry has always been driven by templates, whether it has been histamine or the β-lactam nucleus; the pharmaceutical giants have made many thousands of analogues. The chemical diversity is actually rather narrow. Also, they have made many compounds that are not the right stereoisomers and are therefore inactive. Perhaps naively, I like to think that microbes and other biological systems make these compounds through biosynthetic pathways involving proteins and therefore every one of these secondary metabolites must recognize a protein, somewhere. If I am to use a protein as a method to find a template, I want to maximize the chemical diversity that I present to that receptor or enzyme. I don't think I shall get that from company chemical banks; there are more molecules out there in Nature.

Chater: You are effectively dismissing many people's opinions on the biological role of antibiotics, and that shouldn't go unchallenged! You find that diatomaceous earth is extremely good at mopping up things like antibiotics. But almost no microorganisms will grow on this mineral, whether or not they produce antibiotics or compete with organisms that do. Where they grow is in that organic material, which may be near the diatomaceous earth but not actually in it; it may be separated from it by many μm. That would be quite far enough for the kinds of models that we have been talking about to be applicable. Your explanation may instead provide a reason why it is so difficult to detect antibiotics by extracting soil; because, when you do that, you will be bringing the antibiotics into contact with diatomaceous earth and removing them from any possibility of detection in biological tests.

Nisbet: I may be crediting evolution with more power than it has had time to exercise, but I would have thought that if organisms were to develop defence systems based on antibiosis, they would have developed many more broadly active (Gram-positive and Gram-negative) antibiotics; one might have expected more antibiotics that are active against resistant organisms. Why haven't all the other organisms in Nature developed defence systems to these drugs through evolution? If this was an important way of killing them, surely they would have done?

Chater: The point is that they have; but what you see is the continuing process of evolution; you miss all the compounds to which universal resistance systems have already developed.

Nisbet: I still would have expected more thienamycins around than we have. Why haven't we got many more and why are so many of them non-specific?

Demain: There are many thienamycins (i.e. carbapenems). They are very labile and present in Nature at very low concentrations, but we know about 50 of them already.

Nisbet: But why aren't there more types of blockbusters—that is, antibiotics that will hit Gram-positive and Gram-negative bacteria resistant to β-lactamase?

Demain: There are a lot of those, but you don't hear about them, because they are toxic and are eliminated from further study very early on.

Chater: If they act on such a wide range of organisms and systems, they will have other inconveniences, so they are thrown out early in the screening.

Davies: Can anybody give me evidence that you can actually detect antibiotics in soil?

Demain: Very simply put, if you go to Nebraska and take soil, you will find patulin and all sorts of other mycotoxins. They cause a lot of problems in agriculture, such as afflicting animals and contaminating crops; they have antibiotic activities and they are there—in straw, in soil, and in crops.

Davies: What about antibiotics produced by streptomycetes in soil?

Demain: You can find those too.

Davies: Many published papers have claimed they could never find antibiotics in soil.

Demain: There was great prejudice against finding them! David Gottlieb (1976) asked the following: when an apple falls off a tree, and in the next few days it's filled with patulin, is that a natural situation or an unnatural situation? The fallen apple is injured, it is infected by the fungus, the antibiotically active mycotoxin is produced, and thus almost all apple juice you drink contains some patulin. It is a very good antibiotic.

Vining: Dr Nisbet, I wonder how you explain the persistence of secondary metabolites over long periods of evolution if they don't have biological activity against organisms that interact with the producer? If they are primarily cholecystokinin inhibitors rather than antibiotics, what selective value would they have for a soil microorganism? Why would the pathway for their biosynthesis have evolved and been maintained if the product did not directly benefit the producer?

Nisbet: I'm not suggesting that secondary metabolites are not useful; I am just not convinced that they are useful as antibiotics in the natural environment. I don't see why β-lactams, for example, might not be, at low concentrations, morphogenetic agents that will change cell shape and ability to penetrate, which will affect cell division, sporulation, perhaps; not affecting competitor organisms, but maybe modifying the growth and development of the producer itself.

Demain: Do you believe that when a fungal cell encounters a plant and the plant makes a phytoalexin in response to the fungus, it is not a defence mechanism for fighting the fungus? Then the fungus is induced to make another compound to attack the plant: is that accidental, or not? There are thousands of phytoalexins. What bothers me is that people will never argue against that type of chemical warfare, but are always ready to say that microbial antibiotics are not agents of chemical warfare.

Davies: It's not that the molecules are not antibiotics; but they have other functions which have importance to the biology of the producing organism, for example as sex pheromones. Gentamicin is a good inducer of bacterial conjugation at low concentrations. It also induces translocation in bacteria.

Demain: If you consider aflatoxin, maybe its function is to increase mutation in Nature! The point is that whether we have identified the right function or not, there is a function.

Nisbet: Absolutely! There's no question of there being a function.

Cane: We have to avoid being hung up on the question of a single function, or *the* main function. If you consider the central metabolites (amino acids, sugars, nucleotides), if you can't ascribe a single function to an amino acid, then is it legitimate to require only a single function for a secondary metabolite? After all, proteins have structural, energetic and catalytic roles. Nature is simply opportunistic with the basic biological materials which are available or can be made.

Dr Demain's idea of looking through the chemist's store of synthetic compounds as a way of looking for physiologically active materials (p 227) has set me thinking about the question of exploiting not just the synthetic capacity of microorganisms but their catabolic capacity. A very effective thing to do at present is to look in chemical waste dumps for organisms which will degrade toxic materials, as ways of finding biological agents for the disposal of polychlorinated biphenyls (PCBs) and other chlorocarbons. In principle, we should be able to devise enzymic screens for looking at this potential.

Chater: I wanted to add another dimension to the issue of antibacterial function, namely instances where microorganisms make two biosynthetically independent compounds which act synergistically to give very strong antibacterial action. This is a particularly compelling example of specificity.

Demain: Virginiamycin (produced by *Streptomyces virginiae*), which antagonizes CCK, is made of two separate chemicals which act synergistically against bacteria; are both needed for activity as a CCK antagonist?

Nisbet: No, it requires only one of them, virginiamycin M.

Davies: Do other synergistins work in the same way?

Nisbet: I doubt if they do, because it is a double enzyme blockade in the case of the two different components; they are structurally different.

Vining: As antibiotics the two virginiamycin components (A and B) act synergistically on the 50S subunit of prokaryotic ribosomes to inhibit protein biosynthesis. As you would expect for synergists, they are each active if used alone, but the specific activity is lower.

Davies: There are similar components in every one of the synergistin mixtures; there are as many as six producers, as far as I know. If virginiamycin has these different activities, some of the other synergistins might be expected to work in the same way.

Williams: I favour the idea that at least a good number of antibiotics have evolved to have antibiotic function. For example, consider CC1065; this molecule is a banana shape with exactly the right chirality to go into the minor groove of DNA, which it has been shown to do (Hurley et al 1984). It has a hydrophobic face that is then buried. At one end it has a cyclopropane ring; after binding, N-3 of an adenine residue of the DNA opens that ring. It has been shown by NMR that this is where the molecule attacks. When I see such sophistication, I can't help thinking that it's because that is a selected function for the substance. I am not precluding other functions. The case is very strong if you look at the hydrophilic face as well, and the complementary shape of the fit.

Nisbet: I was struck, though, by Eric Cundliffe's example of novobiocin, which actually does bind to the host's DNA gyrase, the constitutive gyrase, and then the host produces a second gyrase. One wonders what that second one does. Are certain genes transcribed by the second gyrase that wouldn't normally be transcribed, because of the changing conformation of the DNA?

Williams: I wouldn't speak for all cases, but where we know what is going on, it seems to be very sophisticated, and if we reject that sophistication, it's as if you isolate an alcohol dehydrogenase and get its crystal structure; you have a nice mechanism whereby it will reversibly convert acetaldehyde and ethanol, and then you say that it's there, but you don't believe that this is its evolved function!

Cane: What is its function in the horse? Actually, it is to detoxify terpenols. Ethanol is probably an accidental substrate. Horse liver alcohol dehydrogenase is known for its broad substrate specificity.

Williams: But the function is still as an alcohol dehydrogenase. One wouldn't deny that liver alcohol dehydrogenase has a useful function for which it is well evolved in *Homo sapiens*.

Cundliffe: Louis Nisbet referred to Professor Ōmura's work in Japan showing that erythromycin functions as a motilin agonist (Table 6, p 220). It binds to bacterial ribosomes with a dissociation constant of about 10^{-8} M. We are beginning to find evidence that it's making primary contacts with ribosomal RNA, and we are starting to get a feel for which bit of rRNA is involved. Before long we may get something, perhaps not as beautiful as vancomycin docking in on its receptor, but some sort of model that shows how erythromycin binds in a rather neat way to the bacterial ribosome.

Then we discover that it binds much better to some receptor in the gastric wall, and acts as a motilin agonist. One wonders which is the appropriate phrase: that giving erythromycin to your patient has the side effect of causing diarrhoea and stomach cramps; or is it the sadistic physician, who is trying to get the patient out of the prescribing room and doesn't want to see him again, and so gives him something pretty awful, and by chance cures him of Legionnaire's disease or osteomyelitis or an upper respiratory tract complaint? I wonder at what point you feel you have found the 'right' receptor. Anyway, I am not very impressed by dissociation constants of 10^{-8} M. Most antibiotics that I have worked with are very poor in comparison with, for example, the way in which hormones bind to their receptors with dissociation constants several orders of magnitude lower. So I am not particularly impressed by some of these molecules as drugs, in the form in which we know them now. I wonder whether they don't have other functions of which we are totally unaware.

Cane: How tight is tight enough?

Cundliffe: I don't know!

Demain: Another point about motilin is that there is a dissociation of antibacterial activity and motility; the best form of erythromycin, for agonist activity, does not have significant antibacterial activity.

Cundliffe: You mean that it is actually a different molecule doing that?

Davies: It's a molecule synthetically derived from erythromycin, which doesn't bind to the motilin receptor.

Rinehart: In relation to comments on CC1065, the drug that will actually be used is a second- or third-generation version of that; so the idea of looking for the pharmacophore and then improving on it by chemical means is very appropriate.

Nisbet: Yes. I really didn't want to scotch the idea that none of these molecules developed as antibiotics, but I wanted to reinforce the point that you need to look at the business end of these molecules, to understand what the full potential may be within the producing organism, in relation to the environment within which that organism is growing, and then to investigate their potential as new therapies for diseases, beyond their role as antibiotics. It's very important to look on these microbial compounds as sources of pharmacophores.

We have heard a lot about homologies between different proteins, including classes of proteins that were far apart but with a lot of homology of sequences, where David Cane is looking at short sequences and saying there were 1200 hits out of 27 000 or so. This emphasizes that there's a lot of sequence identity between different proteins, and therefore these pharmacophores, which in some situations may exert antibiotic action, may have a different action, a different potency, and we may see a different response in the cell, as a result of activity against another protein.

Cavalier-Smith: Similarity isn't the same as homology, since it can arise by convergence.

Cundliffe: No; it's very important to distinguish between them—there are no degrees of homology; two proteins are either homologous, or they are not.

Zähner: The possible roles of secondary metabolites can be considered in terms of their intrinsic and extrinsic functions. I am sure that the category of intrinsic functions will be enlarged in the near future. Possible candidates for such intrinsic functions are: (a) The inhibition of spore germination in the case of spore-forming producers of secondary metabolites (e.g. germicidin B from *Streptomyces viridochromogenes*; Petersen 1991). (b) Iron transport (siderophores). More than 200 siderophores from microorganisms have been described. Some of these compounds were first screened as antibiotics. For example, nocardamine (desferrioxamine E) was published in 1951 as an antibiotic against mycobacteria, but it is in fact not a weapon against competitors in the ecological niche; it is in reality the major iron transport compound of the producing nocardia strain. (c) The regulation of differentiation. The A-factor from *Streptomyces griseus* is not the only example. I expect there to be many other differentiating factors from the pool of secondary metabolites; some may also be butyrolactones, but other chemical groups can be expected. (d) Regulators of cell wall biosynthesis. If the β-lactams were originally products of actinomycetes or eubacteria, a possible function in the regulation of cell wall or spore wall formation would make even more sense for the producing cells than a possible role as weapons against competitors. (e) Discrimination between potassium and sodium ions. Ionophores such as valinomycin, macrotetralides and polyethers show optimal properties for discriminating among the alkali ions, and they could be part of an ion influx or efflux system. Ionophores are detectable in eubacteria, actinomycetes and fungi; some hundreds have been described.

Vining: I am still attracted by the idea, following Dudley Williams' line of thought, that the biological activity has been beautifully tailored, presumably by opportunistic selection through evolution, to do a specific job. It troubles me that what you, Dr Nisbet, are saying is that secondary metabolites are tailored to do *two* separate, very specific and very good jobs. How does this happen, if the metabolites are produced in a situation where they are not selected to do one of them?

Nisbet: I am not saying that they necessarily do a good job. I am suggesting that we should not presume that we know what they do, just because they kill other cells. It may be that the selection pressure has been to be more successful in sporulation; maybe the compounds regulate the growth and differentiation of the producer at lower concentrations than those at which they kill other cells. That may be a more effective selection pressure than the ability to throw out what must be minute traces of toxins into the local environment of a piece of mycelium growing in soil. One wonders, in fact, just how they achieve the concentrations that we use them at, in such a local environment.

Cavalier-Smith: Why do they need to? All they need is a quantitative reduction in the populations of competitors; it could be a minute reduction. It doesn't need to be killing; just a slowing down of growth rates by a tiny percentage, systematically, would be enough to favour the spread of their genes.

Nisbet: Why would that be a more important selective pressure than, say, sporulation?

Cavalier-Smith: But slowing growth rates in itself is a very strong selective pressure; you are invoking activities for which there is no evidence, in most cases, so it is an unparsimonious hypothesis!

Nisbet: That is not true; the nikkomycins can regulate negatively the production of a zygospore. I am saying that these compounds *can* have other functions; in a number of cases, their activities in relation to those other functions are much more potent than the functions we have ascribed to those molecules.

Davies: As a comment on what Leo Vining said, which is also relevant to Dudley Williams' point, it seems more reasonable to assume that the molecule finds its receptor at a very early stage in evolution, rather than that the molecule evolves to fit the receptor. The receptor subsequently evolves and it is possible to have different receptors which respond to the same type of molecule.

Vining: Julian, if you argue that the ligand was there before the receptor, and that various different receptors have evolved from the original, how would you explain the evolution of a cryptic receptor? Going back to the example when the original receptor is for antibiotic activity, what selection pressure would there be to evolve from this a different kind of receptor associated with cholecystokinin function in an organism where this activity seems irrelevant?

Davies: I don't know what the organisms were exposed to! You can always imagine selective pressure of some kind, especially since this does not require an all-or-nothing situation; a small change is all that may be needed.

Vining: What you are really saying is that the second activity of the ligand is adventitious; it's pure chance that it happens to fit another receptor.

Davies: I am happier with the notion, for example with RNA, that a small molecule recognizes RNA at a very early stage of evolution, and then the RNA subsequently evolves to be part of a ribosomal RNA, or some other RNA-based structure which still recognizes the same small molecule. I don't have evidence for that, but I do have evidence that the same types of small molecule will bind to quite different types of RNA.

Chater: There is some evidence that bacteria do get exposed to antibiotics naturally, in the sense that preserved culture samples from the pre-antibiotic era (i.e., before the 1950s) sometimes have antibiotic resistance plasmids and transposons in them. Some examples of this are cited by Paul Broda in his book *Plasmids* (1979). This suggests that those organisms were exposed to antibiotics before we began their widespread use, and that there was some selective benefit to the ability to be resistant.

Davies: There's a question about some of those experiments. It depends what you mean by the pre-antibiotic era. The best collection of organisms that were taken for these experiments were strains from the Murray collection at the University of Western Ontario. Naomi Datta looked at those strains (Hughes & Datta 1983). She found mercury resistance and sex factor activity, but no antibiotic resistance. This is considered to be a study of the first group of pre-antibiotic era organisms. The studies done subsequently came from Lederberg's collection, and antibiotics were already in use at that time.

Chater: Even if there should be doubts about the incidence of resistance in the 'pre-antibiotic era', I think that for resistance transposons and plasmids to have emerged for the first time early in the antibiotic era would have required remarkably rapid evolution. The resistance genes are perfectly fitted to be expressed in Gram-negative organisms like *E. coli*, and have very different codon usage from the organisms known to produce the relevant antibiotics. It is true that there is no well-attested case for there being close similarity, at the DNA sequence level, between a resistance determinant on a transposon and one in a producing organism.

Davies: How close do you want? Sixty per cent identity? The aminoglycoside phosphotransferase of RSF1010, one of the most promiscuous plasmids, has a high degree of similarity with the phosphotransferase gene from *Streptomyces griseus*.

Chater: That is 40% sequence divergence, but *E. coli* and *Salmonella* have diverged for maybe 100–200 million years, very much more than the five or 10 years that we are talking about here; in that time their sequences have diverged by about 50% at positions that do not affect protein structure (Ochman & Wilson 1987). So 40% divergence in a few years is surely an extraordinary amount of divergence.

Davies: I would agree!

Cane: I always like to think of superoxide dismutase (SOD) as an honorary natural product, because it's a large molecule to which we can ascribe a function, but we don't know why it's really there, because we don't know what it actually does.

Davies: Can we live without it?

Cane: I don't know how to do the experiment! Many people believe that its role is to protect the cell against oxidative damage.

Davies: If you give an organism superoxide dismutase, is it likely to have 'improved' characteristics of growth? The gene has been cloned, so it is possible to introduce the capacity to make superoxide dismutase into any organism.

References

Broda P 1979 Plasmids. WH Freeman, Oxford

Gottlieb D 1976 The production and role of antibiotics in soil. J Antibiot (Tokyo) 29:987–1000

Hughes VM, Datta N 1983 Conjugative plasmids in bacteria of the 'pre-antibiotic' era. Nature (Lond) 302:725–726

Hurley LH, Reynolds VL, Swenson DH, Petzold GL, Scahill TA 1984 Reaction of the antitumor antibiotic CC-1065 with DNA: structure of a DNA adduct with DNA sequence specificity. Science (Wash DC) 226:843–844

Petersen F 1991 Germicidin B, ein autoregulatorischer Keimungshemmstoff aus *Streptomyces viridochromogenes*. Dissertation, University of Tübingen

Ochman H, Wilson AC 1987 Evolutionary history of enteric bacteria. In: Neidhardt FC (editor-in-chief) Escherichia coli and Salmonella typhimurium: cellular and molecular biology. American Society for Microbiology, Washington, DC, vol 2:1649–1654

Secondary metabolites from marine organisms

Kenneth L. Rinehart

Roger Adams Laboratory, University of Illinois, Urbana, IL 61801, USA

Abstract. Marine macroorganisms and microorganisms, like terrestrial species, produce a dizzying array of secondary metabolites, including terpenes, steroids, polyketides, peptides, alkaloids and porphyrins. Most of the marine metabolites are found in terrestrial species as such or have close counterparts in land-based species, but some are sufficiently unusual to constitute a separate class (e.g. marine sterols). Although in many cases the functions of these secondary metabolites in the marine species themselves are unclear, other compounds play well-defined roles—for example as trail markers, sexual attractants, antifouling substances or antifeedants. What is clear is that many of the most interesting marine secondary metabolites have potent activities largely unrelated to their *in situ* roles. Examples abound of antitumour, antiviral, immunosuppressive and antimicrobial agents, as well as neurotoxins, hepatotoxins and cardiac stimulants. Relatively few biosynthetic studies of marine secondary metabolites have been done because of the logistical problems of working under water and the primitive state of techniques for growing marine invertebrates in culture. However, recent studies indicate that many compounds isolated from marine macrospecies (e.g. sponges) may instead be produced by microorganisms (e.g. bacteria or phytoplankton). Studies of these symbionts may facilitate efforts to understand the biosyntheses of these metabolites.

1992 Secondary metabolites: their function and evolution. Wiley, Chichester (Ciba Foundation Symposium 171) p 236–254

Marine natural products have exploded as a field of study in the past 20 years, as attested by extensive annual reviews (Faulkner 1991), series of monographs (Scheuer 1973, 1983), and symposium reports (including Fautin 1988, Jefford et al 1988, Thompson et al 1991).

In part, the explosion is a result of improved physicochemical techniques, especially long-range nuclear magnetic resonance correlations which allow study of more complex structures, and fast atom bombardment and tandem mass spectrometry, which facilitate the study of minute quantities of non-polar compounds. In part, too, the investigation of marine natural products is a result of man's curiosity about the natural world around him, newly enhanced by the ability to penetrate the earth's aquatic environment by SCUBA and submersible techniques. The focus of a number of investigations has been on unique or characteristic aspects of marine life (Scheuer 1991)—for example, on sexual attractants in brown algae (Müller et al 1982), on trail-marking substances from sea hares (Fenical et al 1979), on the food chain from algae (Faulkner 1984) or sponges (Faulkner 1991) to molluscs, on antifouling substances (Keifer et al 1986), and on antifeedants (Tachibana & Gruber 1988). However, in large measure, the impetus to investigate marine natural products stems from a number of observations of pharmacological activity—for example, as antitumour, antiviral, antibacterial, antifungal, anti-inflammatory and cardioactive agents (Scheuer 1991, Fautin 1988, Jefford et al 1988, Thompson et al 1991, Zaborsky & Attaway 1992, Kaul & Sindermann 1978, Rinehart et al 1983) and as toxins (Halstead 1965, 1967, 1970, Hashimoto 1979).

In approaching this brief discussion of marine-derived secondary metabolites, I have chosen to stress macroorganisms (such as tunicates, sponges or seaweeds) rather than microorganisms, but to arrange the discussion according to compound types rather than phyla, with emphasis on novel or characteristic structural features of marine natural products and to use, where appropriate, pharmacologically active materials as examples.

Polyketides, macrolides and polyethers

Although biosynthetic evidence relating to these compounds is still sparse, there are numerous examples of compounds from marine species clearly derived by linear condensation of acetate, propionate or butyrate (Fig. 1). One of the simplest among them is laurencin (**1**) (Moore 1978), derived from *Laurencia glandulifera*, a red alga, which illustrates a characteristic feature of many marine secondary metabolites, the incorporation of halogen (Hager et al 1974) via the intervention of one of the haloperoxidases that are widespread in the marine environment. The cyclization can be envisaged as being triggered by a bromonium ion. The linear polyketide-derived plakortic acid (**2**), an active antifungal compound (Phillipson & Rinehart 1983), illustrates another characteristic of secondary metabolites from marine species, the peroxide group.

Recently, a number of polycyclic polyether toxins have been isolated from marine species. Prominent among the polyethers is bryostatin 1 (**3**) (Pettit et al 1982), currently in Phase I clinical trials as an anticancer agent. Another, okadaic acid (**4**), illustrates one difficulty inherent with marine species—identifying the true source of the compound. It was isolated originally from the sponges *Halichondria okadai* and *H. melanodocia* (Tachibana et al 1981) and most recently from a New England sponge, a *Phakellia* species (Sakai 1991), but it now appears that okadaic acid is produced by a dinoflagellate (perhaps *Dinophysis fortii*), along with the closely related dinophysistoxin and related compounds (Oshima et al 1982), and becomes concentrated in the sponges, which are efficient filter feeders.

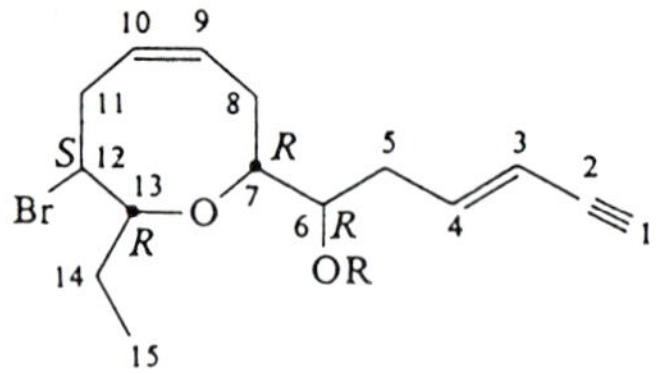

R = Ac

1 Laurencin

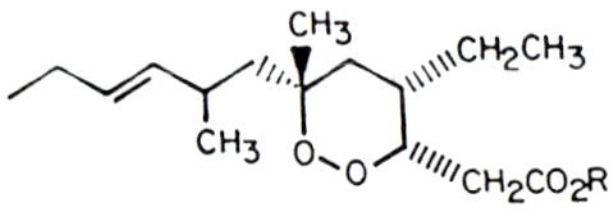

R = H

2 Plakortic acid

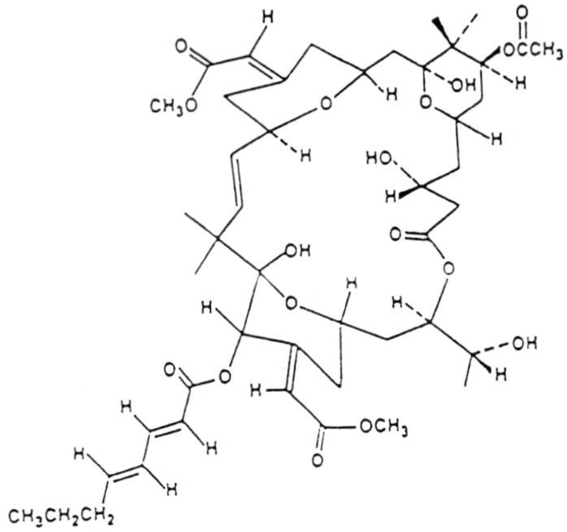

3 Bryostatin 1

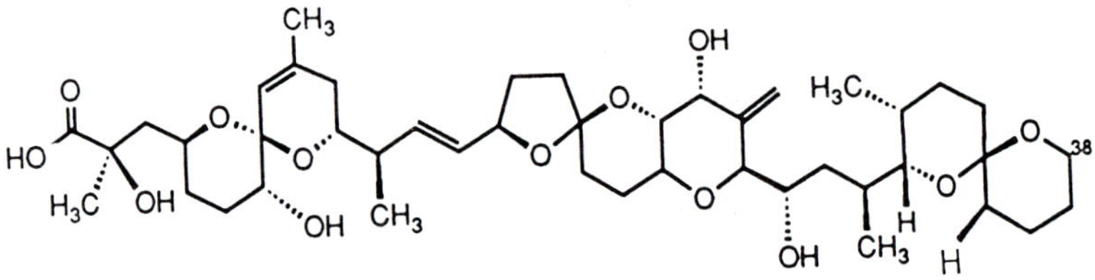

4 Okadaic acid

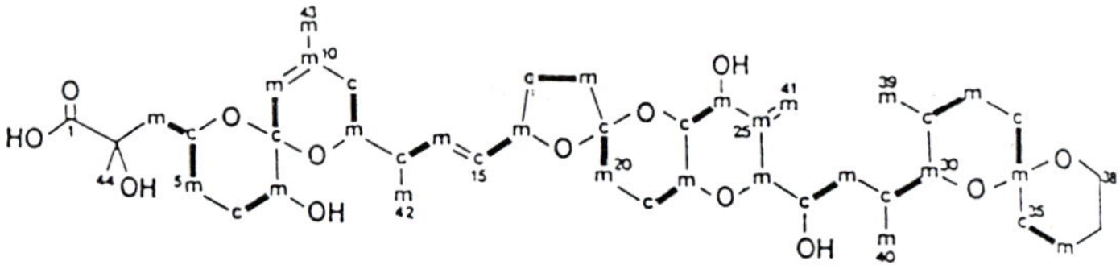

Okadaic acid labeling pattern

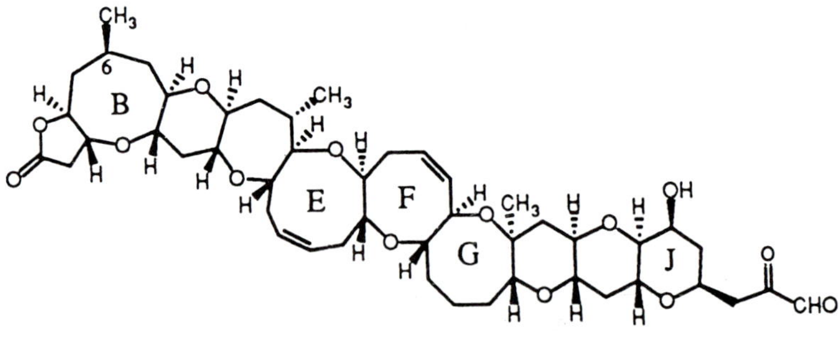

5 Brevetoxin A

FIG. 1. Marine natural products presumed to be derived from a polyketide biosynthetic pathway.

The brevetoxins (e.g. **5**), produced by the dinoflagellate *Gymnodinium breve*, are more complex polyethers. Their biosyntheses have been studied by two groups (Chou & Shimizu 1987, Lee et al 1986) and those of the related toxins okadaic acid and prorocentrolide by a third (Schmitz & Yasumoto 1991, Torigoe & Yasumoto 1991). All agree on a partial polyketide origin of these molecules, but the usual pattern is complicated by scrambling of the label from simple precursors (notably acetate).

Terpenes and steroids

For the most part, terpenes derived from marine organisms resemble closely their terrestrial counterparts (Fig. 2), as illustrated by the sesquiterpenoid antitumour agent crassin acetate (**6**) (Weinheimer et al 1968) and the diterpenoid antifouling agent renillafoulin C (Keifer et al 1986), both isolated from coelenterates. However, some systems (e.g. the capnellenetriol **7**) (Kaisin et al 1974) have been found first or only in marine products, and some functional groups, such as halogen substituents, are also characteristic of marine terpene metabolites. A particular example is laurinterol (**8**), from the red algal genus *Laurencia*, whose activity against Gram-positive bacteria rivals that of streptomycin (Sims et al 1975). It can be noted that the position brominated in laurinterol is that predicted by rules for electrophilic aromatic substitution by a bromonium ion.

Other groups characteristic of marine-derived terpenes are isonitrile, isothiocyanate and thiocyanate substituents, as illustrated in isocyanopupukeanane (**9**) (Karuso & Scheuer 1989). This group has been shown to originate from a cyanide ion.

Other marine-derived terpenes of interest are the diterpenes avarol (**10**) and avarone, which have been reported to be inhibitors of the human immunodeficiency virus (HIV) (Sarin et al 1987). Yet another terpene of interest is manoalide (**11**), one of only three marine natural products to have reached clinical trials (manoalide as an anti-inflammatory agent; Mayer & Jacobs 1988).

A large number of triterpene saponins have been obtained from holothurians (sea cucumbers) and have been found to have antifungal activity (Burnell & ApSimon 1983). These compounds, illustrated by holothurin A (**12**), generally contain a number of sugar substituents and are toxic to mammals as well.

No other class of marine natural products has been as exhaustively and systematically studied as the sterols. The Djerassi group (Kerr & Baker 1991) have done most of the work on their occurrence and biosynthesis. Marine sterols are characterized by complex patterns of alkylation in the side chain, as illustrated by calysterol (**13**), which has branching carbons derived from *S*-adenosylmethionine. Another class of marine steroids consists of saponins, highly oxygenated, carbohydrate-substituted metabolites found mainly in

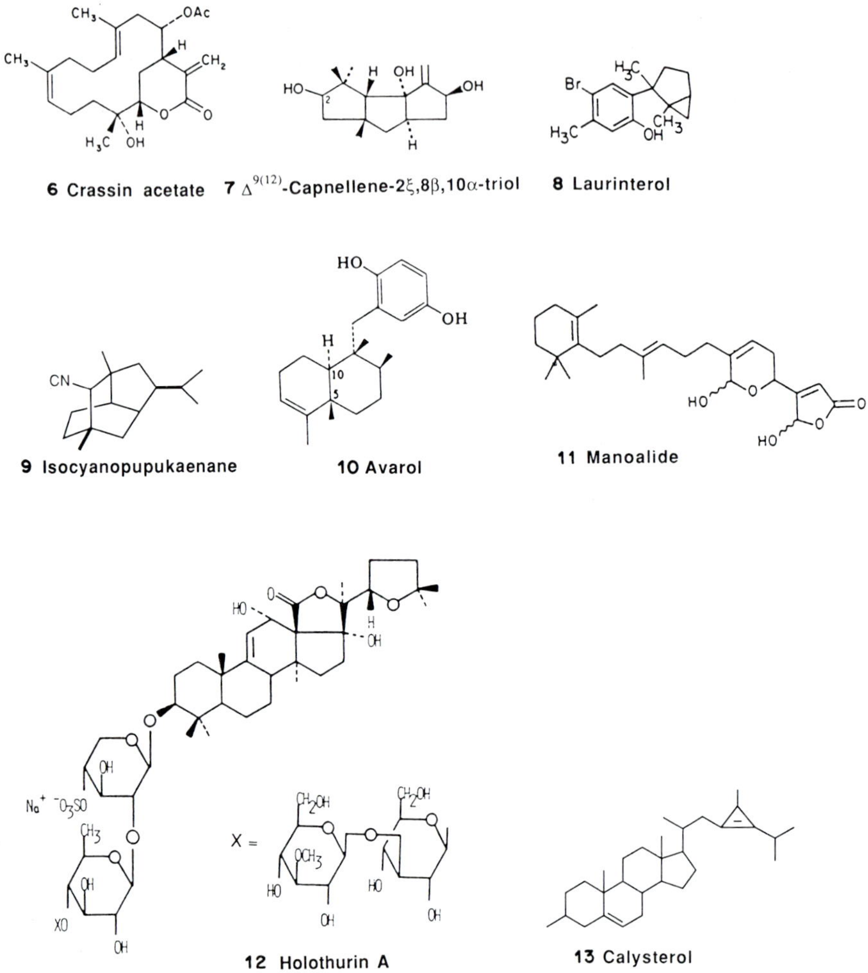

FIG. 2. Characteristic terpenes and steroids isolated from marine sources.

starfish (Burnell & ApSimon 1983). Some marine saponins (e.g. mosesin 1) are produced as antifeedants by different sole species and repel sharks (Tachibana & Gruber 1988), with the effect being enhanced by an accompanying polypeptide.

Linear peptides

A number of linear peptides containing common amino acids have been obtained from marine species. Examples include the sea anemone toxins (Fig. 3)

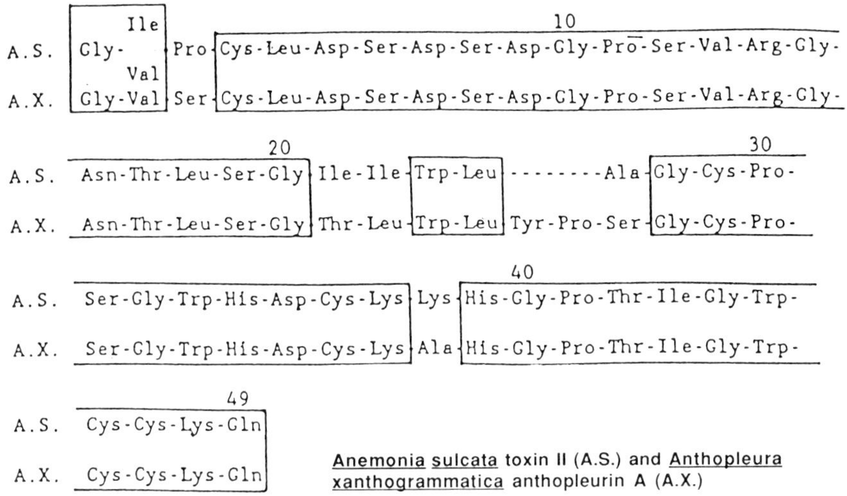

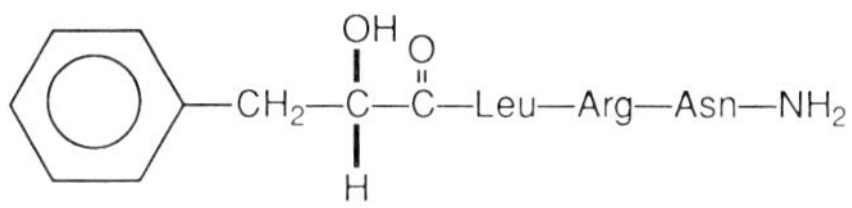

14 Antho–RNamide

FIG. 3. Linear peptides from sea anemones. *Above*: Cardioactive peptides; identical residues are boxed. *Below*: A neuropeptide isolated from *Anthopleura elegantissima*.

(Beress 1978, Norton et al 1978), linear peptides with molecular masses in the range 3000–5000 Da (Fig. 3). These are potent cardiostimulants, and it has been sometimes suggested that they could be replacements for digitalis, but their function in the animal is unknown. Smaller peptides from the anemone *Anthopleura elegantissima*, the hydromedusa *Polyorchis penicillatus*, and the sea pansy *Renilla koellikeri* are all neuropeptides, which stimulate motility by the coelenterates (Grimmelikhuijzen et al 1992). These peptides are capped with an amide group at the C-terminus and blocked at the N-terminus as well. In the case of antho-RNamide (**14**) the N-terminal blocking group is the previously unreported phenyllactyl group. Some of these neuropeptides are intriguing for another reason, being expressed as part of a larger peptide. For example, antho-RFamide, <Glu-Gly-Arg-Phe-NH_2, is expressed 14 times (as -Gln-Gly-Arg-Phe-Gly-) in the precursor protein from *Anthopleura elegantissima* (Grimmelikhuijzen et al 1992).

Cyclic peptides

Probably the most exciting compounds among the cyclic peptides (Fig. 4) are the didemnins (**15**), cyclic depsipeptides isolated from the colonial tunicate *Trididemnum solidum* (Rinehart et al 1981). Didemnin B is currently in Phase II clinical trial as an anticancer agent. All didemnins contain the *N*-methylated amino acids L-Me_2Tyr and D-MeLeu and the novel C_8 amino acid isostatine, which is apparently derived from D-alloisoleucine and acetate or malonate. The function of the didemnins in the tunicate is uncertain but they are potent cytotoxic and thus antifouling agents, killing the coral or sponge that the tunicate grows on (nothing grows on *T. solidum*), providing a competitive advantage over other filter feeders. Nor is it completely certain that the tunicate produces the didemnins, because it grows commensally with a cyanobacterium, *Synechocystis trididemni* (Lafargue & Duclaux 1979).

Another important class of cyclic peptides includes the microcystins (e.g. **16**) (heptapeptides; Botes et al 1985) and nodularin (**17**) (a pentapeptide; Rinehart et al 1988). These compounds are produced by the cyanobacteria *Microcystis aeruginosa* (freshwater) and *Nodularia spumigena* (brackish) and have dramatic hepatotoxic activity in mammals. They are also potent inhibitors of protein phosphatases, but their function in the cyanobacteria themselves is unknown. The most interesting aspect of their structures is the presence of the C_{20} amino acid Adda, which has recently been demonstrated in both of the above cyanobacteria (Moore et al 1991, Choi 1992) to arise from phenylalanine, acetate and methionine.

Modified amino acids

Among the modified amino acids found in marine species some of the most noteworthy are, as with the terpenes, halogenated compounds, which are found in sponges (Fig. 5). Thus, the genus *Aplysina* (formerly *Verongia*) contains a series of brominated conversion products of phenylalanine and, more directly, tyrosine. The most pharmaceutically interesting of these is probably aeroplysinin I (**18**) (Fattorusso et al 1970), which has some *in vivo* antitumour activity, but other more complex brominated tyrosine metabolites are also found. In studies with intact sponges, the conversion of [^{14}C]- and [^{15}N]phenylalanine and [^{14}C]tyrosine to some of the more abundant metabolites has been followed (Tymiak & Rinehart 1981, Carney 1990).

Other amino acid-derived metabolites with similar characteristics are the antiviral sceptrins (**19**) and ageliferins (**20**) (Keifer et al 1991) from the *Agelas* sponge genus, which also have a high degree of bromination and appear to be derived from pyrrolecarboxylic acid and its precursor (proline or hydroxyproline) and, perhaps, homoarginine.

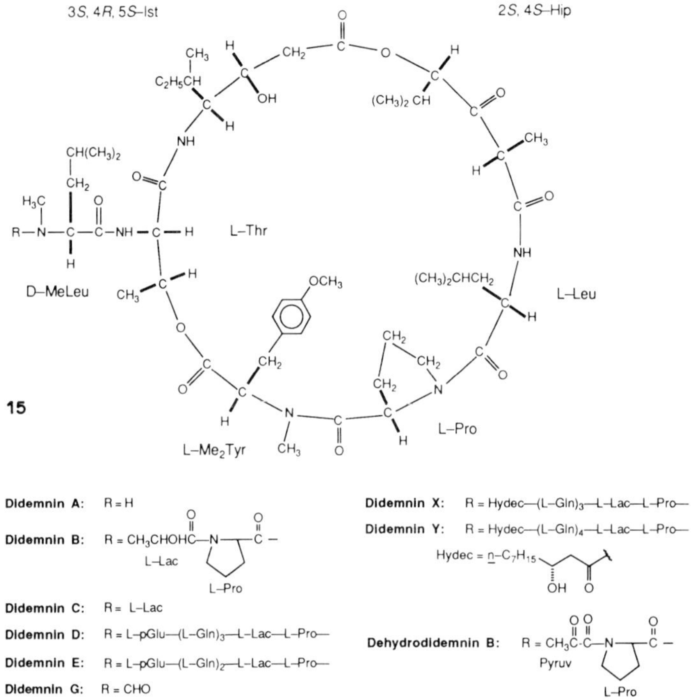

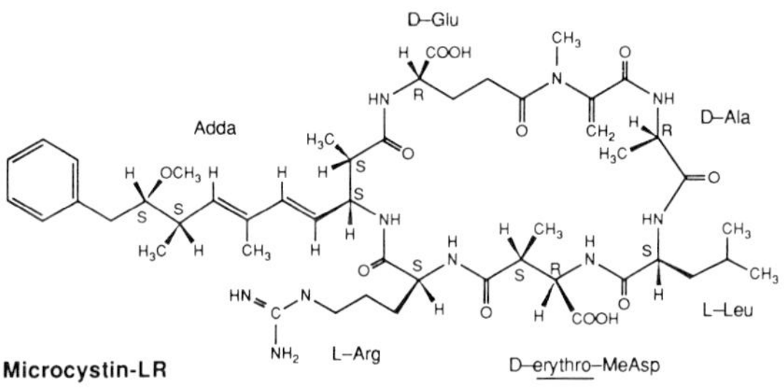

16 Microcystin-LR

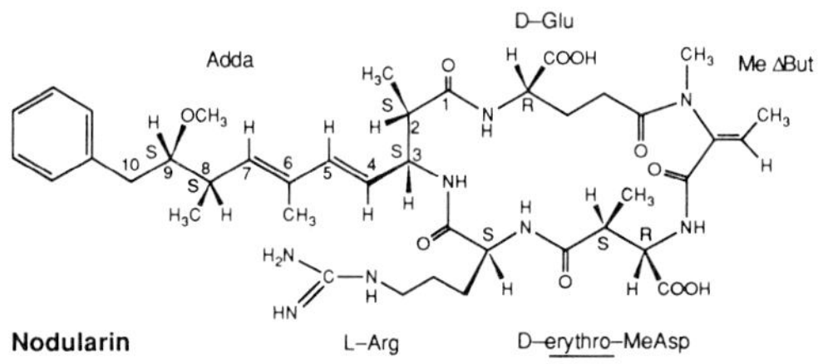

17 Nodularin

FIG. 4. Cyclic peptides from a tunicate and cyanobacteria.

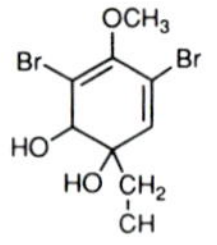

18 Aeroplysinin-I

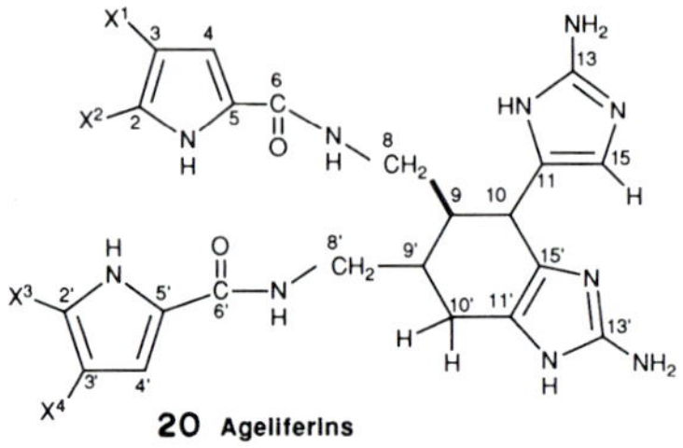

20 Ageliferins

X^1	X^2	X^3	X^4
Br	H	H	Br
Br	H*	Br*	Br
Br	Br	Br	Br

*Substituents may be interchanged.

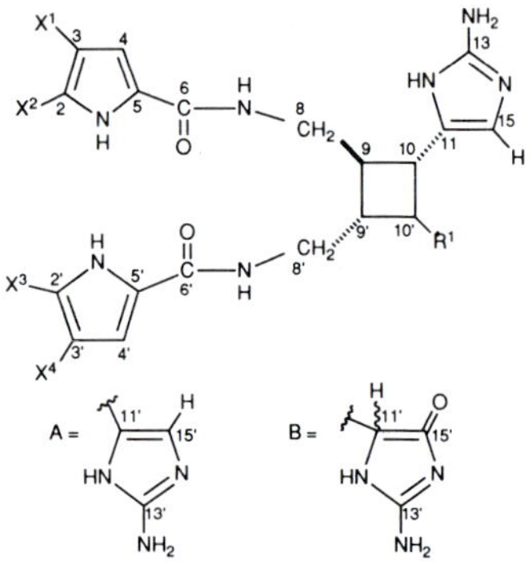

19 Sceptrins

X^1	X^2	X^3	X^4	R^1
Br	H	H	H	A
Br	H	H	Br	A
Br	Br	Br	Br	A

Oxysceptrins

X^1	X^2	X^3	X^4	R^1
Br*	H	H	H*	B
Br	H	H	Br	B

21

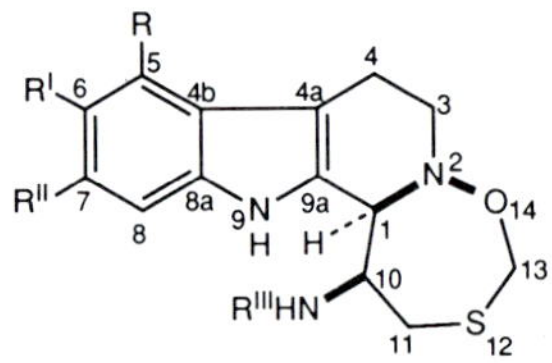

Eudistomin C: $R = R^{III} = H, R^{I} = OH, R^{II} = Br$
E: $R = Br, R^{I} = OH, R^{II} = R^{III} = H$
K: $R = R^{I} = R^{III} = H, R^{II} = Br$
L: $R = R^{II} = R^{III} = H, R^{I} = Br$

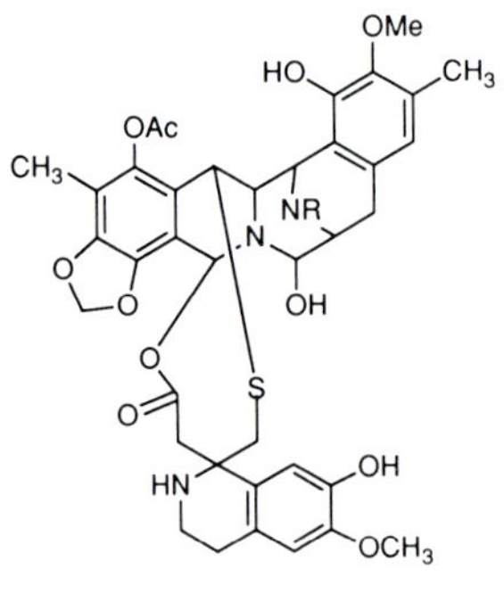

22 Ecteinascidin 729: R = H
Ecteinascidin 743: $R = CH_3$

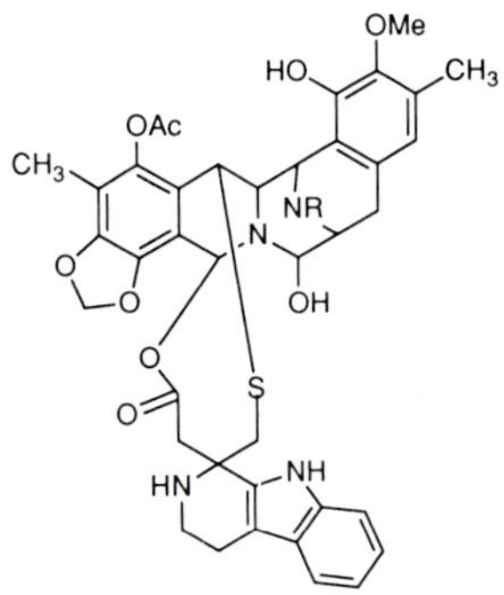

23 Ecteinascidin 722: R = H
Ecteinascidin 736: $R = CH_3$

FIG. 5. Marine natural products derived (or presumed to be) from amino acids.

Alkaloids

Nitrogen-containing heterocycles are among the most interesting marine secondary metabolites (Fig. 5). Like alkaloids produced by terrestrial plants, many are obviously derived by cyclization of amino acid precursors. Thus, the eudistomins (**21**), from the colonial tunicate *Eudistoma olivaceum*, are clearly derived biosynthetically from tryptophan (or tryptamine) and a second amino acid, as are terrestrial β-carbolines (Rinehart et al 1987). In the case of the oxathiazepinotetrahydro-β-carbolines like eudistomins C, E, K and L, their structure indicates that the requisite second amino acid is cysteine, although this has not yet been demonstrated. Their function in the tunicate is unknown but they are among the most active antiviral agents from marine organisms.

Similarly, the ecteinascidins (**22, 23**), from the colonial tunicate *Ecteinascidia turbinata*, are among the most active antitumour agents from marine sources (Rinehart et al 1990, Sakai 1991). Their biosynthesis has not been studied but that of the related *Streptomyces*-derived antibiotic saframycin A has (Mikami et al 1985), confirming the presumptive tyrosine as a precursor of the tetrahydroisoquinoline units. The β-carboline unit of ecteinascidins 722 and 736 (**23**) should be derived from tryptophan. As with most other secondary metabolites from marine organisms, the function of the ecteinascidins is not known. However, the tunicates are almost always found clean of other organisms, so an antifouling activity for these filter feeders can be suspected. The larvae of *E. turbinata*, too, appear to be chemically protected, perhaps also by ecteinascidins.

A class of compounds apparently unique to marine secondary metabolites (usually found in tunicates or sponges) consists of a set of aromatic heterocycles with several condensed rings (Fig. 6), illustrated by amphimedine (**24**) (Schmitz et al 1983). These compounds are usually toxic to mammalian cells but their function in the marine animals which produce them is unknown and their biosyntheses have not been investigated.

Porphyrins

An intriguing example of an apparent symbiosis between two marine organisms is provided by the pigment tunichlorin (Fig. 6, **25**), isolated from the same tunicate (*T. solidum*) that produces the didemnins (Bible et al 1988). This chlorin is related to chlorophyll, but in it magnesium is replaced by nickel (in the only natural nickel-containing chlorin), the vinyl group is replaced by a hydroxymethyl, and the long-chain phytol ester is lacking. It appears that the commensal blue-green alga noted above (*Synechocystis trididemni*) produces phaeophytin a, which is then modified by the tunicate, which presumably has previously concentrated nickel. It is significant that only chlorophyll a is

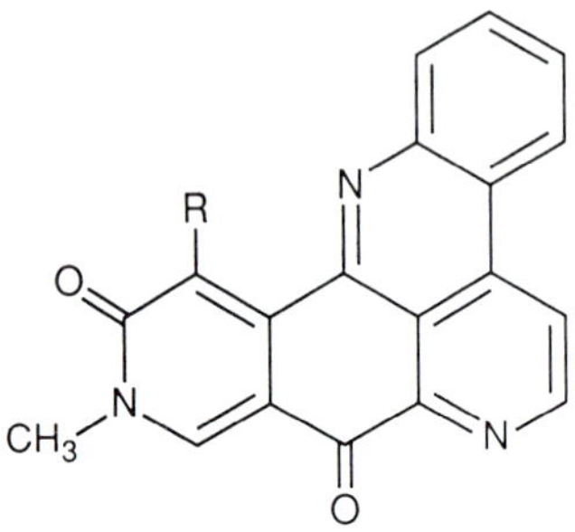

R = H

24 Amphimedine

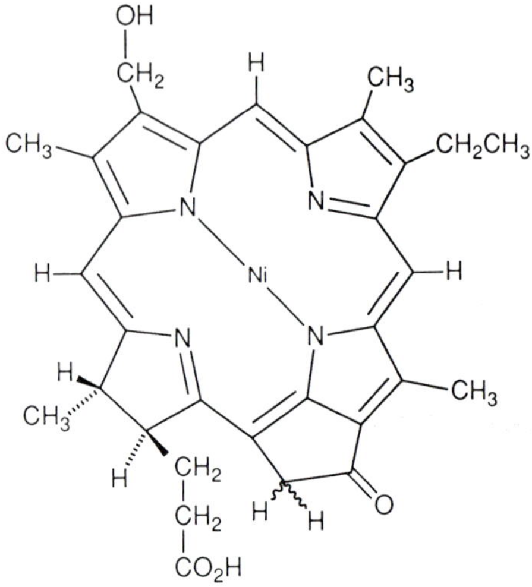

25 Tunichlorin

FIG. 6. Polycyclic aromatic marine natural products.

produced by the microorganism; this characterizes it as a cyanobacterium rather than a prochloron (Lewin 1983), which would have produced both chlorophylls a and b.

Conclusions

Marine-derived secondary metabolites provide a wealth of compounds with intriguing structures. Although most structural features correspond directly to those found in terrestrial organisms, some are quite characteristic of marine organisms, such as halogen, isonitrile, isothiocyanate or peroxy substituents or highly condensed aromatic heterocycles. Many of the compounds have potent bioactivities as toxins or therapeutic agents but relatively little is known about their roles in their natural environment—in inhibiting predation, providing competitive advantage, or marking trails, for example.

Study of the biosynthesis of the secondary metabolites has been inhibited by difficulties in working with marine macroorganisms in the ocean and with culturing them in the laboratory. Moreover, symbiosis and the inevitable food chain confuse the issue of exactly which organism is producing the compound of interest.

The challenges posed by the field make it certain to be a provider of rich scientific rewards in the future.

Acknowledgements

Our research at the University of Illinois on marine natural products has been supported at various times by grants from the National Institute of Allergy and Infectious Diseases (AI 4769, AI 1278), the National Institute of General Medical Sciences (GM 27029), the National Science Foundation, The Upjohn Co, SeaPharm, Inc and PharmaMar, SA.

References

Beress L 1978 Biologically active polypeptides, toxins and proteinase inhibitors from the sea anemones. In: Kaul PN, Sinderman CJ (eds) Drugs and food from the sea. University of Oklahoma, Norman, OK, p 59–72

Bible KC, Buytendorp M, Zierath PD, Rinehart KL 1988 Tunichlorin: a nickel chlorin isolated from the Caribbean tunicate *Trididemnum solidum*. Proc Natl Acad Sci USA 85:4582–4586

Botes DP, Wessels PL, Kruger H et al 1985 Structural studies on cyanoginosins-LR, -YR, -YA, and -YM, peptide toxins from *Microcystis aeruginosa*. J Chem Soc Perkin Trans I, p 2747–2748

Burnell DJ, ApSimon JW 1983 Echinoderm saponins. In: Scheuer PJ (ed) Marine natural products, vol V. Academic Press, New York, p 287–389

Carney JR 1990 The biosynthesis of brominated tyrosine metabolites by *Aplysina fistularis*. PhD thesis, University of Illinois, Urbana, IL

Choi BW 1992 Studies on secondary metabolites produced by actinomycetes and cyanobacteria. PhD thesis, University of Illinois, Urbana, IL

Chou H-N, Shimizu Y 1987 Biosynthesis of brevetoxins. Evidence for the mixed origin of the backbone carbon chain and the possible involvement of dicarboxylic acids. J Am Chem Soc 109:2184–2185

Fattorusso E, Minale L, Sodano G 1970 Aeroplysinin-I, a new bromo-compound from *Aplysina aerophoba*. Chem Commun (J Chem Soc Sect D) p 751–752

Faulkner DJ 1984 Marine natural products: metabolites of marine algae and herbivorous marine molluscs. Nat Prod Rep 1:251–280

Faulkner DJ 1991 Marine natural products. Nat Prod Rep 8:97–147, and earlier reports

Fautin DG (ed) 1988 Biomedical importance of marine organisms. California Academy of Sciences, San Francisco, CA

Fenical W, Sleeper HL, Paul VJ, Stallard MO, Sun HH 1979 Defensive chemistry of *Navanax* and related opisthobranch molluscs. Pure Appl Chem 51:1865–1874

Grimmelikhuijzen CJP, Carstensen K, Darmer D et al 1992 Coelenterate neuropeptides: structure, action and biosynthesis. Am Zool 32:1–12

Hager LP, White RH, Hollenberg PF, Doubek DL, Brusca RC, Guerrero R 1974 A survey of organic halogens in marine organisms. In: Webber HH, Ruggieri GD (eds) Food-drugs from the sea. Marine Technology Society, Washington, DC, p 421–428

Halstead BW 1965 Poisonous and venomous marine animals of the world, vol 1: Invertebrates. Government Printing Office, Washington, DC

Halstead BW 1967 Poisonous and venomous marine animals of the world, vol 2: Vertebrates. Government Printing Office, Washington, DC

Halstead BW 1970 Poisonous and venomous marine animals of the world, vol 2: Vertebrates (contd). Government Printing Office, Washington, DC

Hashimoto Y 1979 Marine toxins and other bioactive marine metabolites. Japan Scientific Societies Press, Tokyo (Translated by Konosu S, Hashimoto K, Onoue Y, Fusetani N)

Jefford CW, Rinehart KL, Shield LS (eds) 1988 Pharmaceuticals and the sea. Technomic, Lancaster, PA

Kaisin M, Sheikh YM, Durham LJ et al 1974 Capnellane—a new tricyclic sesquiterpene skeleton from the soft coral *Capnella imbricata*. Tetrahedron Lett 2239–2242

Karuso P, Scheuer PJ 1989 Biosynthesis of isocyanoterpenes in sponges. J Org Chem 54:2092–2095

Kaul PN, Sindermann CJ (eds) 1978 Drugs and food from the sea. University of Oklahoma Press, Norman, OK, and earlier volumes

Keifer PA, Rinehart KL Jr, Hooper IR 1986 Renillafoulins, antifouling diterpenes from the sea pansy *Renilla reniformis* (octocorallia). J Org Chem 51:4450–4454

Keifer PA, Schwartz RE, Koker MES, Hughes RG Jr, Rittschof D, Rinehart KL 1991 Bioactive bromopyrrole metabolites from the Caribbean sponge *Agelas conifera*. J Org Chem 56:2965–2975

Kerr RG, Baker BJ 1991 Marine sterols. Nat Prod Rep 8:465–497

Lafargue F, Duclaux G 1979 Premier example en Atlantique tropical d'une association symbiotique entre une ascidie didemnidae et une cyanophycée chroococale: *Trididemnum cyanophorum* nov. sp. et *Synechocystis trididemni* nov. sp. Ann Inst Oceanogr Paris 55:163–184

Lee MS, Repeta DJ, Nakanishi K, Zagorski MG 1986 Biosynthetic origins and assignments of ^{13}C nmr peaks of brevetoxin B. J Am Chem Soc 108:7855–7856

Lewin RA 1983 The problems of *Prochloron*. Ann Microbiol (Paris) 134B:37–41

Mayer AMS, Jacobs RS 1988 Manoalide: an antiinflammatory and analgesic marine natural product. In: Fautin DG (ed) Biomedical importance of marine organisms. California Academy of Sciences, San Francisco, CA

Mikami Y, Takahashi K, Yazawa K et al 1985 Biosynthetic studies on saframycin A, a quinone antitumor antibiotic produced by *Streptomyces lavendulae*. J Biol Chem 260:344–348

Moore RE 1978 Algal nonisoprenoids. In: Scheuer PJ (ed) Marine natural products, vol I. Academic Press, New York, p 43–124

Moore RE, Chen JL, Moore BS, Patterson GML, Carmichael WW 1991 Biosynthesis of microcystin-LR. Origin of the carbons in the Adda and Masp units. J Am Chem Soc 113:5083–5084

Müller DG, Gassman G, Marner F-J, Bolard W, Jaenicke L 1982 The sperm attractant of the marine brown alga *Ascophyllum nodosum* (phaeophyceae). Science (Wash DC) 218:1119–1120

Norton TR, Kashiwagi M, Shibata S 1978 Anthopleurin A, B, and C, cardiotonic polypeptides from the sea anemones. In: Kaul PN, Sindermann CJ (eds) Drugs and food from the sea. University of Oklahoma Press, Norman, OK, p 37–50

Oshima Y, Murakami Y, Yasumoto T 1982 Identification of okadaic acid as a toxic component of a marine dinoflagellate *Prorocentrum lima*. Bull Jpn Soc Sci Fish 48:69–72

Pettit GR, Herald CL, Doubek DL, Herald DL, Arnold E, Clardy J 1982 Isolation and structure of bryostatin 1. J Am Chem Soc 104:6846–6848

Phillipson DW, Rinehart KL Jr 1983 Antifungal peroxide-containing acids from two Caribbean sponges. J Am Chem Soc 105:7735–7736

Rinehart KL Jr, Gloer JB, Hughes RG Jr et al 1981 Didemnins: antiviral and antitumor depsipeptides from a Caribbean tunicate. Science (Wash DC) 212:933–935

Rinehart KL Jr, Gloer JB, Wilson GR et al 1983 Antiviral and antitumor compounds from tunicates. Fed Proc 42:87–90 and other papers in the symposium

Rinehart KL Jr, Kobayashi J, Harbour GC et al 1987 Eudistomins A-Q, β-carbolines from the antiviral Caribbean tunicate *Eudistoma olivaceum*. J Am Chem Soc 109:3378–3387

Rinehart KL, Harada K, Namikoshi M et al 1988 Nodularin, microcystin, and the configuration of Adda. J Am Chem Soc 110:8557–8558

Rinehart KL, Holt TG, Fregeau NL et al 1990 Ecteinascidins 729, 743, 745, 759A, 759B, and 770, potent antitumor agents from the Caribbean tunicate *Ecteinascidia turbinata*. J Org Chem 55:4512–4515

Sakai R 1991 Bioactive compounds from tunicates and a sponge. PhD thesis, University of Illinois, Urbana, IL

Sarin PS, Sun D, Thornton A, Müller WEG 1987 Inhibition of replication of the etiologic agent of acquired immune deficiency syndrome (human T-lymphotropic retrovirus/lymphadenopathy-associated virus) by avarol and avarone. J Natl Cancer Inst 78:663–666

Scheuer PJ 1973 Chemistry of marine natural products. Academic Press, New York

Scheuer PJ (ed) 1983 Marine natural products. Chemical and biological perspectives, vol V. Academic Press, New York, and earlier volumes

Scheuer PJ (ed) 1991 Bioorganic marine chemistry, vol IV. Springer-Verlag, Berlin, and earlier volumes

Schmitz FJ, Yasumoto T 1991 The 1990 United States–Japan seminar on bioorganic marine chemistry: meeting report. J Nat Prod 54:1469–1490

Schmitz FJ, Agarwal SK, Gunasekera SP, Schmidt PG, Shoolery JN 1983 Amphimedine, new aromatic alkaloid from a Pacific sponge *Amphimedon* sp. Carbon connectivity determination from natural abundance ^{13}C–^{13}C coupling constants. J Am Chem Soc 105:4835–4836

Sims JJ, Donnell MS, Leary JV, Lacy GH 1975 Antimicrobial agents from marine algae. Antimicrob Agents Chemother 7:320–321

Tachibana K, Gruber SH 1988 Shark repellent lipophilic constituents in the defense secretion of the Moses sole (*Pardachirus marmoratus*). Toxicon 26:839–853

Tachibana K, Scheuer PJ, Tsukitani Y et al 1981 Okadaic acid, a cyclotoxic polyether from two marine sponges of the genus *Halichondria*. J Am Chem Soc 103:2469–2471

Thompson M-F, Sarojini R, Nagabhushanam R (eds) 1991 Bioactive compounds from marine organisms. Oxford & IBH Publishing Co Pvt Ltd, New Delhi, India

Torigoe K, Yasumoto T 1991 Biosynthesis of polyether compounds from dinoflagellates. In: Hayashi Y (ed) 33rd Symposium on the Chemistry of Natural Products, Osaka, Japan: Symposium Papers. Osaka City University, p 525–532

Tymiak AA, Rinehart KL Jr 1981 Biosynthesis of dibromotyrosine-derived antimicrobial compounds by the marine sponge *Aplysina fistularis* (*Verongia aurea*). J Am Chem Soc 103:6763–6765

Weinheimer AJ, Middlebrook RE, Bledsoe JO Jr, Marsico WE, Karns TKB 1968 Eunicin, an oxa-bridged cembranolide of marine orgin. Chem Commun (J Chem Soc Sect D) p 384–385

Zaborsky OR, Attaway DH (eds) 1992 Advances in marine biotechnology, vol 1. Plenum Press, New York

DISCUSSION

Hopwood: I am fascinated by the linear peptides of the sea anemones where a precursor gene with several copies of the coding sequence within it has been found. Is it known how the processing of the primary protein gene product takes place?

Rinehart: Not yet. This work is very recent, by Grimmelikhuijzen et al (1992), who have established that the protein yields more than one copy of the neuropeptide.

Williams: I think the peptide processing must be analogous to the processing found for other neuropeptides. We found in *Xenopus laevis* that there are lytic peptides produced from long precursors. There is a Gly-Arg-Arg sequence in the precursor sequence, and the dibasic residue (Arg-Arg) is recognized as a site of cleavage. The processing cleavage then produces a peptide, as in this case, with glycine at the N-terminus. The glycine becomes hydroxylated at the α-carbon and this product is then cleaved by a retro-aldol reaction. This leaves the nitrogen of the glycine residue at the C-terminus of the peptide as an amide. So instead of a carboxylate, you end up with $CONH_2$, and the nitrogen of that group comes from the glycine that would normally follow in the peptide sequence.

Rinehart: There are specificities at the N-terminus as well. The pyroglutamyl unit comes from a glutamine residue, but generally there is an acidic amino acid to the left of that in a rather well-defined sequence. One thing which we had hoped to find, but didn't, was a phenylalanine as an N-terminus which would be converted to a phenyllactic acid, as in antho-RNamide (**14**) (Fig. 3, p 241). One way to get that is by transamination of a phenylalanine unit to a ketone, and its reduction to a phenyllactyl group.

Davies: Are all the peptides glycine amides?

Rinehart: Yes.

Davies: There is a very strong parallel with a large number of neuropeptide hormones found in mammals.

Williams: Substance P is a case in point.

Rinehart: In fact, Grimmelikhuijzen has pointed out that the coelenterates are probably the phylogenetically oldest organisms which contain neural peptides. He feels there may be a genealogical relationship to mammalian neuropeptides.

Cavalier-Smith: The question has been raised whether any secondary compounds were acquired by endosymbiosis (the intracellular establishment of foreign cells). For saxitoxin, found in toxic dinoflagellates, the compound is known to be made both by dinoflagellates and by a cyanobacterium, so there is a possibility that it might have come into the dinoflagellate through chloroplast endosymbiosis. Is anything known about the enzymes making saxitoxin in either group, and whether they are homologous?

Baldwin: Very little is known about the enzymic biosynthesis, but Shimizu has shown that it is derived from the amino acid arginine (Gupta et al 1989).

Rinehart: The origin of the carbon and nitrogen atoms is indeed known, but the enzymes are not, as far as I know.

Demain: Most of the studies of microorganisms making these secondary products apparently have been made *after* a period of time in which it was thought that the higher marine form was making these compounds. What about the intentional examination of marine microorganisms? I realize there are probably a lot of compounds made, but there must have been some examination of various types of microorganisms for the products that they make. Are they unique products, compared to related terrestrial microorganisms?

Rinehart: I don't think so. At least several classes of compounds are the same. Much work on streptomycetes' production has been done by Okami's group (Takeuchi 1987). The compounds are typical of terrestrial streptomycetes.

Compounds produced by *Vibrio* species are, however, characteristic of marine bacteria. A number of groups are looking at marine bacterially derived material in Japan, in Spain, and in the USA, so far mainly with the goal of finding pharmaceutically useful agents.

Williams: What is known about the mechanism by which puffer fish protect themselves from toxins which they have taken up via the food chain, and to which they are not themselves susceptible?

Rinehart: The storage of tetrodotoxin doesn't seem to bother the fish, but I don't know why. Ciguatoxin has been probably the most elusive of the toxins through the years. Dr T. Tosteson recently told me that as many as 40% of the fish in Puerto Rico sometimes contain this toxin. Ciguatoxin turns out to be produced by another dinoflagellate, *Gambierdiscus toxicus*, as has recently been shown by Yasumoto (Murata et al 1989). They were specifically looking for the microorganism, and cultured it from toxic Moray eels, which are especially prone to eat detritus or other fish, concentrating the ciguatoxin through the food chain. However, I don't know how eels protect themselves either.

Davies: You mentioned the terpene metabolite laurinterol, which has an activity against Gram-positive bacteria approaching that of streptomycin. Is anything known about its toxicity?

Rinehart: It is active against both Gram-positive and Gram-negative organisms *in vitro*, and it is toxic. Many of the halogenated secondary metabolites are acidic phenols, which explains their toxicity.

Brandl: You have reported many interesting metabolites. Have you ever tested them in plant protection applications, perhaps aimed against fungi, weeds, or insects?

Rinehart: We haven't done this, but there are groups looking at such applications. Four marine natural products are currently marketed. Kainic acid is used as an antihelminthic drug in Japan. Padan is in agricultural use, modelled after a marine product. The other two, Ara-A and Ara-C, are also analogues of marine natural products and are in use as antiviral and antitumour agents, respectively; they were modelled after spongouridine and spongothymidine, discovered in a sponge. Thus, of the four compounds marketed at the moment, one is for agricultural use.

Davies: In terms of using these compounds commercially, can you say anything about the problems of trying to grow marine organisms in culture, if that's what one has to do in order to make the product?

Rinehart: I don't think this is necessarily what one needs to do to obtain marine natural products. In some cases it will be possible, I think, to get enough material from the marine environment. This would be feasible, for example, with didemnin B, if that becomes a useful drug, because *Trididemnum solidum* is a nuisance, overgrowing and killing coral. The local naturalists in Bonaire might very much like to have it harvested, therefore. But for ecteinascidins this approach would be difficult, because of the small amounts produced. Since it is present only to the extent of $10^{-5}\%$, one has to harvest an enormous amount of tunicate. Probably it can be cultured. Studies in Puerto Rico suggest that the life cycle of *Ecteinascidia turbinata* is 2–3 months, so three or four harvests per year would be theoretically possible, and useful, if one had big enough aquaria. The culture of *E. turbinata* is being studied at PharmaMar in Madrid and a company in San Diego, California proposed growing *T. solidum* in culture. There are questions of temperature, of salinity, and of trace metals which all have to be answered in these studies. I think *Trididemnum* would be difficult, because of the symbiosis between the tunicate and a cyanobacterium, but *E. turbinata* is quite feasible because it grows very fast.

Escoubas: In the case of antagonistic interactions between organisms, is there any evidence in marine systems for the induction of the production of the compounds, upon an attack by a predator, for instance? In other words, can a parallel be drawn with terrestrial ecosystems where such phenomena have been described, for example in plant–herbivore interactions?

Rinehart: I am not aware of any.

Waterman: You gave us this example of an organism (*Ecteinascidia turbinata*) living on mangrove roots, and speculated that it is producing secondary metabolites that keep it clean in a very dirty environment. How do you think this is happening? Is it slowly releasing compounds into the water?

Rinehart: I presume it is. Tunicates are filter feeders with two holes, for water coming in and going out. We have never measured ecteinascidin in the water leaving the tunicate, but we did that with sponges and demonstrated that up to 2–3 inches from the top of a cylindrical sponge, you can measure antibacterial metabolites.

Waterman: It would have to be a constant production, to maintain suitable inhibitory levels in the immediate environment.

Rinehart: Yes, in *Ecteinascidia* at a very low level, because the ecteinascidins are exceedingly active cytotoxic compounds, at nanogram levels.

Davies: Louis Nisbet spoke earlier about microorganisms as sources of interesting pharmaceutical agents. What is your opinion about marine microorganisms as potential sources?

Nisbet: The fascinating thing here is that we seem to be finding new chemistries; that's the important point for the pharmaceutical industry. The whole subject of so-called biodiversity, and how that relates to chemical diversity, is a fascinating subject. I think that marine organisms are important sources of chemical diversity, but, given the challenges of growing them, there is even a greater pressure on those searching through these chemicals to look for smaller, perhaps synthesizable molecules; this may take us away from some of the novel chemistry that is present in some of the more complex compounds.

Rinehart: Organic chemists are very clever people! One compound that has been of interest for five to eight years now is halichondrin B, which was shown to have potent antitumour activity *in vivo*. Now Y. Kishi's group has synthesized it (Aicher et al 1992).

Baldwin: The recent history of chemical synthesis providing commercially available quantities of complex, natural product-related substances is very bad. The best case is the thienamycin synthesis by Merck, and I think this involves about 22 stages. It's based upon the fact that it can be sold at an enormous cost, as an antibiotic drug. If you are going to do a total synthesis of some of the structures you have shown us, I wouldn't put any time base on when you are going to get the materials! That's the problem.

Hunter: Earlier, we discussed the pre-antibiotic era and the development of drug resistance (p 234). I imagine that marine organisms have been putting these compounds out into the environment for very many years. Is there any incidence of other organisms acquiring protective mechanisms against these compounds, or is it too early to tell?

Rinehart: I don't think there are any studies yet on this. We have been concentrating on macroorganisms; studies on microorganisms haven't been done, so far.

Bu'Lock: To some extent, you could regard the cases where an organism takes in some of these highly active materials, segregates them into specialized organs or organelles, and appropriates them for use as part of their own defence mechanisms, as going one better than simple detoxification or other resistance mechanisms.

Rinehart: Nudibranchs (such as *Dolabella auricularia*) are defenceless molluscs that don't have shells. Some of the most interesting compounds have come from them, but generally those compounds prove in fact to be produced by something else, like a sponge. Frequently, the coloration of the nudibranch depends on the dyes in the sponge that it has been feeding on.

References

Aicher TD, Buszek KR, Fang FG et al 1992 Synthetic studies towards halichondrins: synthesis of the C.27-C.38 segment. Tetrahedron Lett 33:1549–1552

Grimmelikhuijzen CJP, Carstensen K, Darmer D et al 1992 Coelenterate neuropeptides: structure, action and biosynthesis. Am Zool 32:1–12

Gupta S, Norte M, Shimizu Y 1989 Biosynthesis of saxitoxin analogues: the origin and introduction mechanism of the side-chain carbon. J Chem Soc Chem Commun, p 1421–1424

Murata M, Legrand AM, Ishibashi Y, Yasumoto T 1989 Structures of ciguatoxin and its congener. J Am Chem Soc 111:8929–8931

Takeuchi T 1987 Bioactive products from marine microorganisms. In: Takeuchi T (ed) Institute of Microbial Chemistry, 1962–1987. Microbial Chemistry Research Foundation, Tokyo, Japan, p 44–48

Roles for secondary metabolites in plants

Peter G. Waterman

Phytochemistry Research Laboratories, Department of Pharmaceutical Sciences, University of Strathclyde, Glasgow G1 1XW, Scotland

Abstract. More than about 20 000 secondary metabolites have now been identified and their isolation and characterization continues at an undiminishing rate. Although the production of most secondary metabolites is under genetic control, in relatively few cases have convincing arguments been put forward to rationalize their occurrence in terms of primary metabolic functions. Whatever the initial reason for their evolution, secondary metabolites are now an essential part of the armamentaria used by plants in the battle to survive and propagate, to the extent where the expenditure of energy, photosynthate and nutrients for their production can be demonstrated to be 'cost effective' for that. Their role may centre on defence of the producer against predators (herbivores), pathogens or competitors, on aid to pollination or seed dispersal, or on protection against or adaptation to extrinsic abiotic factors, or on combinations of these functions. Various examples are given in support of this argument.

1992 Secondary metabolites: their function and evolution. Wiley, Chichester (Ciba Foundation Symposium 171) p 255–275

Plant biochemists may differ in what they perceive to be secondary metabolites, but all will agree that living organisms are capable of producing a remarkable array of organic compounds with limited distribution. Currently, the number of isolated and chemically defined secondary metabolites must stand in excess of 20 000 (Harborne 1988), and may be considerably greater. Given that techniques for their separation and characterization are still improving and that many species are still only partially investigated or wholly uninvestigated, the complete array of such metabolites must be vast indeed. Of course, that array will include the many minor variations that occur with each skeletal theme (through methylation, glycosylation, etc.) but, because apparently minor chemical changes can lead to appreciable changes in biological activity, it seems valid to consider each structure as having distinct attributes.

The hypothesis supported in this paper is that most of these metabolites, through interaction with the environment, or with other organisms in the environment, enhance the prospects for survival of the producer or its offspring.

In proposing this function it is not necessary to presume that this was the *raison d'être* for the metabolite. Rather, the presumption being made is that their continued presence confers an advantage sufficient to make the expenditure of primary metabolites incurred in their production cost-effective.

Obvious routes through which advantage can be obtained from secondary metabolites concern the defence of the producer against predators, pathogens or competitors. These functions imply bioactivity, but this is not a prerequisite for usefulness; for example, pollination and indication of fruit ripeness are equally important roles. Nor is it necessary to think only in terms of direct interactions with other organisms—secondary metabolites may also function in more subtle ways by influencing the external environment.

Secondary metabolites in defence against predators and pathogens

No one who has ever seen a cow pasture, closely grazed except for a profusion of ragwort (*Senecio jacobaea*), can doubt the feeding-deterrent properties of at least some secondary metabolites. In ragwort the antifeedant compounds are pyrrolizidine alkaloids such as senecionine (**1**), which seem particularly active in deterring mammalian herbivores. The deterrent power of hydrocyanic acid has been demonstrated by comparative studies of cyanogenic and acyanogenic forms of clover (*Trifolium repens*) and *Lotus corniculatus*, which reveal that the acyanogenic form is preferred by slugs and snails (Compton & Jones 1985). The capacity of the cyanogenic clover to deter large herbivores or insects is less well substantiated (Dirzo & Harper 1982, Mowat & Shakell 1989). Alkaloids as a class of secondary metabolite have long been recognized for their high propensity for biological activity. This is even more striking if one remembers that the nitrogen-containing antibiotic products of *Penicillium* and *Streptomyces* are technically alkaloids.

Of course, no defence is unbreachable and it is to be expected that counter-adaptations will occur. Among insects in particular, there are many examples of the use of different groups of secondary metabolite as feeding or egg-laying cues; the relevant plant metabolites are then avoided, detoxified or sequestered, or are even put to use by the herbivore. For example, pyrrolizidine alkaloids

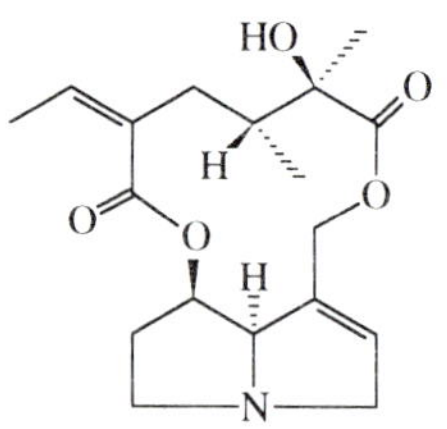

1 senecionine

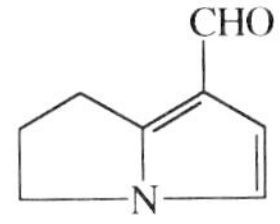

2 danaidal

can be exploited by many insects. Perhaps the most well-explored example is in danaiad butterflies, which convert them into pheromones such as danaidal (**2**) (Boppre 1990). Other examples include the arctiid moth *Creatonotos transiens* which is able to modify pyrrolizidines to act as defence compounds and as pheromones (von Nickisch-Rosenegk et al 1990), and the grasshopper *Zonocerus variegatus*, which can store them, and where they presumably act as a deterrent to predators (Bernays et al 1977).

Some classes of secondary metabolite seem particularly effective as insect antifeedants. Terpene derivatives such as azadirachtin (**3**) and pyrethrin-1 (**4**) have very specific insect toxicity. Saponins, iridoids, sesquiterpene lactones and non-protein amino acids also commonly exhibit insect-deterrent activity. Other triterpene derivatives, like cardenolides (**5**), appear to act specifically in mammals (as heart toxins). Danaiad butterflies are also able to sequester plant-produced cardenolides (Brower et al 1984), this time to be used as a 'chemical defence' against predators. Bufodienolides (**6**), steroidal lactones very similar to cardenolides, are produced by frogs and toads and stored in the skin as a very effective defence against predators. Plants also commonly store active metabolites in their bark: a volatile resin, consisting mainly of long-chain methyl ketones, released from the injured bark of *Commiphora rostrata* (Burseraceae) (McDowell et al 1988), presents a most effective barrier against fungal pathogens (Table 1).

In relation to feeding deterrency, more has been written about phenolic compounds than about any other metabolites. There is no doubt that all

TABLE 1 Antifungal activity of the whole resin and separated components of the resin of *Commiphora rostrata*

	Growth inhibition (%)[a]			
Component	*AF*	*AN*	*AO*	*P*
Whole resin	50.0	78.5	78.9	68.2
2-Decanone	59.4	74.6	84.3	63.5
2-Undecanone	7.4	75.6	71.3	47.1
2-Dodecanone	17.9	53.2	39.8	66.3

AF, *Aspergillus flavus*; AN, *A. niger*; AO, *A. ochraceus*; *P*, *Penicillium* species.
[a]Growth inhibition based on incorporation of the resin or resin components at 5000 p.p.m.
(Taken from McDowell et al 1988 by permission of Pergamon Press.)

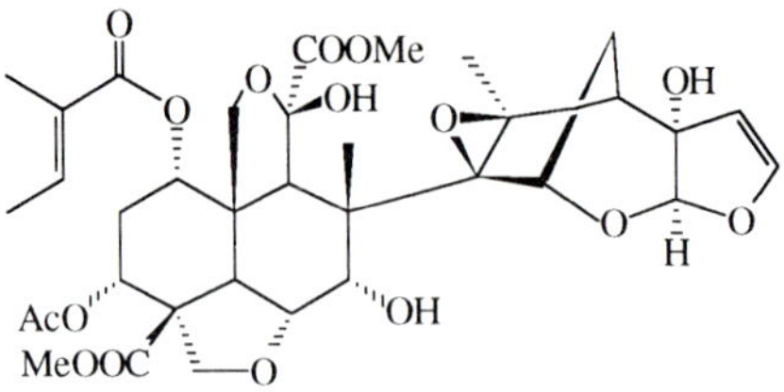

3 azadirachtin

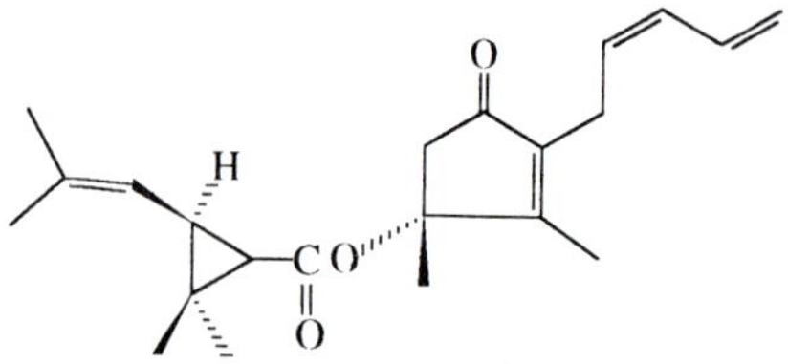

4 pyrethrin-1

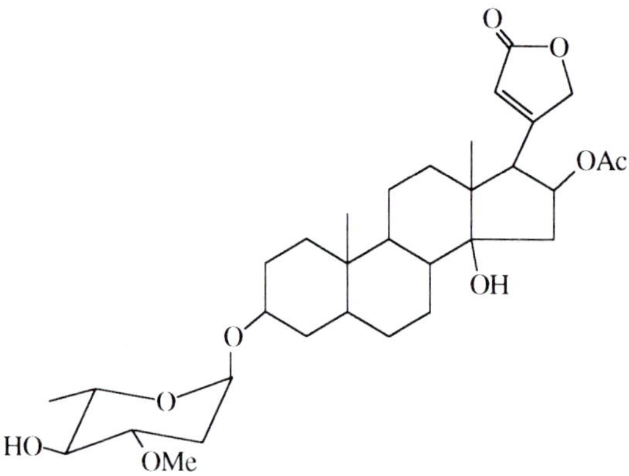

5 oleandrin

herbivores do have to deal with phenolics because of their ubiquitous occurrence in plants, yet there is little evidence to support the general statement that their presence acts as a feeding deterrent. Although there are exceptions, animals appear to have very efficient detoxification procedures for most simple phenolics (Harborne 1988); however, this does not mean that herbivores do not avoid excessive intake, because detoxifying phenols costs both energy and substrate.

Of all phenolic metabolites, greatest emphasis has been placed on the group of compounds known collectively as tannins (condensed tannins, hydrolysable tannins, polymeric phloroglucinols; see Haslam 1989). In the past 16 years we have 'progressed' from a position where tannins were regarded as the ultimate quantitative defence, acting by starving the ingestor of protein through the formation of insoluble and immutable tannin–protein complexes in the gut (Rhoades & Cates 1976), to a position where their capacity to deter at all and their mode(s) of action are being questioned (Bernays et al 1989). Although there is overwhelming evidence, particularly for domestic animals, that excessive levels of tannins in foodstuffs do have a deleterious effect on nitrogen balance,

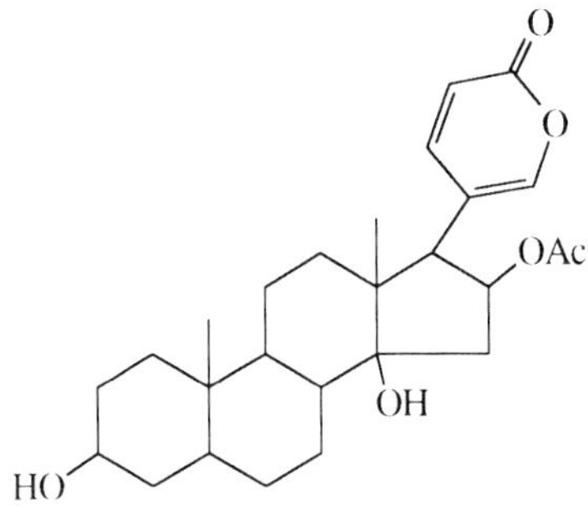

6 bufotalin

TABLE 2 Hydrolysis of bovine serum albumin (BSA) by trypsin in the presence of tannins and the influence of cholic acid

Tannin	Ratio tannin to BSA	Cholate (8 mM)	Free amino acid[a,b] 30 min	Free amino acid[a,b] 60 min
Tannic acid	0.1	–	0.64	0.76
	0.1	+	0.93	0.93
	1.0	–	0.14	0.09
	1.0	+	0.47	0.54
Condensed tannin	0.1	–	0.66	0.74
	0.1	+	0.93	0.91
	1.0	–	0.20	0.16
	1.0	+	0.57	0.64

[a]Free amino acid measured as absorbance reading for ninhydrin assay of free amino acids.
[b]60 min values that are lower than 30 min values are within experimental error.
(Reproduced from Mole & Waterman 1985 by permission of Plenum Press.)

it is now clear that what is 'excessive tannin intake' varies from species to species (Mole & Waterman 1987a, Bernays et al 1989) and that the tannin–protein complexation process is by no means as immutable as it was once thought to be (Bernays et al 1989). For example, the specificity of a tannin for complex formation with different proteins varies widely—proteases are particularly unreactive (Mole & Waterman 1987b). In an *in vitro* model of proteolysis (Mole & Waterman 1985) it is possible, by changing the relative concentrations of substrate protein and tannin, to achieve rates of proteolysis either lower (the expected result) or higher than those measured when no tannin is present (Fig. 1 in Mole & Waterman 1985). The reduction in proteolysis caused by adding tannin can be lessened by adding surfactants such as cholic acid (Table 2). A possible explanation for these various observations is given in Fig. 1.

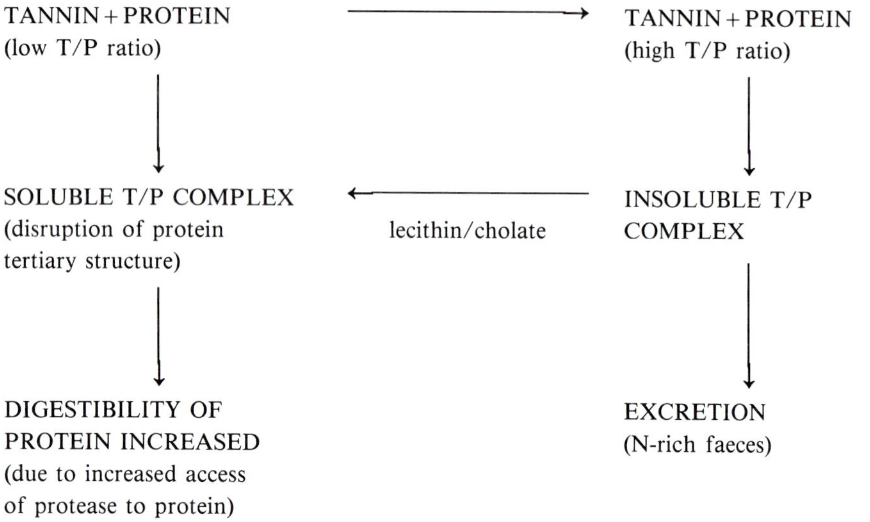

FIG. 1. Variation in the outcome of the tannin–protein interaction. (Adapted from Mole & Waterman 1987a.)

Induced defences

The secondary metabolite producer should not be regarded as a passive participant in its interactions with herbivore or pathogen. One group of natural products that clearly has a defensive role is the phytoalexins. These antifungal compounds (see Fig. 2 for examples derived from a number of different metabolic routes) are formed *de novo* in response to attack by pathogens. This production is thought to be stimulated by compounds termed elicitors, usually oligosaccharides, that are produced as a consequence of the attack (Ebel 1986).

There have recently been a number of reports of an enhanced production of phenolics and alkaloids by plants in response to real herbivory or to experimental procedures designed to mimic herbivory (Baldwin 1988, Karban & Myers 1989, Waterman & Mole 1989). In most cases where phenolic compounds are involved it seems doubtful that this 'induced' production occurs specifically to combat an attack: current ideas would suggest that their production represents a response to metabolic overload or overflow (Haslam 1985, Tuomi et al 1990) and that any benefit accrued is entirely fortuitous. This is not always the case, however. When foliage of bog myrtle (*Myrica gale*) is either protected from or subjected to insect herbivory there is an approximate doubling of the number of volatile oil-containing trichomes on the leaf and an increase in the production of one component in the mixture of leaf phenolics, the flavone glycoside kaempferol-3-(2,3-diacetoxy-4-*p*-coumaroyl)-rhamnoside (7), in plants subjected to herbivory (Carlton et al 1990). Neither oil nor flavone

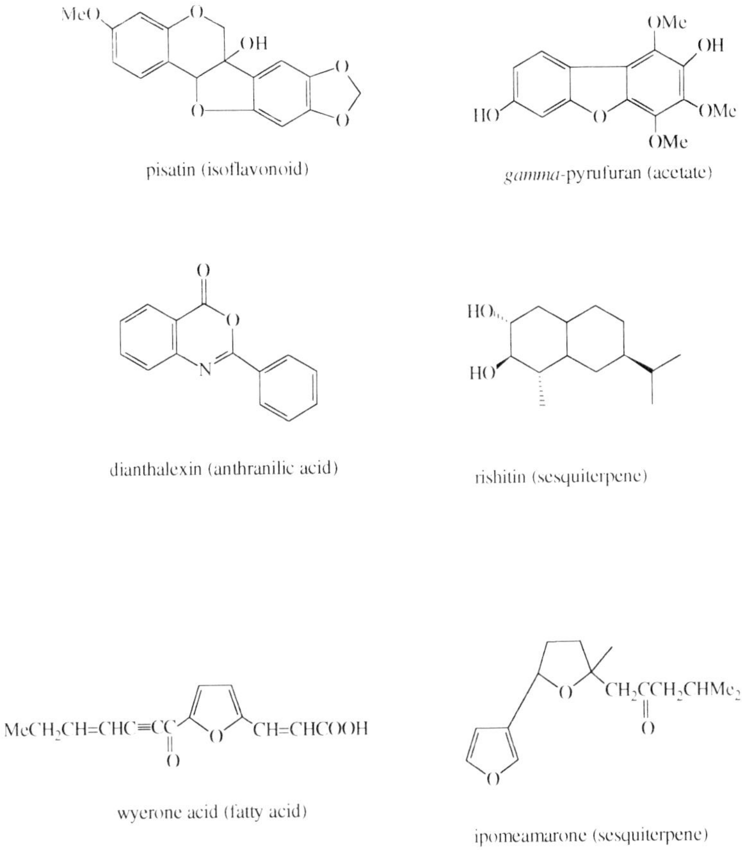

FIG. 2. Examples of phytoalexins derived from different metabolic routes.

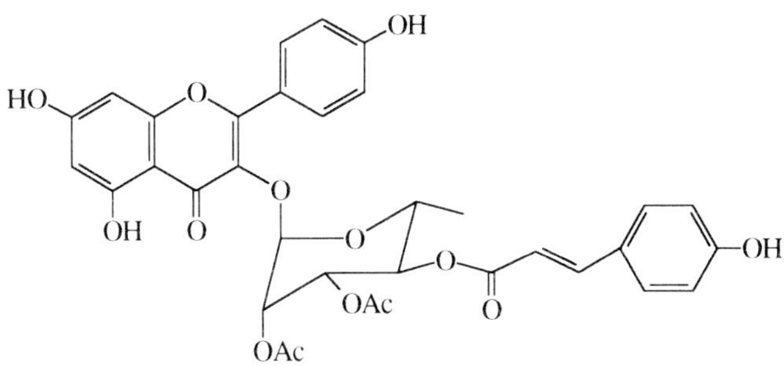

7 kaempferol-3-(2,3-diacetoxy-4-*p*-coumaroyl)-rhamnoside

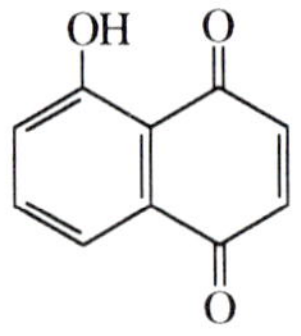

8 juglone

appears to have significant antifeedant activity but both are antifungal, suggesting the intriguing possibility that insect attack leads to a strengthening of antifungal defences to counter the invasion of pathogens through the wound (Carlton 1991).

Allelopathy and nutrient cycling

In the laboratory, many secondary metabolites have been shown to inhibit the growth of seedlings and to be toxic to growing plants. It is argued that these compounds are leached from the leaf or root into the soil where they benefit the producer by preventing the growth of competitors (a process termed allelopathy). Among metabolites often cited as allelopathic are a range of simple benzoic and cinnamic acid derivatives which are widely distributed in plants. Their clear activity in the laboratory does not prove that they play a useful role in the field, because they may be rapidly absorbed in the soil, or detoxified by microorganisms or by the plants themselves.

The most often quoted example of a proven allelopathic activity is that for the walnut, *Juglans nigra*, which produces the naphthoquinone juglone (**8**), derivatives of which are germination inhibitors and phytotoxins (Harborne 1988). Juglone or its reduced precursor leached from the aerial parts of the walnut is certainly capable of preventing competition by other plants through direct phytotoxic activity. However, recent studies by Schmidt (1988) show that microorganisms in the soil under walnut trees can rapidly degrade juglone so that the allelopathic effect will be short lived.

Another potential outcome of the deposition of phenolic metabolites on the soil is an effect on nutrient turnover, resulting from the inhibition or deterrence of detritivores, root symbionts and other microorganisms (Horner et al 1988, Kuiters 1990). Polyphenolic compounds in particular have been implicated in the acidification of soils, which tends to diminish the availability of nutrients, particularly nitrogen (Kuiters 1990). The resultant increase in soil acidity is thought to lead to metabolic imbalance in which carbon fixation outstrips nutrient supply and as a consequence there is a greater production of phenolics (Coley et al 1985, Baas 1989). This has been suggested to lead, in extreme cases, to severely nutrient-deficient environments (Janzen 1974). Whatever the cause, it is certainly true that the level of production of phenolics, particularly condensed tannins, does vary widely between different rain forests (Waterman & McKey 1989).

The impact of secondary metabolites on soil chemistry may also have beneficial effects. One intriguing possibility is that polyphenols and some other metabolites are actually important to the success of a plant species in nutritionally poor environments because, by slowing the rate of decomposition and nutrient release, they maximize the efficiency of nutrient cycling within the ecosystem (Horner et al 1988).

There is one other aspect of allelopathy and nutrient cycling where volatile secondary metabolites might play an important role. Low molecular mass terpenes, first absorbed into soils and then slowly released, have been implicated in allelopathy in desert environments (Harborne 1988). In the Californian chaparral these volatile compounds are eliminated from the soil during the regular bush fires that are critical in nutrient cycling and plant succession. As far as I am aware, all ecosystems which rely on regular fire cycles to regulate succession and to convert vegetable litter into nutrients are characterized by vegetation which has a high capacity for producing volatile oils. Low molecular mass terpenes burn with a relatively low heat, so that a fire relying on these for fuel will have the effect of turning the detritus to ash with a minimum of damage to viable elements in the vegetation (such as the tree cover).

Pollinator attraction and seed dispersal

Visual stimuli are used by plants to attract a vast range of potential pollinators and seed dispersal agents. These may be obvious to the casual human observer in a bright red flower or in a fruit turning from green to red as it ripens. In reality, the true nature of flower colour, as seen by a pollinator, can be much more complex: many insects, notably bees, are able to detect electromagnetic radiation well into the ultraviolet spectrum. Different groups of pollinator are known to be preferentially attracted to particular flower colours (Faegri & van der Pijl 1979); butterflies tend toward strong purples and reds, bees prefer intense yellows and blues, birds are particularly sensitive to reds and flies prefer dull browns, greens and purples.

The most common metabolites used to produce flower colours are the flavonoids. In general orange, red and purple flower coloration is caused by anthocyanins (**9**); flavonols (**10**), aurones (**11**) and chalcones (**12**) lead to orange or yellow colours and flavones (**13**) and flavonols (**10**) also give white and cream colours. Carotenes (**14**), alone or in combination with flavonoids, yield yellows and whites. In the families of the Centrospermae, highly coloured flowers are produced by betacyanins (**15**) and betaxanthins, which replace the anthocyanins for this function. The colour of green flowers is due to chlorophyll.

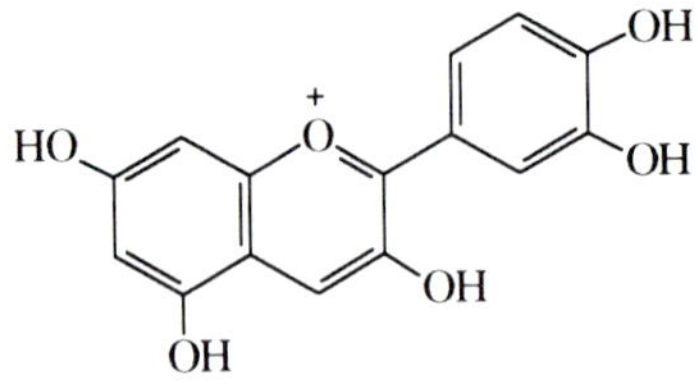

9 anthocyanidin (λ_{max} 525 nm)

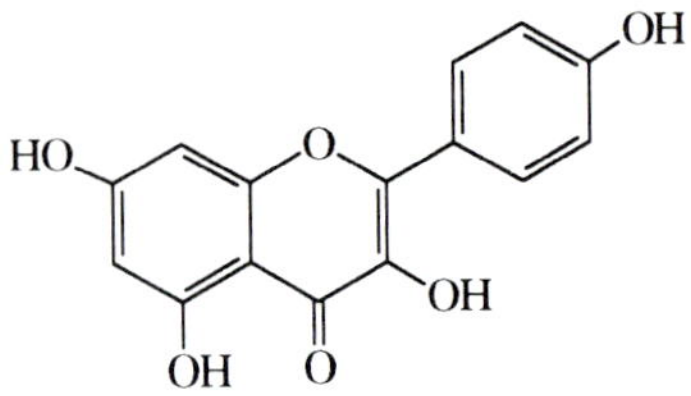

10 flavonol (λ_{max} 370 nm)

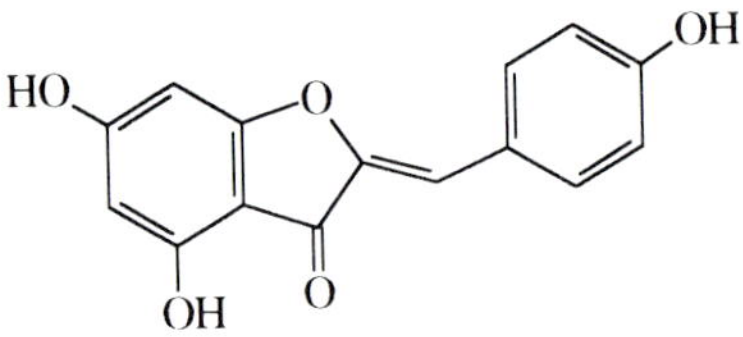

11 aurone (λ_{max} 390 nm)

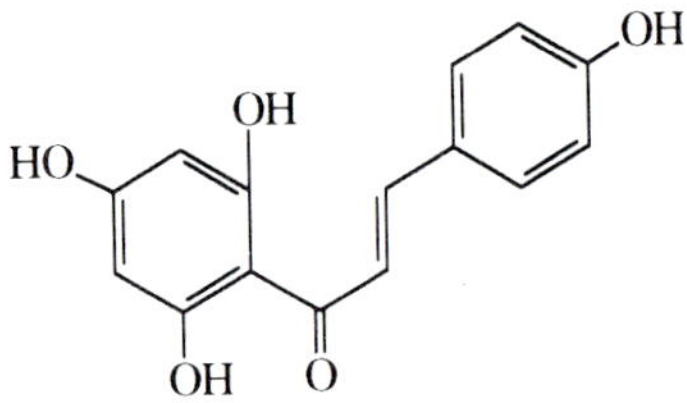

12 chalcone (λ_{max} 375 nm)

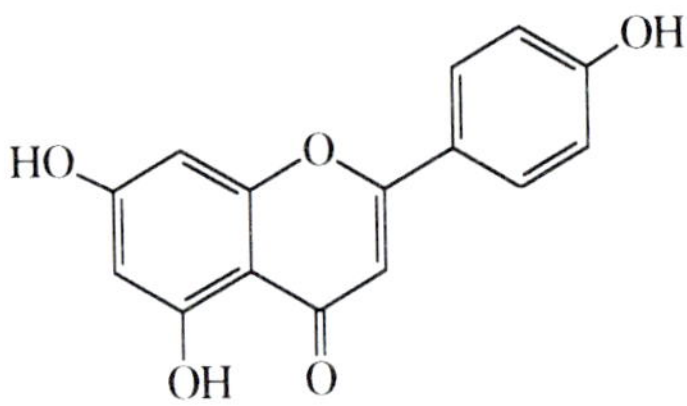

13 flavone (λ_{max} 335 nm)

Subtle differences in the final colour are achieved by modification in structures through oxidation or reduction, methylation and/or glycosylation, by the use of co-pigments such as chelating metal ions (aluminium, magnesium), or by the arrangement of co-pigments within the coloured organ. Differential deposition of various pigments in various parts of the organ can cause patterns (honey guides) which become apparent only if viewed in UV light (Harborne & Nash 1983, Harborne 1988).

The other main secondary metabolite-mediated route to attraction used by plants is olfactory. Flower scents are commonly produced by low molecular mass terpenes (mono- and sesquiterpenes) which often occur in very complex mixtures in flowers. Other 'pleasant' scents are produced by simple benzoic and cinnamic acid derivatives and long-chain alkanes, alkanols and alkanals.

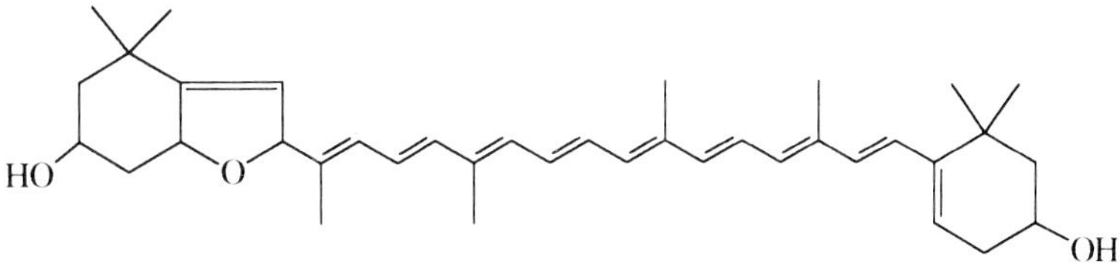

14 carotene (λ_{max} 440-460 nm)

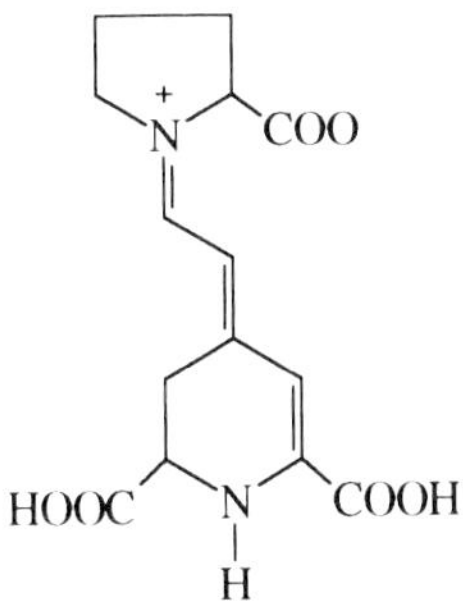

15 betaxanthine (λ_{max} 480 nm)

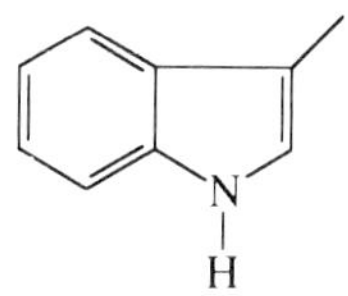

16 skatole

In some cases the structures of the odiferous compounds produced by flowers appear to mimic those of the mating pheromones produced by the potential pollinator.

At the more unsavoury end of the odour spectrum (for humans) are plants that rely on flies or beetles for pollination. These often produce odours suggestive of decaying meat, generated by mixtures of amines, diamines and simple cyclic compounds like skatole (**16**). These are most often used to attract pollinators, but such 'carrion-mimicking' compounds can also lure prey to insectivorous plants where they are trapped and consumed.

A third chemical strategy for attracting pollinators is the production of nectar, which consists primarily of simple sugars and, sometimes, amino acids and/or lipids. There is evidence that secondary metabolites harmless to nectar-feeders but toxic to other animals may occur in some nectars (Harborne 1988). One of the most common manifestations of this is the problem of 'toxic' honey.

The chemical strategies used to attract pollinators to flowers can also be exploited to signal to seed-dispersal agents that the fruit is now edible. Given the wide variety of fruits that do signal ripeness through colour change, this must be a successful strategy, although such signals are equally visible to seed predators. Our understanding of the complex changes in chemistry that occur during the ripening of an animal-dispersed seed is still fragmentary. Because the 'nutrient reward' has to become available to the disperser, while the seed must remain unattractive to the disperser and predator, the use of secondary metabolites will often differ radically in fruit and seed. One long-standing problem has been to understand the fate of the tannins responsible, through their binding with salivary proteins and mucopolysaccharides, for the astringency of many unripe fruit, which disappears as the fruit ripens. Ozawa et al (1987) suggest that the tannin actually remains unchanged, but that alterations in the structure of carbohydrates in the fruits may produce small oligosaccharides which, in the mouth, compete with salivary proteins and mucopolysaccharides for tannin binding.

Secondary metabolites and response to abiotic factors

The production of secondary metabolites may be enhanced or reduced by many abiotic factors, including seasonality, levels of macro- and micronutrients, osmotic stresses (drought and salinity) and light intensity (Waterman & Mole 1989). The relationship between the production of polyphenols and soil 'quality' has already been discussed in the section on allelopathy.

Much has been written on the variation in the production of phenolic compounds as a function of incident light intensity (reviewed in Waterman & Mole 1989). There are now numerous examples of a positive correlation between light and levels of phenolics, both between and within individual plants (Table 3; Mole et al 1988). These variations may well have an impact on a plant's susceptibility to herbivores (Bryant 1987, Mole & Waterman 1988, Karban & Myers 1989). These variations have been rationalized in terms of carbon/nutrient balance (Coley et al 1985).

Another area where secondary metabolites may be important is in protection against the harmful environmental effects of excessive incident UV light and drought. Aromatic compounds occurring on the leaf surface, notably simple lipophilic flavones, are capable of screening out harmful wavelengths, whereas other groups of compounds have the capability to act as anti-oxidants that will nullify harmful free radicals arising through the effects of UV radiation. Hydrophobic surface resins and waxes may present an important mechanical barrier to desiccation.

TABLE 3 Comparison of levels of total phenolics, condensed tannins, protein and tannin : protein ratios in young, mature and senescent 'sun' and 'shade' leaves of *Trema guineensis*

Measure	*Leaf age*	*Sun-growing leaves*	*Shade-growing leaves*
Total phenolics[a]	Young	2.72 (0.39)	1.42 (0.46)
	Mature	2.32 (0.22)	1.75 (0.57)
	Old	2.03 (0.43)	1.49 (0.20)
Condensed tannins[b]	Young	14.11 (2.73)	5.96 (1.72)
	Mature	9.56 (1.30)	5.54 (2.04)
	Old	7.99 (2.01)	6.19 (2.05)
Protein[c]	Young	18.59 (2.21)	18.03 (1.47)
	Mature	14.46 (1.23)	15.30 (2.54)
	Old	11.10 (1.53)	12.55 (1.84)
Tannin : protein ratio[d]	Young	0.76 (0.36–1.06)	0.33 (0.15–0.50)
	Mature	0.67 (0.30–0.82)	0.36 (0.09–0.62)
	Old	0.71 (0.34–1.31)	0.49 (0.19–0.70)

[a]Total phenolics: % dry weight, measured by Folin-Denis method.
[b]Condensed tannins: % dry weight, measured by butanol : HCl method.
[c]Protein measured by Kjeldahl protein nitrogen assay.
[d]Tannin : protein ratio based on assay of functional tannin obtained by precipitation with protein.
Figures in parentheses are standard deviations.
(Data taken from Mole & Waterman 1988 and Mole et al 1988, by permission of Plenum Press.)

Concluding comments

It seems abundantly clear that secondary metabolites do have a considerable impact on the success of the producer and, in some cases, can be instrumental in actually shaping the structure and productivity of the ecosystem in which they are produced. It is probably only occasionally correct to view the presence of these compounds as part of a grand 'co-evolutionary' design. Their synthesis may originally have arisen through evolutionary (or environmental) pressures to improve on an already existing functional metabolite, but in many cases it seems far more likely to have arisen through metabolic accidents or overloads. Whichever scenario reflects the origins of secondary metabolites is hardly relevant to current function(s), because their value lies in the fact that they provide a selective advantage (or advantages) to the producer that outweigh their cost. What is more, even if production was initially by design, it is possible (even probable) that since that event the metabolite has acquired new values. Take as an analogy the simple drug, aspirin: originally synthesized as a painkiller, its most important use today is in the prevention of blood platelet aggregation in patients recovering from a heart attack.

Acknowledgements

The author extends his thanks to the Ciba Foundation and to the organizers of this symposium for their kind invitation to participate.

References

Baas WJ 1989 Secondary plant compounds, their ecological significance and consequences for the carbon balance. In: Lambers H (ed) Causes and consequences of variation in growth rate and productivity of higher plants. SPB Academic Publishing, The Hague, p 313–340

Baldwin IT 1988 The alkaloidal responses of wild tobacco to real and simulated herbivory. Oecologia (Berl) 77:378–381

Bernays EA, Edgar JA, Rothschild M 1977 Pyrrolizidine alkaloids sequestered and stored by the aposematic grasshopper *Zonocerus variegatus*. J Zool (Lond) 182:85–87

Bernays EA, Cooper-Driver G, Bilgener M 1989 Herbivores and plant tannins. Adv Ecol Res 19:263–302

Boppre M 1990 Lepidoptera and pyrrolizidine alkaloids. J Chem Ecol 16:165–185

Brower LP, Seiber JN, Nelson CJ, Lynch SP, Hoggard MP, Cohen JA 1984 Plant determined variation in cardenolide content and thin layer chromatography profiles of monarch butterfly, *Danaus plexippus*, reared on milkweed plants in California. J Chem Ecol 10:1823–1857

Bryant JP 1987 Feltleaf willow–snowshoe hare interactions: plant carbon/nutrient balance and floodplain succession. Ecology 68:1319–1327

Carlton RR 1991 An investigation into the rapidly induced chemical response of *Myrica gale* to insect herbivory. PhD thesis, University of Strathclyde

Carlton RR, Gray AI, Lavaud C, Massiot G, Waterman PG 1990 Kaempferol-3-(2,3-diacetoxy-4-*p*-coumaroyl)-rhamnoside from leaves of *Myrica gale*. Phytochemistry 29:2369–2371

Coley PD, Bryant JP, Chapin FS 1985 Resource availability and plant antiherbivore defense. Science (Wash DC) 230:895–899

Compton SG, Jones DA 1985 An investigation of the responses of herbivores to cyanogenesis in *Lotus corniculatus* L. Biol J Linn Soc 26:21–38

Dirzo R, Harper JL 1982 Experimental studies on slug–plant interactions. III. The performance of cyanogenic and acyanogenic morphs in *Trifolium repens* in the field. J Ecol 70:119–138

Ebel J 1986 Phytoalexin synthesis: the biochemical analysis of the induction process. Annu Rev Phytopathol 24:235–264

Faegri K, van der Pijl L 1979 Principles of pollination ecology, 3rd edn. Pergamon Press, Oxford

Harborne JB 1988 Introduction to ecological biochemistry, 3rd edn. Academic Press, London

Harborne JB, Nash RJ 1983 Flavonoid pigments responsible for ultraviolet patterning in the genus *Potentilla*. Biochem Syst Ecol 12:315–319

Haslam E 1985 Metabolites and metabolism. Clarendon Press, London

Haslam E 1989 Plant polyphenols. Cambridge University Press, Cambridge

Horner JD, Gosz JR, Cates RG 1988 The role of carbon-based plant secondary metabolites in decomposition in terrestrial ecosystems. Am Nat 132:869–883

Janzen DH 1974 Tropical blackwater rivers, animals and mast fruiting by the Dipterocarpaceae. Biotropica 6:69–103

Karban R, Myers JH 1989 Induced plant responses to herbivory. Annu Rev Ecol Syst 20:331–348

Kuiters AT 1990 Role of phenolic substances from decomposing forest litter in plant–soil interactions. Acta Bot Neerl 39:329–348

McDowell PG, Lwande W, Deans SG, Waterman PG 1988 Volatile resin exudate from stem bark of *Commiphora rostrata*: potential role in plant defence. Phytochemistry 27:2519–2521

Mole S, Waterman PG 1985 Stimulatory effects of tannins and cholic acid on tryptic hydrolysis of proteins. J Chem Ecol 11:1323–1332

Mole S, Waterman PG 1987a Tannins as antifeedants to mammalian herbivores—still an open question? In: Waller GR (ed) Allelochemicals: role in agriculture and forestry. Am Chem Soc Symp Ser vol 330:572–587

Mole S, Waterman PG 1987b Tannic acid and proteolytic enzymes: enzyme inhibition or substrate deprivation? Phytochemistry 26:99–102

Mole S, Waterman PG 1988 Light induced variation in phenolic levels in foliage of rain forest plants. II. Potential significance to herbivores. J Chem Ecol 14:23–32

Mole S, Ross JAM, Waterman PG 1988 Light induced variation in phenolic levels in foliage of rain forest plants. I. Chemical changes. J Chem Ecol 14:1–21

Mowat DJ, Shakell MA 1989 The effect of different cultivars of clover on numbers of, and leaf damage by, some invertebrate species. Grass Forage Sci 44:11–18

Ozawa T, Lilley TH, Haslam E 1987 Polyphenol interactions: astringency and the loss of astringency of ripening fruit. Phytochemistry 26:2937–2942

Rhoades DF, Cates RG 1976 Toward a general theory of plant antiherbivore chemistry. Recent Adv Phytochem 10:168–213

Schmidt SK 1988 Degradation of juglone by soil bacteria. J Chem Ecol 14:1561–1571

Tuomi J, Niemala P, Siren S 1990 The panglossian paradigm and delayed inducible accumulation of foliar phenolics in mountain birch. Oikos 59:399–410

von Nickisch-Rosenegk E, Schneider D, Wink MN 1990 Time-course of pyrrolizidine alkaloid processing in the alkaloid exploiting arctiid moth, *Creatonotos transiens*. Z Naturforsch Sect C Biosc 45:881–894

Waterman PG, McKey DB 1989 Herbivory and secondary compounds in rain forest plants. In: Lieth H, Werger MJA (eds) Tropical rain forest ecosystems. Elsevier, Amsterdam

Waterman PG, Mole S 1989 Extrinsic factors influencing production of secondary metabolites in plants. In: Bernays EA (ed) Insect–plant interactions. CRC Press, Boca Raton, FL, vol 1:107–134

DISCUSSION

Demain: Many of us spend much time arguing about the functions of secondary metabolites in microorganisms, but most of us know nothing about ecology. We are proposing a number of these as ecologically important compounds. What do you think of the arguments that you hear among microbiologists and chemists about antibiotics being agents of chemical warfare, in Nature?

Waterman: The immediate problem here is that almost all your experiments are of necessity done in unreal conditions. In my paper I referred to the impact of compounds leached into the soil from microorganisms. This is an area that microbiologists have to move into, if they want to get to grips with the 'real world' ecological impact of these compounds. However, working at that level, with the yields that you appear to get from microorganisms in real-world situations, will be exceedingly difficult.

Davies: Of the phytoalexins that are already known, can one in most cases put one's finger on an actual role for these molecules, with respect to the plant that produces them?

Waterman: The presumed role for the phytoalexins is an inhibitory effect against the invading pathogen that has triggered their production.

Davies: Can that be shown in *all* cases?

Waterman: Certainly in the well-documented cases, this has been adequately established.

Cane: Some of this has been well studied. For example, Rod Croteau studied the production of resins and terpenes in species of pine and has shown by the simple model of wounding with a razor blade that you can follow not only the production of these compounds, but also the induction of particular cyclases, and can relate cyclase levels to types of wounding (Lewinsohn et al 1991, 1992, Gijzen et al 1991).

Phytoalexins produced by solanacaeous plants (e.g. capsidiol) are quite easy to study because they can be produced in tissue culture, where there is a whole variety of elicitors. You can follow mRNA levels, for example (Vögeli & Chappell 1990). Compared to the gross production of plant metabolites, the production of defence substances is quite well understood at the genetic and molecular levels. If one doesn't necessarily know the details of the transduction of the signals, one knows a lot about what the signals are and the effect of these compounds on the species that provoke their production.

Bu'Lock: On the other hand, we now have the nice example of the bog myrtle, where the 'deterrent effect', if any, is exercised upon something quite different to what is actually eliciting the production of the deterrent.

Waterman: Yes. The initial effect of the herbivorous insect on the plant is the production of compounds to respond to the almost inevitable subsequent attack by fungal pathogens, because a wound has been produced; this particular insect, the capsid *Lygocoms spinolai*, is a cell sucker and, when feeding, destroys pockets of cells, in which you can later observe fungal growth. Spread of the fungus has been stopped, every time, by the production of the induced metabolite.

Turner: Another example is the fungus *Nectria haematococca*, which is a pathogen of pea plants. Peas produce a phytoalexin called pisatin, an inhibitor of fungal growth, in response to infection. This is not a powerful inhibitor, but it slows the growth of the fungus on laboratory media. More virulent strains of *Nectria* produce pisatin demethylase, which confers some degree of resistance to pisatin (van Etten et al 1989).

Waterman: This is an important point. High levels of activity need not equate with effectiveness. Although in this case the induced compound may not be a powerful inhibitor, the plant is producing it in large quantities but very localized, and it is effectively doing the job.

Wächtershäuser: What is the difference between the secondary metabolite repertoires of lower plants and of flowering plants?

Waterman: There are differences in emphasis. The fungi in particular are good at producing acetate-derived compounds, probably better overall than higher plants. If you take the number of organisms studied and the range of compounds found, the shikimate pathway has probably reached its highest level of diversity in ferns, gymnosperms and more 'primitive' taxa of flowering plants. There is an interesting paper touching on this matter by Kubitzki & Gottlieb (1984).

Wächtershäuser: I wonder if the potential of a chemical class of compounds ever becomes exhausted? There was a theory that dinosaurs died out because flowering plants invented poisons.

Waterman: Tony Swain was feeding tortoises at London Zoo with alkaloids and found that these reptiles were less able to detect the toxins than were mammals; and he extrapolated from there (Swain 1976). This is not a well-founded hypothesis, but there is some merit in the idea that a group of compounds develop, evolve, and then gradually, as time passes, because of increasing resistance among the organisms they are supposed to defend against, they are supplanted by other chemical groups. Among the alkaloids of the flowering plants (the angiosperms), the benzylisoquinoline group is probably the oldest, if the botanists are correct about what are primitive and what are relatively modern groups of angiosperms. The alkaloids in the more primitive angiosperms tend to be derived from tyrosine. Then you see a switch-over from tyrosine to tryptophan and the changeover from a non-alkaloidal part of the molecule which is also tyrosine-derived to a non-alkaloidal part of the molecule which is terpene-derived. Again I refer you to Kubitzki & Gottlieb (1984).

Haslam: You can construct a satisfying ecological series of arguments and a framework in which to place plant products, but it doesn't give you any clue as to why these compounds are probably made—the metabolic events. For example, the hypothesis about polyphenolic repellants could be made more compelling by saying that polyphenols bind to proteins which are very rich in proline, often by four orders of magnitude more than other types of protein. And the salivary proteins in the mouth are very rich in proline: 30–40% of the amino acids in salivary proteins are proline or hydroxyproline. In that sense, this is a defence mechanism. On the other hand, if you look at the phenols in the plant, very often the plant accumulates phenolic materials in which it has substantially desensitized the ability of the polyphenol to bind to the protein, compared to intermediates earlier in the biosynthetic pathway; so, taking this viewpoint, the plant must be making them for some other reason.

A problem which concerns me more is the question of the metabolic events which cause the plants to make these substances. One can try to draw analogies between microorganisms, which have been very well studied, and plants. There are considerable similarities. Plants secrete secondary substances, and also sequester them. Eric Cundliffe drew our attention earlier to those two features of the self-defence of the organism itself. And plants make secondary metabolites from essentially the same substances as do microorganisms, from acetic acid and certain amino acids. In plants, there is a dominance of the shikimate pathway and the amino acids derived from it, and these are the most expensive amino acids to make, in terms of ATP equivalents. It has always struck me that the plant is trying to get back something from amino acid production which has gone awry.

Cane: One thing that troubles me about many evolutionary ecology arguments is that we can identify currently occurring interactions, where there is still the pitched battle, or dynamic tension, between different organisms; but if we want to ask where these defence mechanisms come from, we need to identify the organisms which are now extinct because they could not adapt to the defence put up by their normal food source, or those organisms which learn to eat something else, because the plant they normally ate changed—as distinguished from learning some way of overcoming the defence. This information is not there in any gross way.

When Dr Haslam talks about the amount of ATP necessary to make shikimate metabolites, these metabolites come from 4- and 3-carbon sugars which are generated by the pentose shunt cycle, which in turn is fed by photosynthesis. In these plants, this is their primary carbon source. I don't know if that is why they make shikimates, but they certainly have no shortage of the 'feedstuff' (carbon dioxide and sunlight) with which to manufacture the starting materials. Why they put them in that direction is beyond me, but they certainly don't have to work very hard to acquire the starting materials.

Waterman: Not all plants expend photosynthate on condensed tannin production. The occurrence of condensed tannins is correlated, to a large degree, with the arboreal habit. But just about all plants do metabolize secondary metabolites.

Cane: Tannins are one expression of shikimate metabolism, as are aromatic amino acids.

Waterman: They are a peculiar expression of shikimate metabolism, in that I don't know of any evidence that any condensed tannins can be reclaimed. When looking at costs, as Professor Haslam was doing, it is important to remember that most secondary metabolites are in a constant turnover situation.

Haslam: May I comment on the matter of expense? I meant that the ATP equivalents from the carbohydrate and intermediary metabolism pool are highest when aromatic amino acids are being synthesized; it's something like 60 or 70 ATP equivalents, whereas for amino acids which are near to the Krebs cycle, it's a very small ATP equivalent (about 30–40). This is what I mean by saying that in making aromatic amino acids, the plant is expending a lot of ATP, which it has to generate in some way.

Davies: We talk about the energy cost to plants and microbes, but I don't know that we can make this kind of argument with any conviction.

Cane: In the USA we burn more oil per capita than any other country; that doesn't necessarily prove that it's economical, but it also doesn't prove that there's no reason to do it: we want to keep warm!

Williams: I was struck by the speculation that trees that are very high terpene producers might be such high producers that there is occasionally a conflagration. The level at which the uninjured trees, for example the eucalyptuses, can put out these things is amazing. In many parts of the world there are so-called 'blue mountains', named because of the enormous concentrations of terpenes in the air, enough to change the colour of the local environment; we are talking of hundreds of thousands of tonnes of these materials. An enormous amount of biosynthetic energy is going into this.

Orgel: When a compound is made by a rather long synthetic pathway, the intermediates must have been useful in some way, because the organism couldn't have evolved all the enzymes simultaneously. Are there well-documented cases in which each of the intermediates is still represented in some surviving species, thus giving us a kind of 'fossil' record of the biosynthetic pathway?

Baldwin: One case could be the fungi *Penicillium* and *Cephalosporium acremonium* (*Acremonium chrysogenum*). The former make penicillin by basically two steps, whereas the cephalosporin producers modify that intermediate penicillin to give the cephalosporins, which I presume is an advantage, because the cephalosporins are more resistant to β-lactamases in 'target' organisms.

Bu'Lock: A striking example is the progressive evolution to greater and greater toxicity. Virtually every stage in the elaboration of aflatoxins by *Aspergillus flavus* also exists as a terminal metabolite of some other organism, but if you string the whole thing together you end up with the most toxic of all these metabolites, and the frequency with which you end a biosynthetic series with the most potent compound in that series is something quite striking and not entirely explained, where we cannot see any actual advantage of the toxic effects.

Wächtershäuser: It is typical of secondary metabolism that all the intermediates have functionality. A good example are the bile acids, for which a biochemical phylogeny has been established. Organisms that are lower in the tree make precursors of the bile acids of the higher organisms. Another example is retinal in the eye. This has been progressively modified; every 70 million years, a new group has been added!

Waterman: You can see the steps of pathway evolution of secondary metabolites in many cases. For example, in the Solanaceae with tropane alkaloids it is often possible to detect trace amounts of each intermediate, from the first addition of an acetate unit to the amino acid to the final accumulated metabolite.

Vining: I was interested in the connection between the antifungal activity of certain plant leaves, their effect on the ruminant digestive system, and the learning or deterrent effect on the animal. How is that actually established?

Waterman: I have been involved in studies of a number of colobid monkeys. Behaviourally, these animals do an awful lot of sampling; they wander through their territory, taking a leaf here and a leaf there, and just sampling the one, maybe eat it or maybe spit it out, and then go on and find a feeding tree and feed. It seems as though they are reinforcing some perception they already have of what is good to eat and what is not good. This is purely observational, so far.

Vining: Is there a learned response based on what the effect is—stomach pains, or whatever?

Waterman: There must be, but what their taste buds are telling them is clearly not the same as what our taste buds tell us, from personal observation from sampling their foods!

Vining: There could be an indirect deterrence resulting from effects on intestinal microorganisms. Most digestive systems in animals have a microbial component, so there could be quite a wide-ranging effect on the microbial flora.

Nisbet: A recent article in *The Economist* (Anon 1992) referred to Ugandan chimpanzees eating a particular plant, *Aspilia*, when they are infected with nematode worms. It is postulated that this is an example of ethnobotany at the level of the monkey, and that it actually cured the infection.

Waterman: There is a growing literature on this subject (known as zoopharmacognosy), and there are a number of cited examples of behaviour of this nature. The first example was indeed *Aspilia* (Rodriguez et al 1985), and the behaviour of the chimpanzees is certainly very striking. They fold the leaf up, roll it and swallow it. The leaf passes through the animal intact. What has not yet been done is to show satisfactorily what has been extracted from the leaf.

Baldwin: I think they have shown that parts of the leaf are removed during the passage through the animal.

Waterman: It would be surprising if it went through without something happening.

References

Anon 1992 Chimp's choice. The Economist, 15 February 1992

Gijzen M, Lewinsohn E, Croteau R 1991 Characterization of the constitutive and wound-inducible monoterpene cyclases of grand fir (*Abies grandis*). Arch Biochem Biophys 289:267–273

Kubitzki K, Gottlieb OR 1984 Phytochemical aspects of angiosperm origin and evolution. Acta Bot Neerl 33:357–368

Lewinsohn E, Gijzen M, Savage TJ, Croteau R 1991 Defense mechanisms of conifers–relationship of monoterpene cyclase activity to anatomical specialization and oleoresin monoterpene content. Plant Physiol 96:38–43

Lewinsohn E, Gijzen M, Croteau R 1992 Wound-inducible pinene cyclase from grand fir—purification, characterization, and renaturation after SDS-PAGE. Arch Biochem Biophys 293:167–173

Rodriguez E, Aregullin M, Nishida T et al 1985 Thiorubin-A, a bioactive constituent of *Aspilia* (Asteraceae) consumed by wild chimpanzees. Experientia 41:419–420

Swain T 1976 Angiosperm–reptile co-evolution. In: d'A Bellairs A, Cox CB (eds) Morphology and biology of reptiles. Linn Soc Symp Ser no. 3, p 107–122

van Etten HD, Matthews DE, Matthews PS 1989 Phytoalexin detoxification: importance for pathogenicity and practical implications. Annu Rev Phytopathol 27:143–164

Vögeli U, Chappell J 1990 Regulation of a sesquiterpene cyclase in cellulase-treated tobacco cell suspension cultures. Plant Physiol 94:1860–1866

Defensins: endogenous antibiotic peptides from human leukocytes

Robert I. Lehrer* and Tomas Ganz†

*Department of Medicine and †Will Rogers Pulmonary Research Laboratory, University of California at Los Angeles, 10833 LeConte Avenue, Los Angeles, CA 90024, USA

Abstract. A variety of endogenous antimicrobial peptides equip mammals, amphibians, insects and plants to defend themselves against microbial pathogens. Defensins are small peptides of mammalian cells that contain 29–35 amino acid residues, including six invariant cysteines that form three intramolecular disulphide bonds. They are produced by the sequential proteolysis of precursors that contain approximately 95 amino acids and are synthesized by several types of cells, especially the bone marrow precursors of blood neutrophils. Defensins constitute 5–7% of the total protein in human neutrophils and are present in high concentration in the azurophil granules and phagocytic vacuoles. The production of defensins by human neutrophils amounts to approximately 10 mg/kg body weight per day. In certain mammalian species lung macrophages and specialized epithelial (Paneth) cells in the small intestine also produce defensins. Defensins are complexly folded, amphipathic, rich in antiparallel β-sheet but devoid of α-helical domains. Their unusually broad antimicrobial spectrum encompasses Gram-positive and Gram-negative bacteria, many fungi, mycobacteria, spirochetes and several enveloped viruses. The antimicrobial properties of defensins result from their insertion into target cell membranes and the formation of voltage-sensitive channels. Given their abundance and broad spectrum of antimicrobial activity, defensins may contribute substantially to innate resistance to infection.

1992 Secondary metabolites: their function and evolution. Wiley, Chichester (Ciba Foundation Symposium 171) p 276–293

Not only are we outnumbered, surrounded and colonized by microbes, each breath introduces bacteria into our lungs and each mouthful allows microbes to enter our intestinal tract. Fortunately, most bacteria are inept pathogens and the components of our innate immune system maintain the sterility of our lungs and ensure that our small intestine normally contains few bacteria. Although many molecular mechanisms contribute to the constitutive resistance of animals to infection, recent evidence suggests that host peptides and proteins with antimicrobial properties may play significant roles. Lysozyme, the discovery of which by Fleming preceded his observation of penicillin by only a few years, provides an historical link between endogenous antimicrobial molecules and the more 'orthodox' antibiotics covered elsewhere in this volume.

Defensins are small cysteine-rich peptides with antimicrobial activity. They are most prominent in polymorphonuclear neutrophils (PMNs) of mammals, motile white blood cells (leukocytes) whose cytoplasm is filled with granules of several types. PMNs ingest microorganisms and sequester them within vacuoles, wherein they are exposed to a mixture of molecules that can kill microorganisms via oxidative and non-oxidative mechanisms. Prominent among the strong oxidants are chemicals that have the general character of disinfectants, including hydrogen peroxide, hypochlorous acid and chloramines. Phagocytic PMNs initiate the generation of such oxidants by assembling a multicomponent NADPH oxidase on their plasma and vacuolar membranes. Various deficiencies in the activation or assembly of NADPH oxidase have been recognized, and are classified under the rubric of 'chronic granulomatous disease' (CGD). PMNs from patients with CGD have an impaired ability to produce microbicidal oxidants, and most affected persons are highly susceptible to infection by microbes, including bacteria and fungi whose intrinsic pathogenic potential is limited (Smith & Curnutte 1991).

Defensins (Ganz et al 1985) and most of the other proteins responsible for non-oxidative microbicidal activity by PMNs (Spitznagel 1990, Lehrer & Ganz 1990) are contained within azurophil granules (Gabay et al 1989), cytoplasmic organelles whose contents are secreted into phagocytic vacuoles or the external medium by a process known as degranulation. Human neutrophils contain four different defensins. Three of these (human neutrophil peptide [HNP] 1, 2 and 3) have high sequence identity and constitute 5–7% of the total protein of the human PMN and 30–50% of the protein in their azurophil granules. The concentration of human defensins in phagocytic vacuoles may exceed 10 mg/ml. The fourth human defensin, HNP-4, differs considerably from HNPs 1, 2 and 3 in its primary structure and constitutes <2% of the total defensin content (Wilde et al 1989). The relative abundance of HNP-1, 2 and 3 in human neutrophils is readily apparent when extracts of their granules are subjected to electrophoresis (Fig. 1). Six defensins, NP-1, 2, 3a, 3b, 4 and 5, constitute over 15% of the total protein in rabbit PMNs and correspond to the 'lysosomal cationic proteins' initially described by Zeya & Spitznagel (1969). The lesser cationicity of the human defensins undoubtedly delayed their recognition in human PMNs by many years.

Figure 2 shows the primary structures of 15 defensins that have been purified from mammalian white blood cells, including four from human PMNs (Selsted et al 1985a, Wilde et al 1989), six from rabbit PMNs (Selsted et al 1985b), four from rat PMNs (Eisenhauer et al 1989) and one from guinea pig PMNs (Selsted & Harwig 1987). Two of the rabbit defensins, NP-1 and NP-2, were initially purified from rabbit lung macrophages (Patterson-Delafield et al 1980). In contrast to rabbit lung macrophages, where expression of defensins is subject

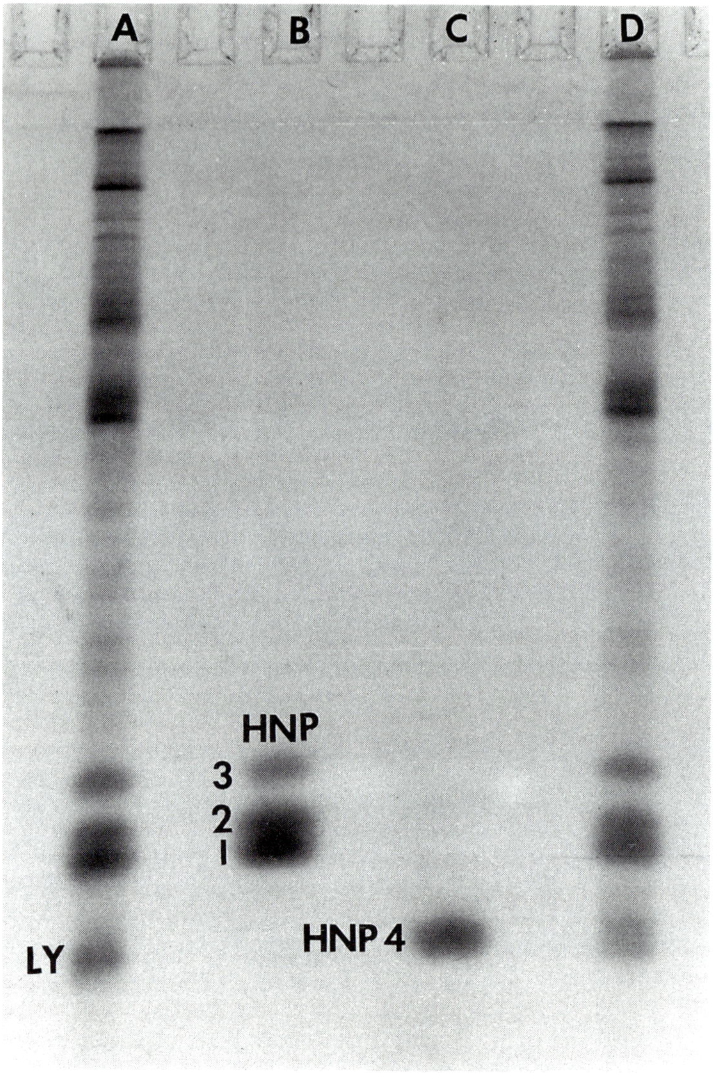

FIG. 1. Coomassie blue-stained acid–urea polyacrylamide gel. Lanes A and D contained an acid extract of human neutrophil granules. The position of lysozyme (LY) has been marked in Lane A. Lane B contained 1 μg each of purified human neutrophil defensins HNP-1, 2 and 3. Lane C contained 1 μg of purified HNP-4, whose migration closely corresponds to that of lysozyme.

	1–10	11–20	21–30	31–
Human:				
HNP1	A C Y C R I P A C	I A G E R R Y G T C	I Y Q G R L W A F C	C
HNP2	C Y C R I P A C	I A G E R R Y G T C	I Y Q G R L W A F C	C
HNP3	D C Y C R I P A C	I A G E R R Y G T C	I Y Q G R L W A F C	C
HNP4	A C S C R L V F C	R R T E L R V G N C	L I G G V S F T Y C	C T R Y
Rabbit:				
NP1	V V C A C R R A L C	L P R E R R A G F C	R I R G R I H P L C	C R R
NP2	V V C A C R R A L C	L P L E R R A G F C	R I R G R I H P L C	C R R
NP3a	G I C A C R R R F C	P N S E R F S G Y C	R V N G A R Y V R C	C S R R
NP3b	G R C V C R K QLL C	S Y R E R R I G D C	K I R G V R F P F C	C P R
NP4	V S C T C R R F S C	G F G E R A S G S C	T V N G V R H T L C	C R R
NP5	V F C T C R G F L C	G S G E R A S G S C	T I N G V R H T L C	C R R
Rat:				
RtNP1	T C Y C R R T R C	G F R E R L S G A C	G Y R G R I Y R L C	C R
RtNP2	V T C Y C R S T R C	G F R E R L S G A C	G Y R G R I Y R L C	C R
RtNP3	C S C R T S S C	R F G E R L S G A C	R L N G R I Y R L C	C
RtNP4	A C Y C R I G A C	V S G E R L T G A C	G L N G R I Y R L C	C R
Mouse:				
CRYPT	V C Y C R S R G C	K G R E R M N G T C	R K G H L L Y T L C	C R
Guinea pig:				
GNP	R R C I C T T R T C	R F P Y R R L G T C	I F Q N R V Y T F C	C

FIG. 2. Amino acid sequences of defensins. All sequences, with the exception of mouse intestinal cryptdin (CRYPT), whose sequence was inferred from published cDNA sequences (Ouellette et al 1989), were obtained by direct analysis of purified peptides recovered from neutrophils (PMNs). The standard one-letter amino acid code is used. The positions of the disulphide bridges are shown.

to developmental and immunological regulation, in PMNs the expression of defensins is believed to be constitutive. Figure 2 also shows the inferred sequence of a mouse defensin, cryptdin, whose mRNA was cloned from the small intestine. Cryptdin originates in Paneth cells (Ouellette et al 1989), which are located at the base of small intestinal crypts, subjacent to the zone of rapidly replicating cells that is responsible for epithelial cell renewal. Paneth cells also contain lysozyme and may participate in control of the small intestine's bacterial flora. The production of cryptdin by mouse Paneth cells resembles defensin production by rabbit lung macrophages in that it is developmentally regulated (Ouellette et al 1989), and does not reach adult levels until the third or fourth week of life.

Although all mammalian neutrophils are likely to contain antimicrobial proteins and peptides, there are notable intraspecific differences in their expression and composition. Romeo and his colleagues have isolated three antimicrobial peptides ('bactenecins') from the granules of bovine neutrophils that are without known counterparts in human cells. One of them (Romeo et al 1988) was a cationic, amphiphilic dodecapeptide containing two cysteine residues which formed an intramolecular disulphide bond. The other two, named Bac5 and Bac7, were larger, approximately 5 kDa and 7 kDa (Frank et al 1990). Both were cationic and unusually proline-rich, but lacked cysteine residues. A cysteine-rich antimicrobial peptide that contained 39 amino acids, including six cysteines, has also been isolated from bovine tracheal epithelium and cloned (Diamond et al 1991). Although its unique cysteine motif distinguishes the peptide, TAP (tracheal antimicrobial peptide), from classical defensins, it is likely to function as an endogenous antibiotic in the respiratory tract of cattle. The human respiratory tract equivalent(s) of TAP remain to be identified.

Studies with genomic clones for rabbit defensins NP-1 and NP-2 indicate that the genes encoding these two highly similar peptides evolved by gene duplication (Ganz et al 1989). The human defensin gene, *DEF*, is located on human chromosome 8p23 (Sparkes et al 1989) and the mouse gene for cryptdin has been localized to a corresponding (syntenic) portion of mouse chromosome 8. HNP-1 and HNP-3 are initially synthesized as 94 amino acid precursors that contain an N-terminal 19 amino acid signal sequence, followed by a 45 amino acid propiece and the 30 amino acid mature defensin peptide (Daher et al 1988). The primary structures of eight defensin precursors are shown in Fig. 3. Although the signal sequences of the human and rabbit neutrophil defensins show remarkable conservation, this was not maintained by mouse cryptdin, the sole defensin known to be produced in Paneth cells. The most interesting feature of the propiece is the presence of five to six conserved acidic residues that impart a net negative charge to this segment.

The post-translational processing of defensins was studied in HL-60 human promyelocytic leukaemia cells and in chronic myeloid leukaemia cells by Valore & Ganz (1992). The initial defensin intermediate, some of which was secreted,

```
                     SIGNAL SEQUENCE                               PROPIECE

            1                 10                 19           20               29
HNP1        M R T L A I L A A I L L V A L Q A Q A             E P L Q A R A D E V
HNP3        M R T L A I L A A I L L V A L Q A Q A             E P L Q A R A D E V
NP1         M R T L A L L A A I L L V A L Q A Q A             E H V S V S I D E V
NP2         M R T L A L L A A I L L V A L Q A Q A             E H I S V S I D E V
NP3A        M R T L I L L A A I L L A A L Q A Q A             E L F S V N V D E V
NP4         M R T L A L L A A I L L V T L Q A Q A             E L H S G M A D D G
NP5         M R T L A L L A A I L L V T L Q A Q A             E L H S G M A D D G
cryptdin    M K K L V L L F A L V L G F - Q V Q A             D S I O N T D E E T

                                       PROPIECE
            30                 40                  50                    60
HNP1        A A A P E Q I A A D I P E V V V S L A W D E S L A P K H P G S R K N M
HNP3        A A A P E Q I A A D I P E V V V S L A W D E S L A P K H P G S R K N M
NP1         V D Q Q P P Q A E D Q - D V A I Y V K E H E S S A L E A L G V - K A G
NP2         V D Q Q P P Q A E D Q - D V A I Y V K E H E S S A L E A L G V - K A G
NP3A        L D Q Q Q P - G S D Q - D L V I H L T G E E S S A L Q V P D T - K
NP4         V D Q Q Q P R A Q D L - D V A V Y I K Q D E T S P L E V L G A - K A G
NP5         V D Q Q Q P R A Q D L - D V A V Y I K Q D E T S P L E V L G A - K A G
Cryptdin    K T E E Q P G E E D Q - A V S V S F G D P E G T S L Q E E S
```

FIG. 3. Amino acid sequences of defensin signal sequences (19 amino acids) and 39–45 amino acid propieces, which lie on the C-terminal side of the 29–35 amino acids of the mature defensin peptides. In the signal sequences, all residues that are identical to those of HNP-1 and HNP-3 have been boxed. The conserved acidic residues in the propieces have also been boxed.

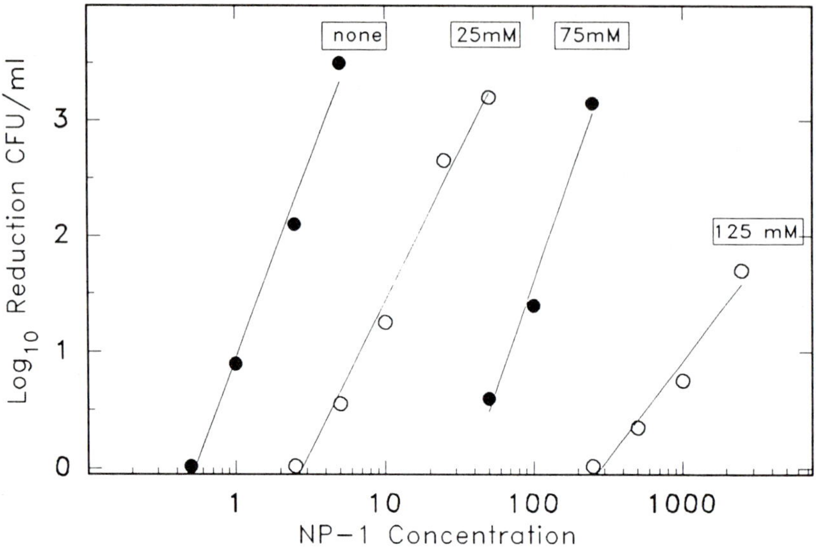

FIG. 4. Effect of ionic strength on the fungicidal activity of rabbit NP-1. *Candida albicans* was suspended at 1×10^6 colony-forming units (CFU)/ml in 10 mM sodium phosphate buffer (pH 7.4) that contained 0, 25, 75 or 125 mM NaCl, and was exposed for 120 min to various concentrations (μg/ml) of NP-1. Fungicidal activity was measured by counting of *C. albicans* colonies and is expressed as $\log_{10}$ reduction in CFU/ml.

was generated by removal of the signal sequence and contained 75 amino acid residues. Its subsequent processing to mature (29–30 amino acid) defensins occurred via a 56 amino acid intermediate. Mature human neutrophils also contained minor amounts of defensin precursors with 39, 34 and 32 amino acids. The acidic propiece removed during defensin processing may mask the positively charged arginine residues in the mature defensin domain, thereby preventing the extraneous binding of nascent defensins (for example, to nucleic acids) or initiation of autocytotoxic processes pending the safe packaging of defensins in azurophil granules (Michaelson et al 1992).

Defensins can kill or inactivate an impressively broad spectrum of microorganisms *in vitro*, including Gram-positive and Gram-negative bacteria (Ganz et al 1985), fungi (Lehrer et al 1988) and certain enveloped viruses (Daher et al 1986). The antimicrobial activity of human defensins is optimal against metabolically active targets (Lehrer et al 1988, 1989), although this property is less apparent with very cationic defensins, such as rabbit NP-1. The antimicrobial effects of defensins are antagonized by increasing ionic strength, as illustrated in Fig. 4, which shows the effect of several salt concentrations on the fungicidal efficacy of NP-1. Note that whereas 1 μg/ml of NP-1 was sufficient to kill approximately 90% (1 $\log_{10}$) of yeasts (*Candida albicans*) suspended in 10 mM phosphate buffer, about 10 or 100 μg/ml NP-1 was required for a similar level of fungicidal activity when the buffer was supplemented with 25 mM or 75 mM NaCl, respectively. Such findings suggest that defensins may be 'designed' to work intracellularly rather than extracellularly. Although the ionic composition of phagocytic vacuoles is unknown, high local concentrations of defensins may render defensin-mediated intravacuolar microbicidal processes relatively independent of local ionic conditions.

We have used *Escherichia coli* ML-35 to probe the bactericidal mechanism of human defensins. For these studies we transformed *E. coli* ML-35, a smooth strain that constitutively expresses β-galactosidase but is deficient in lactose permease, with pBR322, to endow it with a periplasmic β-lactamase. We devised an assay based on the concurrent spectrophotometric measurement of the hydrolysis of two compounds by defensin-treated *E. coli*; PADAC, a coloured β-lactamase substrate, and *o*-nitrophenyl-β-D-galactoside, a substrate for β-galactosidase. This assay allowed us to assess the integrity of the bacterial membranes and follow this over time. Under conditions that supported bactericidal activity, HNP-1 caused sequential permeabilization of the outer and inner membranes. Coincident with these events, bacterial synthesis of macromolecules ceased and colony-forming potential was lost (Lehrer et al 1989). Loss of viability correlated with permeabilization of the inner membrane, which occurred after a lag of approximately 50 minutes if we used stationary phase cells and after 10–15 minutes if we used mid-logarithmic phase organisms. Although HNP-4 has been reported to be substantially more potent than HNP 1–3 against *E. coli* (Wilde et al 1989), in our hands its potency resembles that of HNP-1 (unpublished results), although it acts after a somewhat shorter lag and, unlike HNP-1, its activity is enhanced by the non-ionic surfactant that Wilde et al used in their studies.

Whereas the more cationic rabbit defensins NP-1 and NP-2 killed bacteria effectively even in simple buffer solutions, human defensins were effective only when nutrients, such as dilute trypticase soy broth, were present during the incubation with defensins. We could protect *E. coli* ML-35 from the lethal effects of defensins by supplementing the medium with CCCP (carbonyl cyanide-*m*-chlorophenyl hydrazone), a proton ionophore, with chloramphenicol, a bacteriostatic antibiotic, or with millimolar concentrations of calcium ions. Each of these treatments prevented outer and inner membrane permeabilization of the target cells. Although *E. coli* ML-35 was not killed by HNP-1 when incubated under anaerobic conditions in our usual nutrient-containing buffer, the addition of nitrite, which supports anaerobic aspiration, allowed defensin-mediated microbicidal activity to proceed in the absence of oxygen (unpublished results). The uptake of human defensins by *E. coli* was biphasic. The initial phase was energy dependent and did not occur when incubations were in nutrient-free buffer. This was followed by a larger, energy-independent phase of uptake whose onset coincided with membrane permeabilization (unpublished results).

Immunogold electron microscopy revealed that this secondary uptake allowed defensins to enter the bacterial cytoplasm, as well as to accumulate on the outer membrane and in the periplasmic space (A. Oren, T. Ganz & R. I. Lehrer, unpublished).

We have used *C. albicans*, a human commensal organism and an increasingly important opportunistic pathogen, to examine the fungicidal activity of human defensins also (Lehrer et al 1988). Although HNP-1 killed this organism effectively, it was only about one-tenth as potent as the rabbit defensin NP-1. HNP-3 did not kill *C. albicans*, even though its primary sequence is virtually identical to that of HNP-1 (Fig. 2), the only difference being that its N-terminal residue is aspartic acid rather than alanine. In our hands, HNP-4 and HNP-1 displayed approximately equal potency against *C. albicans* (unpublished work). *C. albicans*, an obligate aerobe, was not killed by HNP-1 under anaerobic conditions. It also survived under aerobic conditions when the incubation medium contained inhibitors of mitochondrial metabolism, such as CCCP or a high concentration of sodium azide. The ability of HNP-1 to kill *C. albicans* was substantially reduced by relatively low concentrations of calcium (Fig. 5) or magnesium ions. Figure 5 also shows that the binding of HNP-1 to *C. albicans* was biphasic, and that the second phase coincided with the organism's loss of viability. In the presence of 0.5 mM calcium, defensin binding was monophasic

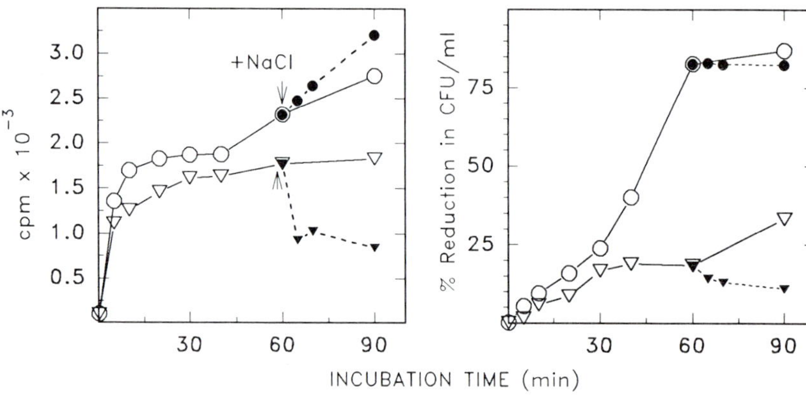

FIG. 5. Effect of Ca^{2+} and NaCl on defensin binding and defensin-mediated activity against *Candida albicans*. *C. albicans* (5×10^6 cells/ml) was incubated with 50 μg/ml of ^{125}I-labelled HNP-1 (specific activity 1870 c.p.m./μg) in 10 mM sodium phosphate buffer, pH 7.4 (open circles) or 10 mM sodium phosphate buffer + 0.5 mM $CaCl_2$ (open triangles). After 60 min (arrow) NaCl (final concentration 100 mM) was added to some tubes (small filled symbols with broken lines) and the incubation was continued for an additional 30 min. At the indicated intervals, samples were removed to determine (*left*) binding of ^{125}I-HNP-1, after centrifugation through silicone oil (expressed in c.p.m., mean from triplicate samples) and fungicidal activity, by colony counting (*right*) (expressed as the percentage reduction from input CFU/ml).

and a substantial fraction of the bound HNP-1 could be displaced by adding salt. In contrast, HNP-1 that bound to *C. albicans* under conditions that supported fungicidal activity was not displaced when NaCl was added.

When tested against a variety of tumour and non-tumour targets in tissue culture, defensins exerted cytotoxic activity that appeared to require elements of target cell metabolism (Lichtenstein et al 1988). Defensins also bound to K562 target cells with biphasic kinetics. Their binding was inhibited when assays were done at low temperature or in the presence of serum. Agents such as azide and 2-deoxyglucose, which interfere with target cell metabolism, antagonized cytotoxicity, as did cytochalasin B and certain other inhibitors. Human defensins are bound by several serum proteins, most notably the activated form of α_2-macroglobulin (Panyutich & Ganz 1991). The inhibitory effects of serum on defensin-mediated cytotoxicity may protect host cells from being damaged by defensins released from leukocytes at sites of infection and inflammation.

Although many defensins are antimicrobial or cytotoxic, their relative potency varies appreciably. The net cationicity of defensins generally (but imperfectly) correlates with their ability to kill *E. coli* and *C. albicans*. Very arginine-rich defensins, such as rabbit NP-1 or NP-2, are about 10-fold more potent in this regard than human HNP-1, 2 or 3. In contrast, human HNP-1 is substantially more effective than rabbit NP-1 against *Mycobacterium fortuitum*. During the past decade, we have obtained neutrophils from several hundred normal human blood donors and subjected acid extracts of their granules to analysis by acid–urea PAGE. Although we have not yet found any subjects to be deficient in HNP-1 or 2, HNP-3 deficiency occurs in over 10% of subjects. Whether the neutrophils of such patients lack HNP-3 because they fail to synthesize it or because they convert it all to HNP-2 remains to be determined. Patients with an exceedingly rare disorder called specific granule deficiency have neutrophils that lack defensins as well as several other granule components.

The effects of human and rabbit defensins on artificial planar 'black lipid' membranes (Kagan et al 1990) have helped to illuminate their mechanisms of action. Both NP-1 and HNP-1 formed voltage-dependent channels in membranes produced from phosphatidylethanolamine, phosphatidylcholine, phosphatidylserine or mixtures of these lipids. The channels showed a rather steep voltage dependency, and could be made to open and close repeatedly as a function of the transmembrane electrical potential. Plotting conductance against defensin concentration on a log–log plot yielded lines with slopes of 2–4, suggesting that a multimer of NP-1 was likely to be the channel-forming species. Because formation of multimers would be favoured when defensins were present at high local concentrations, such as within the azurophil granules or phagocytic vacuoles, this might be another adaptation suiting them to an intracellular site of action. The steep voltage dependency of defensin-mediated channel formation could account for the greater susceptibility of metabolically active *E. coli* to

TABLE 1 Selected endogenous antimicrobial peptides of animals

Peptide	*Source*	*Origin*	*Cysteine-rich*	*No. of amino acids*	*Reference*
Defensins	Many mammals	PMN, mac, int	Yes	29–35	
'Insect defensins'	Insects	Haemolymph	Yes	40–51	
Sapecin	Flesh fly (*Sarcophaga*)	Haemolymph	Yes	40	Matsuyama & Natori 1988
Phormicin	Black blowfly (*Phormia*)	Haemolymph	Yes	40	Lambert et al 1989
Royalisin	Honey bee (*Apis*)	Royal jelly	Yes	51	Fujiwara et al 1990
Tracheal antimicrobial peptide (TAP)	Cattle	Epithelial cells	Yes	38	Diamond et al 1991
Tachyplesin	Horseshoe crab	Haemocytes	Yes	17	Nakamura et al 1988
Bactenecins	Cattle	PMN	Yes		Romeo et al 1988
Dodecapeptide	Cattle	PMN	Yes	12	
Bac5	Cattle	PMN	No	42	Frank et al 1990
Bac7	Cattle	PMN	No	59	
Cecropins	Many insects, [pig]	Haemolymph, [int]	No	35–37 [31]	Bowman & Hultmark 1987 [Lee et al 1989]
Andropin	*Drosophila*	Seminal fluid	No	30	Samakovlis et al 1991
Magainins	Frogs	Skin	No	23	Zasloff 1987
Seminalplasmin	Cattle	Seminal fluid	No	47	Shivaji 1988

PMN, neutrophils; mac, macrophages; int, intestines.

defensins, and for the protective effects of membrane-depolarizing agents such as CCCP (Kagan et al 1990).

The six invariant cysteine residues of defensins form three intramolecular double bonds whose connectivity, established chemically by Selsted & Harwig (1989), is shown in Fig. 1. The structure of a human defensin, HNP-3, was recently obtained by high resolution X-ray crystallography (Hill et al 1991). Unlike magainins or cecropins (other membrane-permeabilizing antibiotic peptides), which are amphipathic α-helices that lack β-sheet or disulphide bonds, HNP-3 contained no α-helical domains. According to Hill et al (1991), HNP-3 crystallized as a dimer that resembled a basket with an apolar base, a polar top, and a positively charged equatorial ring composed of six arginine residues. As the three dimensional structure of the HNP-3 monomer was dominated by a three-stranded antiparallel β-sheet containing 60% of its residues, assembly of the dimer resulted in a six-stranded sheet. Hill et al have proposed several non-exclusive models for the interactions between defensins and target cell membranes. The common features of these models include an initial electrostatic binding of the defensin's arginine residues to negatively charged polar head groups of the target cell's lipid envelope, followed by hydrophobic interactions between defensins and cryptic target cell moieties within the lipid bilayer. The latter were abetted by the transmembrane potential of the target cell, which effectively pulled the defensins into the bilayer. These interpretations are consistent with previous observations that binding of defensins to microbes is biphasic and that membrane-depolarizing agents protect microbes from defensin-induced death.

In addition to their antibiotic and cytotoxic properties, certain defensins (NP-3a and HNP-4) have been reported to block the production of steroids ('corticostasis') by rat adrenal cells that were exposed to adrenocorticotropic hormone (ACTH) (Zhu et al 1987, Singh et al 1988). Rabbit NP-3a was shown to bind specifically and reversibly to the ACTH receptor and to act as a competitive inhibitor of ACTH (Tominaga et al 1990). The specific or promiscuous interaction of defensins with receptors on other cells may allow defensins to perform regulatory as well as effector roles.

Of necessity (or at least by invitation), this discussion has mainly been about defensins. Yet, as interesting as defensins are, they constitute only one family of a much larger array of endogenous antimicrobial peptides produced by animal cells. Table 1 lists some of the other animal-derived antibiotic peptides, most of which have been discovered quite recently. Like defensins, many of these molecules are cysteine-rich and some have been called 'insect defensins'— although they more probably represent instances of convergent molecular evolution rather than descendants of a common primordial defensin gene. The quickening pace of studies in this area is exciting. Not only is such work likely to identify many 'new' molecules that are responsible for the inherent resistance of animals to microbial infection, but it may also provide interesting molecular prototypes for industrial and therapeutic development.

Acknowledgements

Our work on defensins during the past decade has been a group effort involving the efforts of many talented co-investigators. We thank Ann Barton, Kathleen Daher, Robert DeLange, Patricia Eisenhauer, Sylvia Harwig, Bruce Kagan, Mitchell Kronenberg, Rose Linzmeier, Lide Liu, Ami Oren, John Rayner, Ralph Martinez, David Michaelson, Kennet Miyasaki, Judith Patterson-Delafield, Michael Selsted, Dorothy Szklarek and Erica Valore, whose contributions made this brief report possible. Various aspects of this work were supported by research grants from the National Institutes of Health, the University of California and the Will Rogers Pulmonary Research Institute at UCLA.

References

Boman HG, Hultmark D 1987 Cell free immunity in insects. Annu Rev Microbiol 41:103–126

Daher KA, Selsted ME, Lehrer RI 1986 Direct inactivation of viruses by human granulocyte defensins. J Virol 60:1068–1074

Daher KA, Lehrer RI, Ganz T, Kronenberg M 1988 Isolation and characterization of human defensin cDNA clones. Proc Natl Acad Sci USA 85:7327–7331

Diamond G, Zasloff M, Eck H, Brasseur M, Maloy WL, Bevins CL 1991 Tracheal antimicrobial peptide: a cysteine-rich peptide from mammalian tracheal mucosa: peptide isolation and cloning of a cDNA. Proc Natl Acad Sci USA 88:3952–3956

Eisenhauer PB, Harwig SSL, Szklarek D, Ganz T, Selsted ME, Lehrer RI 1989 Purification and properties of three defensins from rat neutrophils. Infect Immun 57:2021–2027

Frank RW, Gennaro R, Schneider K, Przybylski M, Romeo D 1990 Amino acid sequences of two proline-rich bactenecins. Antimicrobial peptides of bovine neutrophils. J Biol Chem 265:18871–18874

Fujiwara S, Imai J, Fujiwara M, Yaeshima T, Kawashima T, Kumpei K 1990 A potent antimicrobial protein in royal jelly. Purification and determination of the primary structure of royalisin. J Biol Chem 265:11333–11337

Gabay JE, Scott RW, Campanelli D et al 1989 Antibiotic proteins of human polymorphonuclear leukocytes. Proc Natl Acad Sci USA 86:5610–5614

Ganz T, Selsted ME, Szklarek D et al 1985 Defensins: natural peptide antibiotics of human neutrophils. J Clin Invest 76:1427–1435

Ganz T, Rayner JR, Valore EV, Tumolo A, Talmadge K, Fuller F 1989 The structure of rabbit defensin genes and their organ-specific expression. J Immunol 143:1358–1365

Hill CP, Yee J, Selsted ME, Eisenberg D 1991 Crystal structure of defensin HNP-3, an amphiphilic dimer: mechanisms of membrane permeabilization. Science (Wash DC) 251:1481–1485

Kagan BL, Selsted ME, Ganz T, Lehrer RI 1990 Neutrophil antimicrobial peptides (defensins) form voltage-dependent ionic channels in planar lipid bilayer membranes. Proc Natl Acad Sci USA 87:210–214

Lambert J, Keppi E, Dimarcq JL et al 1989 Insect immunity: isolation from the immune blood of the dipteran *Phormia terranovae* of two insect antibacterial peptides with sequence homology to rabbit lung bactericidal peptides. Proc Natl Acad Sci USA 86:262–266

Lee J-Y, Boman A, Chuanxin S et al 1989 Antibacterial peptides from pig intestine: isolation of a mammalian cecropin. Proc Natl Acad Sci USA 86:9159–9162

Lehrer RI, Ganz T 1990 Antimicrobial polypeptides of human neutrophils. Blood 76:2169–2181

Lehrer RI, Ganz T, Szklarek D, Selsted ME 1988 Modulation of the in vitro candidacidal activity of human neutrophil defensins by target cell metabolism and divalent cation. J Clin Invest 81:1829–1835

Lehrer RI, Barton A, Daher KA, Harwig SSL, Ganz T, Selsted ME 1989 Interaction of human defensins with *Escherichia coli*: mechanism of bactericidal activity. J Clin Invest 84:553–561

Lichtenstein AK, Ganz T, Nguyen TM, Selsted ME, Lehrer RI 1988 Mechanism of target cytolysis by peptide defensins. Target cell metabolic activities, possibly involving endocytosis, are crucial for expression of cytotoxicity. J Immunol 145:2686–2694

Matsuyama K, Natori S 1988 Purification of three antibacterial proteins from the culture medium of NIH-Sape-4, an embryonic cell line of Sarcophaga peregrina. J Biol Chem 263:11712–11716

Michaelson D, Rayner J, Couto M, Ganz T 1992 Cationic defensins arise from charge-neutralized propeptides: a potential mechanism for avoiding leukocyte autocytotoxicity. J Leukocyte Biol 51:634–639

Nakamura T, Furunaka H, Miyata T et al 1988 Tachyplesin, a class of antimicrobial peptide from the hemocytes of the horseshoe crab (Tachypleus tridentatus). Isolation and chemical structure. J Biol Chem 263:16709–16713

Ouellette AJ, Greco RM, James M, Frederick D, Naftilan J, Fallon JT 1989 Developmental regulation of cryptdin, a corticostatin/defensin precursor mRNA in mouse small intestinal crypt epithelium. J Cell Biol 108:1687–1695

Panyutich A, Ganz T 1991 Activated α-2 macroglobulin is a principal defensin-binding protein. Am J Respir Cell Mol Biol 5:101–106

Patterson-Delafield J, Szklarek D, Martinez RJ, Lehrer RI 1980 Microbicidal cationic proteins in rabbit alveolar macrophages: a potential host defense mechanism. Infect Immun 30:180–192

Romeo D, Skerlavaj B, Bolognesi M, Gennaro R 1988 Structure and bactericidal activity of an antibiotic dodecapeptide purified from bovine neutrophils. J Biol Chem 263:9573–9575

Samakovlis C, Kylsten P, Kimbrell DA, Engstrom A, Hultmark D 1991 The andropin gene and its product, a male-specific antibacterial peptide in Drosophila melanogaster. EMBO (Eur Mol Biol Organ) J 10:163–169

Selsted ME, Harwig SL 1987 Purification, primary structure and antimicrobial activities of a guinea pig neutrophil defensin. Infect Immun 55:2281–2286

Selsted ME, Harwig SSL 1989 Determination of the disulfide array in the human defensin HNP-2, a covalently cyclized peptide. J Biol Chem 264:4003–4007

Selsted ME, Harwig SL, Ganz T, Schilling JW, Lehrer RI 1985a Primary structure of three human neutrophil defensins. J Clin Invest 76:1436–1439

Selsted ME, Brown DM, DeLange RJ, Harwig SL, Lehrer RI 1985b Primary structures of six antimicrobial peptides of rabbit peritoneal neutrophils. J Biol Chem 240:4579–4584

Shivaji S 1988 Seminalplasmin: a protein with many biological properties. Biosci Rep 8:609–618

Singh A, Bateman A, Zhu Q, Shimasaki S, Esch F, Solomon S 1988 Structure of a novel human granulocyte peptide with anti-ACTH activity. Biochem Biophys Res Commun 155:524–529

Smith RM, Curnutte JT 1991 Molecular basis of chronic granulomatous disease. Blood 77:673–686

Sparkes RS, Kronenberg M, Heintzmann C et al 1989 Assignment of the defensin gene to chromosome 8p23. Genomics 5:240–244

Spitznagel JK 1990 Antibiotic proteins of human neutrophils. J Clin Invest 86:1381–1386

Tominaga T, Fukata J, Naito Y et al 1990 Effects of corticostatin-I on rat adrenal cells in vitro. J Endocrinol 125:287–292
Valore EV, Ganz T 1992 Posttranslational processing of defensins in immature myeloid cells. Blood 79:1538–1544
Wilde CG, Griffith JE, Marra MN, Snable JL, Scott RW 1989 Purification and characterization of human neutrophil peptide 4, a novel member of the defensin family. J Biol Chem 264:11200–11203
Zeya H, Spitznagel JK 1969 Arginine-rich proteins of polymorphonuclear leukocyte lysosomes. Anti-microbial specificity and biochemical heterogeneity. J Exp Med 127:927–941
Zasloff M 1987 Magainins, a class of antimicrobial peptides from Xenopus skin: isolation, characterization of two active forms, and partial cDNA sequence of a precursor. Proc Natl Acad Sci USA 84:5440–5553
Zhu Q, Singh AV, Bateman A, Esch F, Solomon S 1987 The corticostatic (anti-ACTH) and cytotoxic activity of peptides isolated from fetal, adult and tumor-bearing lung. J Steroid Biochem 27:1017–1022

DISCUSSION

Turner: One point that comes to mind when considering the effects of defensins on a wide range of organisms is that if there is no distinction between eukaryotes and prokaryotes, and defensins go for the membrane and open it up, even artificial membranes, why don't they have an effect on animal cell membranes also?

Lehrer: One reason is that a remarkable drug delivery system has been evolved for the defensins. One of the things protecting bystander cells is that the peptides are sequestered in a contained space, a phagocytic vacuole. A relatively small amount of these compounds does escape from the cell. There are also binding proteins in the blood, α_2-macroglobulin and γ-globulin, which soak up defensins and thereby protect bystander cells (Panyutich & Ganz 1991).

Circulating neutrophils are very short-lived cells that disappear from the blood with a half-disappearance time ($T_{½}$) of 6–7 hours (Bishop et al 1971). I am not sure why they have been designed that way. Another fascinating aspect of neutrophil physiology is that one of the first responses of a neutrophil to phagocytosis is to depolarize its own membrane. I suspect that this would help to protect it from some of the noxious effects of antibiotic peptides, which are dependent on a transmembrane potential. So it could be any or all of these things.

You can exert cytotoxic effects with these peptides by exposing target cells to them in the absence of serum components, so they are absolutely cytotoxic under those conditions. As soon as you revert to a more physiological state, in the presence of serum proteins, there is no cytotoxicity.

Turner: The neutrophils also deliver superoxide to bacteria?

Lehrer: Yes, neutrophils have a variety of other antimicrobial mechanisms, both oxidative and non-oxidative. In addition to the defensins, of molecular weight under 4000, neutrophils contain lysozyme (muramidase) which has antimicrobial properties. Neutrophils also contain a group of proteins of M_r 25 000–29 000, which are serine proteases, or serine protease congeners. Several of these are also antimicrobial (Lehrer & Ganz 1990). In addition, they contain a 57 000 M_r protein, BPI (bactericidal permeability increasing protein), with a very narrow anti-Gram-negative organism spectrum (Weiss et al 1978). Moreover, the neutrophil's cytosol (the non-granular component) contains a large amount of a curious heterodimeric protein complex, most recently called calprotectin, which has also been reported to possess antimicrobial properties (Steinbakk et al 1990). So probably 50% of the total composition of the human neutrophil is devoted to antimicrobial components.

Davies: Will the defensins permit the entry of other molecules? For example, could you treat *E. coli* with a defensin and find that erythromycin will now get into the cells?

Lehrer: I think the answer would have to be yes. Defensins permeabilized the outer membrane to cephalosporins (Lehrer et al 1989), and we have done experiments with lysozyme and defensins, and found that defensins can permit lysozyme to enter, which means that the 'hole' they make in the outer membrane is large enough for a molecule of 14 000 daltons to get in. Smaller molecules, such as erythromycin ($M_r < 1000$), should pose no problem.

Davies: There have traditionally been problems with the ML series of strains of *E. coli*; they are lysogenic and colicinogenic. They contain about five phages and a number of colicins. Some of the effects seen with the ML strains arise from the fact that the production of colicins is induced. ML-35 is a nice strain to use, but it is very sensitive to membrane-active agents.

Lehrer: We picked this strain because it gives us excellent information on the integrity of the outer and inner membranes, but we have used other strains, and obtained similar results.

Cavalier-Smith: In the mouse intestine, where are the microbes located at the time they are being killed by the defensin analogue?

Lehrer: The defensin analogue (cryptdin) is produced by Paneth cells, which contain granules that are also rich in lysozyme. Paneth cells occur near the base of the crypts, immediately below the zones where small-intestinal cell renewal takes place. It is known that mouse Paneth cells release their granules into the lumen of the crypt when bacteria enter the small intestine. One can also use pharmacological mediators, such as pilocarpine, to degranulate Paneth cells. Although it is unproven, the hypothesis that Paneth cells can control the small intestinal flora is an attractive one.

Cavalier-Smith: But cryptdin doesn't protect against *Giardia*, which is common in mice and would cover a large part of the intestinal wall.

Lehrer: In humans, *Giardia* is typically duodenal. And in species that have crypts and Paneth cells, such as ourselves and mice, the duodenum has relatively few of these structures; more distally, in the jejeunum and ileum, the crypts become considerably more prominent. My view is that the defensins, and the specific intestinal defensins of mice, may also be markers for additional, as yet uncharacterized, small-intestinal antibiotic molecules. It may be that the mouse does not have a small-intestinal antibiotic polypeptide that works against *Giardia*. What fascinates me is the extent to which mammalian species differ in which endogenous antibiotics they possess. This is also interesting in terms of why different host species have different pathogens.

Cane: In a controlled situation, what is the minimum number of molecules of the defensin which will kill a bacterial cell? And, typically, how many cells does a single neutrophil kill during its short lifetime? In other words, if defensin works by essentially sequestering one bacterium and then blasting it, the neutrophil will do this a few times, but if it does this one bacterial cell at a time, that requires a lot of neutrophils.

Lehrer: One can only make guesses. There are about 5×10^6 neutrophils/ml of human blood, and the concentration of bacteria in the blood during bacteraemia is lower by orders of magnitude. Consequently, at least in the blood, neutrophils greatly outnumber the microorganisms. The average phagocytic neutrophil will need to kill relatively few bacteria in its lifetime. In the test-tube it may ingest 50–100 organisms, but I doubt it often engages in such gluttony in the average infection.

Neutrophils are replaced very rapidly, in normal circumstances; and in the presence of infection, they are put out from the bone marrow in greater number and with more rapidity. So these are not cells that are designed to work repetitively. There is another type of phagocyte in mammals, the macrophage, that can live for days, weeks or months in tissues. Macrophages appear to have a completely different set of antimicrobial equipment.

Cane: What about the multiplicity of killing at the molecular level? How many defensin molecules are needed to kill an *E. coli* cell?

Lehrer: We haven't tried very hard to answer that.

Cane: The multiplicity for colicin is one, I believe?

Lehrer: Yes, and in some ways the action of defensin is rather colicin-like, although many more defensin molecules than one are necessary to kill a target cell.

Rinehart: Your minimum inhibitory concentrations are relatively high, about 50 μg per ml, compared with most antibiotics, but you have the advantage that the defensin is right there.

Lehrer: Defensins are different from conventional antibiotics. I suspect that these peptides are designed to work in the milieu of the cell, and *not* to work outside the cell. This may be very important for minimizing side effects, such as tissue damage and inflammation.

Rinehart: The 'black lipid' membrane description is reminiscent of some other antibiotics, for example the gramicidins and alamethicins, which stretch across the membrane. They have been investigated by several pharmaceutical companies and don't do much in the way of killing bacteria, but they are very effective in transporting ions across the membrane by forming channels. They are linear molecules, not convoluted like the defensin; they form dimers, and then the ions slide along the dimers.

Interestingly, special cells containing halogenated compounds have been found in red algae.

Lehrer: Are they phagocytic cells?

Rinehart: They are called halophilic cells, and they perform the halogenation.

Lehrer: This is fascinating, because neutrophils also contain myeloperoxidase, a chloroperoxidase-like enzyme.

Demain: Is there any therapeutic potential of defensins?

Lehrer: This is hard to answer, because the necessary experiments remain to be performed.

Demain: What about chemical synthesis of the peptides?

Lehrer: This has been done in several laboratories and could eventually provide answers to your previous question.

References

Bishop CR, Rothstein G, Ashenbrucker HE, Athens JW 1971 Leukokinetic studies. XIV. Blood neutrophil kinetics in chronic, steady-state neutropenia. J Clin Invest 50:1678–1689

Lehrer RI, Ganz T 1990 Antimicrobial polypeptides of human neutrophils. Blood 76:2169–2181

Lehrer RI, Barton A, Daher KA, Harwig SSL, Ganz T, Selsted ME 1989 Interactions of human defensins with *Escherichia coli*: mechanism of bactericidal activity. J Clin Invest 84:553–561

Panyutich A, Ganz T 1991 Activated α-2 macroglobulin is a principal defensin-binding protein. Am J Respir Cell Mol Biol 5:101–106

Steinbakk M, Andresen CFN, Lingaas E, Dale I, Brandtzaeg P, Fagerhol MK 1990 Antimicrobial actions of calcium binding leucocyte L1 protein, calprotectin. Lancet 336:763–765

Weiss J, Elsbach P, Olsson I, Odeberg H 1978 Purification and characterization of a potent bactericidal and membrane-active protein from the granules of human polymorphonuclear leukocytes. J Biol Chem 253:2664–2672

Final discussion

Davies: In this final discussion of the symposium, I propose that we hear two short presentations, from Wolfgang Piepersberg and John Bu'Lock. Both have considered the questions of the evolution of secondary metabolism and the relationship between this and the 'normal' (intermediary) metabolism of cells, and their notions will be valuable to our understanding of the subject.

Metabolism and cell individualization

Piepersberg: I should like to present some ideas based on our recent experimental observations. The question I am posing is whether secondary metabolism, in actinomycetes, and in general, is part of the ubiquitous and essential process of the 'individualization' of all living cells, in terms of metabolism. By this, I mean that the products of this compartment of the overall metabolism are designed to individualize cell lines for a variety of purposes, especially against other cells, whether differently differentiated cells of the same organism or the cells of other organisms in their environment. This, in turn, is just a facet of the ubiquitous 'individual(ization)' metabolism, as opposed to the 'central' metabolism (encompassing primary, intermediary and essential macromolecular metabolism), which is immanent to and distinctive of all living cells—for example, the differentiated cells of the human body. Most of the products of this metabolism are designed to have a function in the communication with other cells or organisms, for which individualized recognition of the type of receptor/target site specificity is a prerequisite. The products of this type of metabolism we would call 'individualites', and microbial secondary metabolites would represent just a special subgroup of excreted, diffusible and low molecular mass individualites. Cell surface-attached substances and structures involved in specific recognition processes would also be part of the wide range of 'individualites'.

This view would predict that there should be strong metabolic and/or regulatory relationships between the pathways leading to either cell surface or excreted and diffusible individualites. A nice example of this prediction was revealed recently, when we compared streptomycin production genes (*str*) or enzymes (Str) with those for the formation of other compounds containing modified sugars.

We are working on aminoglycosides and other sugar-containing antibiotics and recently concentrated on the 6-deoxyhexose (6DOH) pathways, which in the case of streptomycin biosynthesis in *Streptomyces griseus* and *S. glaucescens* yield dTDP-L-dihydrostreptose as an intermediate (Fig. 1). The activation is started by dTTP (deoxythymidine triphosphate) and glucose-1-phosphate, catalysed by the StrD gene product, followed by a dehydration (StrE), a 3,5-epimerization (StrM), and a reductive synthase (StrL) step (see Fig. 1; Pissowotzki et al 1991). This pathway is very similar to the dTDP-L-rhamnose pathway, the genes (*rfb*) for which are part of the gene cluster for the biosynthesis of the O-antigen (the variable lipopolysaccharide outer chains). These were recently analysed from a *Salmonella* serovar (Jiang et al 1991). The equivalent Str and Rfb enzymes of both pathways (cf. Fig. 1) are very similar in primary structures, even those for the last steps, which differ only in the fact that the StrL gene product of *S. griseus* catalyses branching of the sugar backbone in addition to the stereospecific reduction of the keto group at position 4 (see Fig. 1).

If we extend this consideration of the 6DOH pathways, we see considerable variation in this pathway, with the variable use of a few genetic modules of common origin in the biosynthesis of both types of extracellular substances.

(1) In bacterial or plant secondary metabolites, more than seventy 6DOH derivatives, probably most of them made via the dTDP-hexose pathway, can be found. Some examples are given in Fig. 1. Recently, we have shown (Stockmann & Piepersberg 1992) that the initial steps for probably all the actinomycete 6DOH pathways involved in antibiotic formation are encoded by conserved genes, where the *strM* counterpart is lacking whenever D-6DOH derivatives are made, but is present when the L-forms are synthesized (cf. Fig. 1). Also, these genes are clustered with other production genes in all cases studied more intensively.

(2) A second 6DOH pathway is involved in O-antigen formation, in this case via CDP-activated hexose intermediates, yielding the 3,6-dideoxyhexoses abequose (see Fig. 1) in serotype B, or tyvelose and paratose (not shown) in serotypes D and A salmonellae, respectively. Similarly, the gene products for the first and last steps, RfbF and RfbJ, of the abequose branch of the pathway have more closely related counterparts among the Str proteins, namely StrQ and StrEL. However, these do not have same function in this case, since StrQ and StrP are putative gene products for the production of the third subunit of streptomycin, N-methyl-L-glucosamine, which is made via the UDP-D-glucosamine pathway.

The relationships between the individual *str* and *rfb* genes are outlined in Fig. 2. All the oxidoreductases like StrE, StrL, StrP, RfbB, RfbD, RfbE, RfbG and RfbS are either closer or more distantly related members of the GalE (UDP-glucose 4-epimerase) family of enzymes (Verma & Reeves 1989, Wyk & Reeves 1989, Jiang et al 1991, Pissowotzki et al 1991). In addition, the NDP-hexose synthases, StrD, StrQ, RfbA and RfbF, are members of a still more distantly related protein family.

(3) Even more surprising was the finding, already mentioned, that the probable regulators, StrS and Rfb-Orf10.4 (Distler et al 1992, Jiang et al 1991; the *orf10.4*

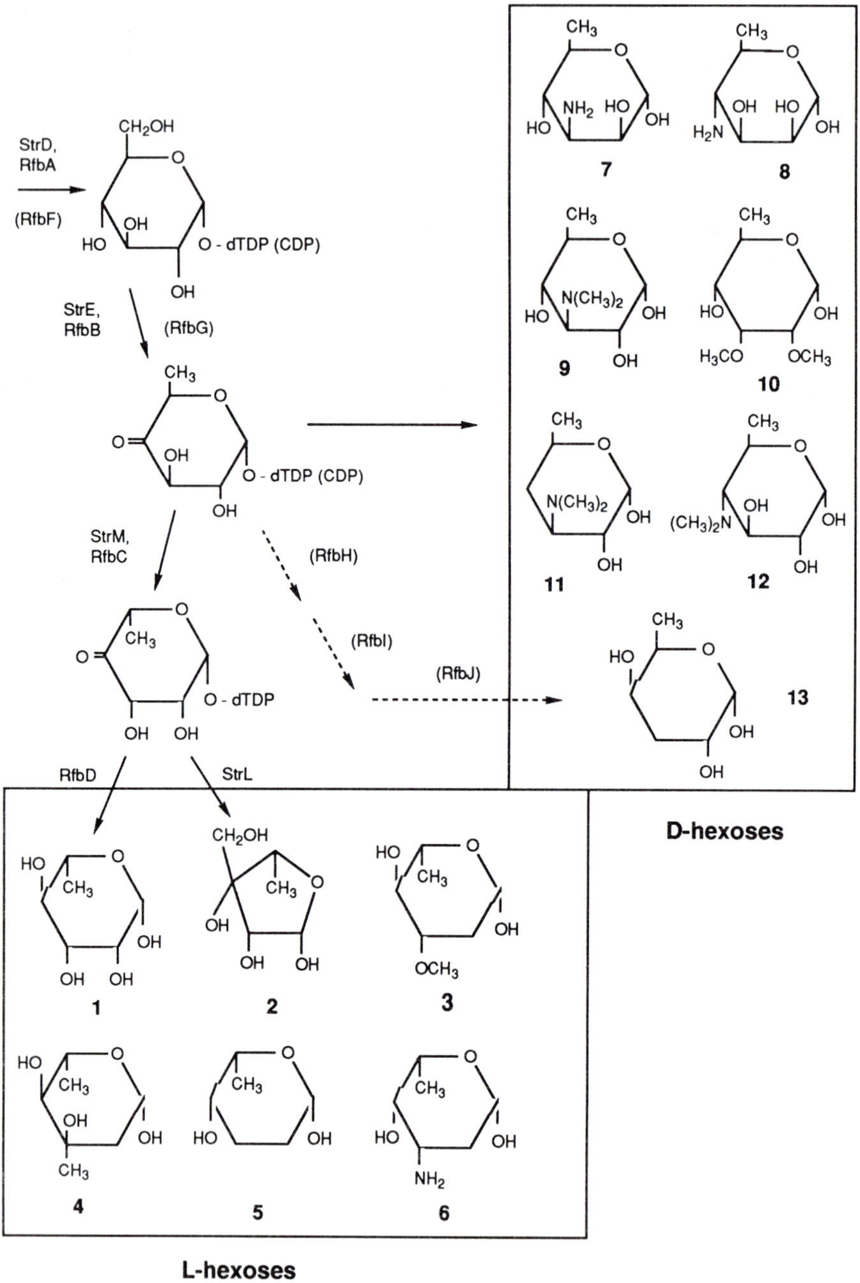
StrD,
RfbA
(RfbF)
StrE,
RfbB
(RfbG)
O - dTDP (CDP)
StrM,
RfbC
(RfbH)
(RfbI)
(RfbJ)
O - dTDP
RfbD
StrL
1
2
3
4
5
6
7
8
9
10
11
12
13
D-hexoses
L-hexoses

gene is called 'rfbR' in Fig. 2), which are of very similar primary structure (see discussion of the paper by Keith Chater, p 158), are present in these two different gene clusters. This type of regulator has previously been observed only in connection with the production of such extracellular, but diverse, 'individualites' as extracellular hydrolases (DegT from *Bacillus stearothermophilus*; Takagi et al 1990), with lipopolysaccharide-O-chain biosynthesis, and with the production of various different antibiotics, such as aminoglycosides, nucleosides, anthracyclines and macrolides (Stutzman-Engwall et al 1992, Lacalle et al 1992).

Taken together, if we ask what is similar between all the extracellular substances specifically made and excreted by bacterial and other cells, we find as the sole unifying theme their use in the specific 'cross-talk' of the producing cell with its environment. In this view, cells (organisms) only find their specific ecological niche, and only survive as a cell line, if they have some means for an individualized response, either agonistic or antagonistic. If we further generalize this view, we could argue that there are two main categories of substances which 'individualize' cells externally: the polymeric compounds attached to and coating the cell surface (which we refer to as category A), and the soluble and therefore diffusibly excreted substances, of both low and high molecular mass (category B). Also, since in general they are used for a similar purpose and are subject to the same evolutionary requirement, namely to deliver cell line-specific (*not* species-specific!) chemical variation to the host cell, we have to expect that very similar gene pools are used for the production of both categories of end products, at least on the regulatory level. Consequently, we would have to look further for common metabolic features, for example in the formation of cell wall materials (all layers could be involved, such as membranes, heteropolymeric substances, glycoprotein layers, sheaths, or mucous capsules), and of category B or so-called secondary metabolites.

Several examples can be found which support this hypothesis. (1) A very nice example consists of the Nod factors of rhizobiae—charged glycolipids based on tetrachitosane oligomers modified by sulphation and by the attachment of unusual highly unsaturated fatty acids, which actually could be polyketide derivatives (Caetano-Anollés & Gresshoff 1991, Downie 1991, Brewin 1991). These have clear relationships in basic structure and probable precursor pools to the peptidoglycans and membrane lipids. Even ATP-driven export proteins, like those involved in the export of antibiotics, occur among the Nod gene

---

FIG. 1. (*Piepersberg*) Examples and basic formation of 6-deoxyhexoses. For the involvement of streptomycin (Str) and O-antigen (Rfb) biosynthetic enzymes, see text. The L-hexose derivatives are: (1) L-rhamnose, (2) dihydrostreptose, (3) oleandrose, (4) mycarose, (5) amicetose and (6) daunosamine. The D-hexose derivatives are: (7) mycosamine, (8) perosamine, (9) mycaminose, (10) mycinose, (11) desosamine, (12) amosamine and (13) abequose.

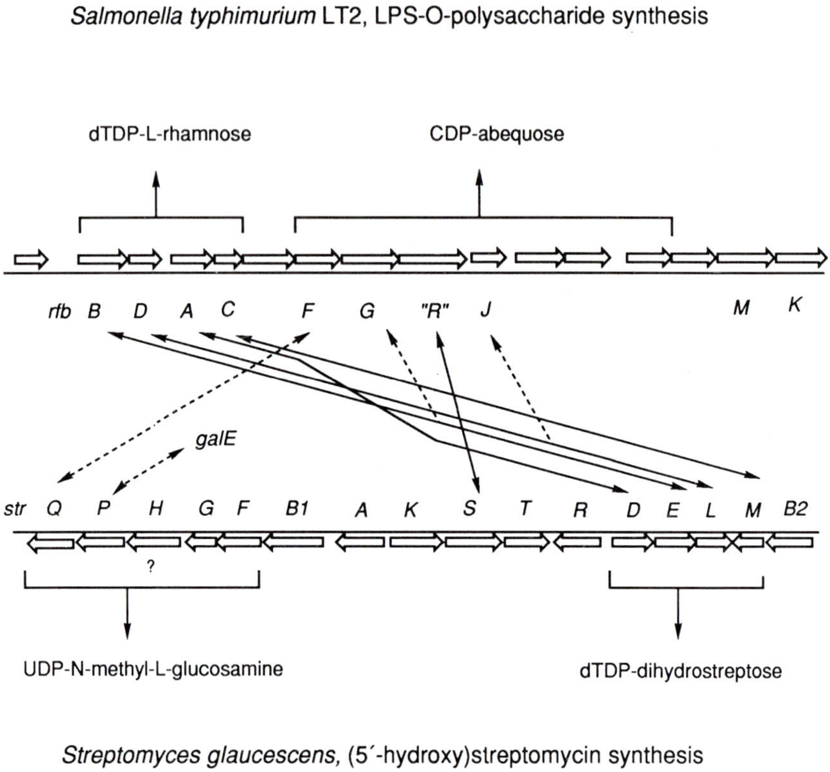

FIG. 2. (*Piepersberg*) Relationships between *rfb* and *str* genes and their products. The gene clusters for the biosyntheses of O-antigen and 5′-hydroxystreptomycin by *S. typhimurium* LT2 and *S. glaucescens* GLA0, respectively, are given. Continuous double-arrows denote genes with (putatively) equal functions and close primary structure resemblance; broken double arrows indicate genes encoding enzymes for similar activities on different substrates and/or co-substrates. For further details, see text.

products (Downie 1991). (2) The well-known relationships between substructures of the peptide side chains of the peptidoglycans and the β-lactams might be another example. (3) Many more similarities will probably be found between enzymic systems involved in the formation of the two categories of individualites. The currently most extensively studied example is the common origin of fatty acid synthases and polyketide synthases (Hopwood & Sherman 1990).

Organisms having either highly differentiated cells, such as higher animals, or a sessile lifestyle with a large and exposed surface and no immune system, such as higher plants or filamentous fungi and bacteria, need a much bigger gene pool for individualite production, especially of category B substances, than, say, single-celled and diffusible organisms, which do not have the same requirement for long-distance interactive communication systems. The latter can reach their agonistic targets or can evade antagonistic situations either by actively moving themselves or by being moved passively. Therefore, for them, category A substances would generally fit their purpose, either in adhesion (as in the example of the O-antigen of salmonellae), in aggression, or for protection. In contrast, in the actinomycetes or specialized symbionts, very similar to the animals and plants, we find diffusible category B compounds subserving intraspecific (hormone-like, immunomodulators, etc.) and interspecific (repellant-type, e.g. antibiotics, or attractant-type, e.g. Nod factors) signalling and cross-talk, for all of which we have had examples in this symposium. Therefore, the sizes of the overall gene pools of any group of organisms should reflect which portion could be used for 'individualization' metabolism and its continuous variation. For example, the differences in genome structure and size between the streptomycetes and other bacteria, such as *Escherichia coli* or corynebacteria, which have just about half or one-third, respectively, the genome size of streptomycetes, suggest that single-celled free-living organisms have a smaller amount of genetic material reserved for the production of 'individualites'.

In conclusion, I propose that we avoid the use of the terms 'secondary metabolism' and 'secondary metabolites'. I believe that there is a specific role for these highly strain- or cell line-specific metabolic events in individual cellular differentiation. I suggest that we instead use the terms 'individual(ization) metabolism' and 'individualites'.

Origins of secondary metabolism

Bu'Lock: After more than 30 years of thinking about secondary metabolism, I am really no nearer than I was to understanding its origin. Nor, I suspect, is anyone else, but for the present occasion I have drawn together some of my more general, and I hope less superficial, ideas about the very pertinent question of the *function and evolution* of secondary metabolites. I make no apology for using the term 'secondary metabolism'; plant physiologists and phytochemists have understood it fruitfully and without serious ambiguities for many years, and despite the elegantly argued position taken by Bennett and Bentley (1989), I continue to find it the most convenient and essentially *neutral* designation for the analogous microbial products.

Whether we use Darwinian or neo-Darwinian modes of thought, we are compelled to ask such questions at least partly in terms of selection pressure, even though it seems that this can only explain the persistence and continued elaboration of secondary metabolism and not any supposed 'origin'. Only if, like so many backsliders and heretics, we allow ourselves to slip into Lamarckian or even Platonic attitudes, can we ask fundamentally silly questions like 'why does the phenomenon exist?' or 'what is it for?'

I think it is clear, and generally acknowledged, that the workings of natural selection—in the specific context of maximizing the selective exploitation of a resource or habitat niche—will very adequately explain the variety of uses which have evolved for specific natural products. But let us keep the horse and cart in their correct order—the fact that *uses* have evolved under natural selection really tells us nothing about how the *products* themselves originated.

In 1966, with microbial metabolites foremost in my mind, and seeking to add to the already discussed roles of secondary metabolites in microbial ecology, I noted one such category of uses, that of secondary metabolites of microorganisms being used as specialist materials (structural, protective, attractive) in the elaboration of differentiated structures (Bu'Lock 1967). At that time a phytochemist, preoccupied with plant products, would similarly have singled out their roles in systems of communication, especially those involving insect vectors. For both viewpoints, the lapse of time and the increasing sophistication of observation and experiment have yielded a host of further and more detailed examples. Secondary metabolites (and their receptors) in microorganisms, plants and insects are seen as key mediators in ecological interactions, as inter- or intraspecific messenger substances, and as both effectors and elements of differentiation. Nevertheless, the wealth and intrinsic interest of these instances, showing how secondary metabolism has provided materials on which natural selection can work, does little or nothing to illuminate the more fundamental problem of how these materials come to be there, *in such variety*, in the first place. Yet, whether we are faced with the awe-inspiring macroscopic variety of the rain-forest flora or with the molecular-scale variety of microbial and plant secondary products, the question of the origin and significance of biological variety has to be faced. Having set out his subject as *The Origin of Species*, Darwin despaired because the essence of the concept of species (which is intrinsic to the question of biological variety) was, to him, both 'undiscovered and undiscoverable'.

In microbiology the extent of natural variety is exemplified not only in the plethora of genetically isolated species but (as a matter of everyday experience to some microbiologists, but not all) in the manifest genetic heterogeneity encountered whenever natural isolates are 'plated out'. In other words, there is a natural accumulation of genetic heterogeneity within the species, which in terms of selection pressure must necessarily be seen as essentially neutral.

Mechanisms for such variations are now well known; they largely depend upon random mutation, particularly within duplicated genes. Microorganisms are all the more prone to this because substantial parts of their genome are expressed only in specific circumstances; there is an obvious survival advantage in not expressing, during competitive vegetative growth, those parts of the genome concerned with processes such as sporulation. It is important to note that variations in an unexpressed genome cannot be subject to natural selection at all!

Thus the combined effect of random mutations in duplicated genes not subject to selection pressure is to create a constant drift towards a genetic diversity within the species; it is by the genetic segregation of such diversities in natural populations that new species are formed, as Darwin realized. However, prior to, or even without, such segregation of populations, the diversity continues to accumulate.

When this diversity comes to be expressed, natural selection can be brought to bear upon it. For example, a characteristic which is selection-neutral during vegetative replicatory growth may be found advantageous during spore development. The fact that variant characters can thus evolve most easily within a functionally segregated portion of the genome most directly explains one of the most well-known and often discussed features of secondary metabolism in microorganisms, namely that it is often minimally expressed during replicatory vegetative growth and is frequently associated (though not necessarily functionally linked) with sporulation and similar processes. Hence the introduction of terms such as *trophophase* and *idiophase*. Note also that this kind of segregation is also a genetically determined and hence variable feature. Conversely, it is well known that where vegetative growth occurs in a competitive situation, secondary metabolic activity may be a feature for negative selection; thus in antibiotic or amino acid manufacture, high-yielding strains are notoriously liable to be outgrown by low-yielding mutants in circumstances favouring vegetative growth. This happens in culture collections too!

A particular feature of this diversity that is especially prominent in microorganisms is one that non-microbiologists have found hardest to assimilate; this is the possibility of *horizontal* transmission of characters between species. So much of biological thinking is based on the presumption that characters are only transmitted vertically, through successive generations, that the possibility of 'horizontal' transmission from one species to another has scarcely been admitted into evolutionary thinking. Yet, in considering the chemotaxonomy of microorganisms, the likelihood of horizontal transmission is almost forced upon us. In fact, of course, all the mechanisms which we use today to transfer and express heterologous genes have existed for millenia 'out there' in the microbial world, and it is almost unthinkable that they have never operated. We are now just beginning to see direct confirmatory evidence of this from gene sequence studies (and see below).

As an illustrative example (Bu'Lock 1973), in the development of the mucoraceous fungi we can postulate a purely vegetative mycelium as the primitive state. Some natural isolates display only this—they are not easily ascribed to any genuine species. To this state, a mechanism for the accumulation of carotenoids under non-vegetative conditions confers an advantage in that it allows carotene to be used as a structural component (sporopollenin) in the elaboration of resistant spores. In order to attain this condition it has probably been necessary to duplicate the rate-limiting genes (e.g. that for hydroxymethylglutaryl [HMG]-CoA reductase) and to transfer one set into the non-vegetative genome. This has permitted positive selection for species also possessing the enzymes necessary to transform β-carotene into sporopollenin.

The carotene accumulators can also acquire enzymes for breaking down β-carotene through retinal; there is no immediately obvious selective advantage in doing this. However, one group of possible breakdown products is the trisporic acids. Segregation of the enzymes for trisporic acid formation between otherwise very similar strains became useful when receptors for these substances were evolved and when these were directly linked to morphogenetic processes. Then the trisporic acids became effectors of the entire process of sexual reproduction of these fungi, permitting wholesale exchange of chromosomal variety between otherwise segregated strains.

Because the entire mechanism is found, with minor variations, throughout the Mucorales, we must conclude that this set of processes was evolved very early in the evolutionary history of phycomycetes. On the other hand, since many isolates (and acknowledged species) do not show any sexual mechanisms whatsoever, we can also conclude that, in the particular ecological niches which these fungi occupy, sexuality itself is not strongly selected for—indeed, we could almost call sexual reproduction in these fungi a secondary metabolic characteristic!

An important contribution to these questions is the article 'the networks of biological creation' by Chris Heron (1992). Writing primarily from the viewpoint of the characteristics of higher plants, he discusses the importance and implications of *lateral* gene transfer in plant evolution. I and others have already noted the natural existence of the mechanisms for such transfers into and between higher plants and microorganisms. He states that: 'most cases of lateral transfer are likely to involve genes not essential to the immediate survival of individual cells—though they may be highly influential to the host organism as a whole. . . . Those small differences that do exist between species tend to be found in parts of the DNA that are functionally "neutral", and are due to point mutations that accumulate at similar rates in diverging lineages.' Unlike most botanists, Heron allows for considerable 'neutral' evolution in plants, noting that such characteristics as leaf or petal shape are genetically determined but have little effect on fitness for survival. Dealing with higher plants, he includes interspecific hybridization, and 'hybridization chains', as mechanisms for establishing variety. He distinguishes between groups of organisms in which parts of the genome are rather 'open' to lateral transfer, in which there can develop innovative, rapidly evolving and highly adaptive species, and groups with relatively 'closed' genomes which can only evolve, and adapt, more slowly and will in general contain many fewer distinguishable species. Heron also adumbrates the complications of phylogenetic 'trees' into which connections by way of lateral gene transfer are introduced.

Much of this thinking is directly applicable to the question of variations in microorganisms of precisely the kind we have to consider when discussing the enormous variety of secondary metabolic characteristics they display. It encourages such ideas as that the Streptomycetes form a genetic continuum; it explains the divergence between the taxonomic results of morphological or nucleic acid-based comparative studies of microorganisms and the results of chemotaxonomy based on secondary metabolite structures. And it has the merit of providing explanations based on the biological world as it exists, without invoking hypotheses about the remote proto-biological past.

Davies: I have very little to add at this stage. It seems that in one respect we have ended up where we started—trying to find the words that best describe the production of large numbers of complex molecules, that have no essential role in maintaining the living process of the producing organism, by cells. On the other hand, over the past few days we have, I believe, probed more deeply into the questions of what secondary metabolites do, and how they are made, than has previously been attempted. We have not come up with many answers to our questions, and we end this exciting symposium with the realization that an understanding of the 'whys' and 'wherefores' of the vast domain of secondary metabolism is still beyond our grasp. However, this symposium was, and will be, of enormous value to all people interested in the bizarre world of secondary metabolism—I know that I am not supposed to use this description! I say this with the conviction that, in the future, studies of these molecules and their biosynthetic pathways will become increasingly of interest in fundamental questions of evolution, gene regulation and cell development on one hand, and in the commercial exploitation of the biological activities of the molecules, on the other.

The symposium organizers are well satisfied with the product of this meeting and the intellectual spirit that it manifested. Much of the success of the symposium is due to the efforts of Derek Chadwick and his excellent and genial staff; we thank them for providing us with such an enjoyable and valuable meeting.

References

Bennett JW, Bentley R 1989 What's in a name?—Microbial Secondary Metabolism. Adv Appl Microbiol 34:1–28

Brewin NJ 1991 Development of the legume root nodule. Annu Rev Cell Biol 7:191–226

Bu'Lock JD 1967 Essays in biosynthesis and microbial development. Wiley, New York

Bu'Lock JD 1973 Comparative and functional aspects of the isoprene pathway in fungi. Pure Appl Chem 34:435–461

Caetano-Anollés G, Gresshoff PM 1991 Plant genetic control of nodulation. Annu Rev Microbiol 45:345–382

Distler J, Mansouri K, Mayer G, Stockmann M, Piepersberg W 1992 Streptomycin biosynthesis and its regulation in Streptomycetes. Gene (Amst) 115:105–111

Downie A 1991 A *nod* of recognition. Curr Biol 1:382–384

Heron C 1992 The networks of biological creation. New Scientist 133 (No. 1807):40–44 (8 February 1992)

Hopwood DA, Sherman DH 1990 Molecular genetics of polyketides and its comparison to fatty acid biosynthesis. Annu Rev Genet 24:37–66

Jiang X-M, Neal B, Santiago F, Lee SJ, Romana LK, Reeves PR 1991 Structure and sequence of the *rfb* (O antigen) gene cluster of *Salmonella* serovar typhimurium (strain LT2). Mol Microbiol 5:695–713

Lacalle RA, Tercero JA, Jimenez A 1992 Cloning of the complete biosynthesis gene cluster for an aminonucleoside antibiotic, puromycin, and its regulated expression in heterologous hosts. Mol Microbiol 11:785–792

Pissowotzki K, Mansouri K, Piepersberg W 1991 Genetics of streptomycin in *Streptomyces griseus*: molecular structure and putative function of genes *strELMB2N*. Mol & Gen Genet 231:113–123

Stockmann M, Piepersberg W 1992 Gene probes for the detection of 6-deoxyhexose metabolism in secondary metabolite-producing streptomycetes. FEMS (Fed Eur Microbiol Soc) Microbiol Lett 90:185–190

Stutzman-Engwall K, Otten S, Hutchinson CR 1992 Regulation of secondary metabolism in *Streptomyces* spp. and overproduction of daunorubicin in *Streptomyces peucetius*. J Bacteriol 174:144–154

Takagi M, Takada H, Imanaka T 1990 Nucleotide sequence and cloning in *Bacillus subtilis* of the *Bacillus stearothermophilus* pleiotropic regulatory gene *degT*. J Bacteriol 172:411–418

Verma N, Reeves P 1989 Identification and sequence of *rfbS* and *rfbE*, which determine antigenic specificity of group A and group D salmonellae. J Bacteriol 171:5694–5701

Wyk P, Reeves P 1989 Identification and sequence of the gene for abequose synthase, which confers antigenic specificity on group B salmonellae: homology with galactose epimerase. J Bacteriol 171:5687–5693

Index of contributors

Non-participating co-authors are indicated by asterisks. Entries in bold type indicate papers; other entries refer to discussion contributions

Indexes compiled by John Rivers

Subject index